THE RUNAWAY BRAIN

OTHER BOOKS BY CHRISTOPHER WILLS

The Wisdom of the Genes

Exons, Introns, and Talking Genes

THE RUNAWAY BRAIN

THE EVOLUTION OF HUMAN UNIQUENESS

Christopher Wills

BasicBooks
A Division of HarperCollins*Publishers*

Picture credits can be found on page 343.

Library of Congress Cataloging-in-Publication Data
Wills, Christopher.
The runaway brain : the evolution of human uniqueness / Christopher Wills.
p. cm.
Includes bibliographical references and index.
ISBN 0-465-03131-5
1. Brain—Evolution. 2. Human evolution. I. Title
QP376.W56 1993
573.2—dc20 92-55029
CIP

Designed by Ellen Levine
Map on pages xiv and xv by Craig Winer

93 94 95 96 ♦/HC 9 8 7 6 5 4 3 2 1

To all the mute inglorious Miltons
who have made us what we are.

Perhaps in this neglected spot is laid
Some heart once pregnant with celestial fire;
Hands, that the rod of empire might have sway'd,
Or waked to ecstasy the living lyre.

—Thomas Gray,
Elegy Written in a Country Churchyard

CONTENTS

III
THE GENES

IV
THE BRAIN

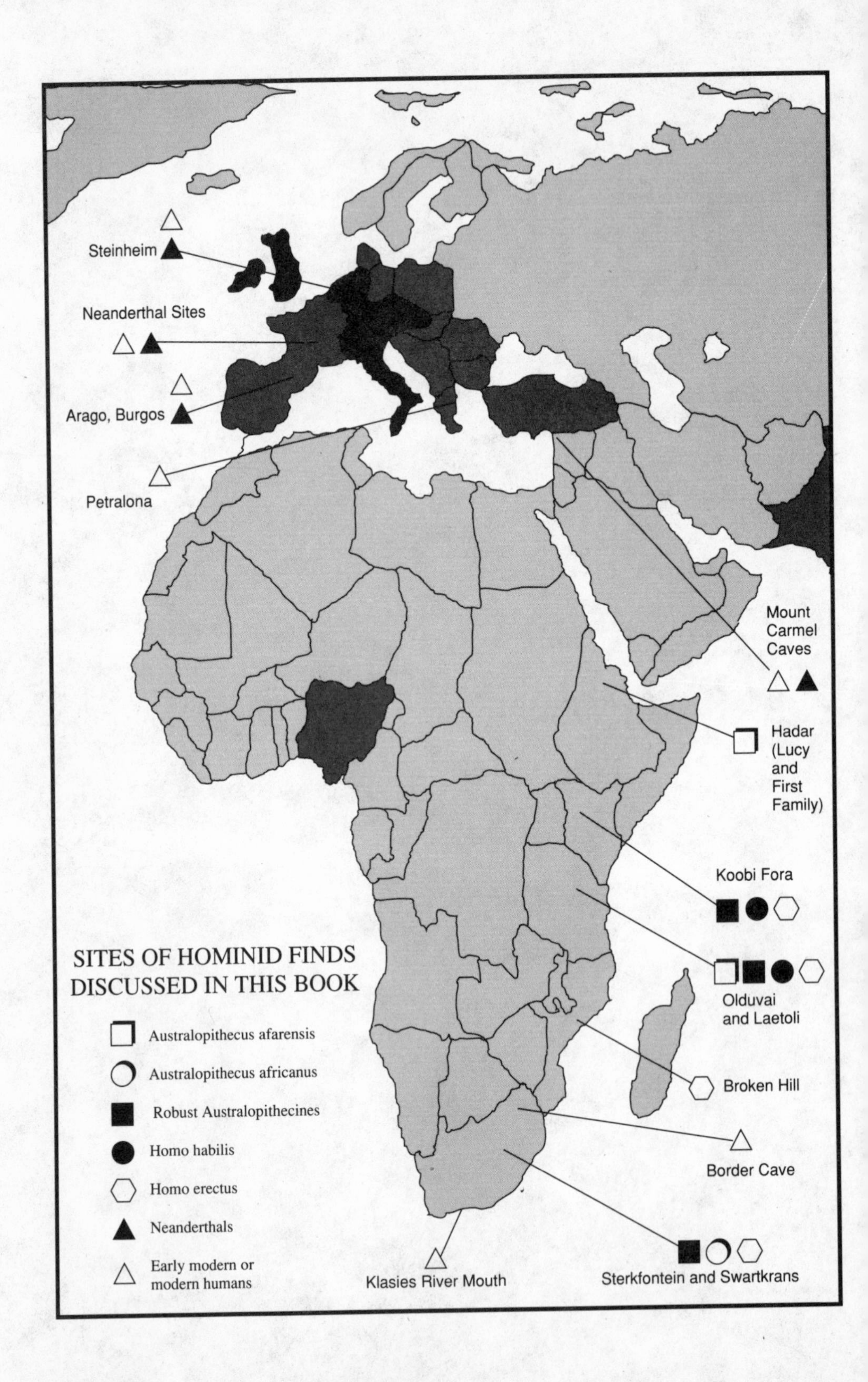
Steinheim
Neanderthal Sites
Arago, Burgos
Petralona
Mount Carmel Caves
Hadar (Lucy and First Family)
Koobi Fora
Olduvai and Laetoli
Broken Hill
Border Cave
Sterkfontein and Swartkrans
Klasies River Mouth
SITES OF HOMINID FINDS DISCUSSED IN THIS BOOK
Australopithecus afarensis
Australopithecus africanus
Robust Australopithecines
Homo habilis
Homo erectus
Neanderthals
Early modern or modern humans

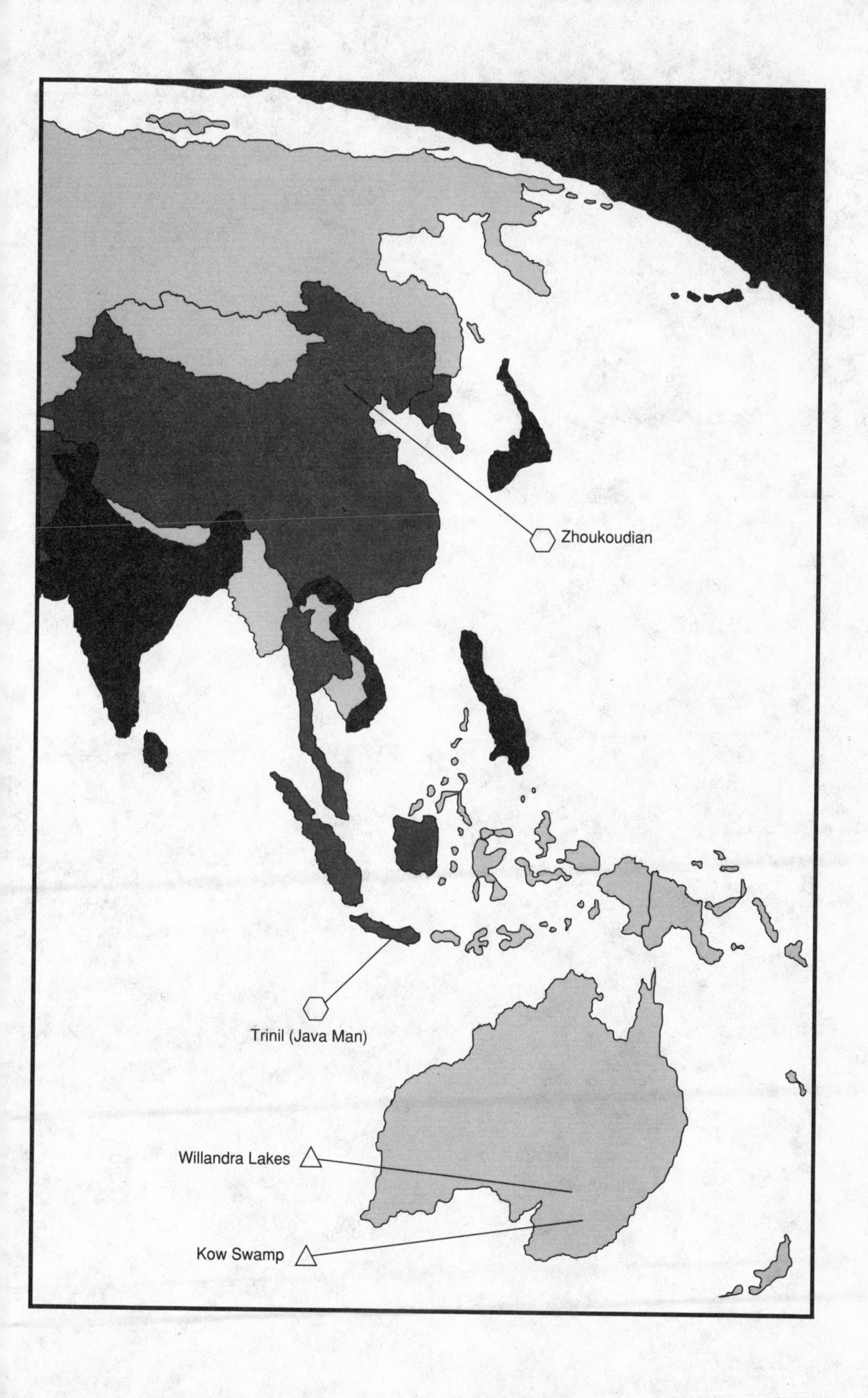
Zhoukoudian
Trinil (Java Man)
Willandra Lakes
Kow Swamp

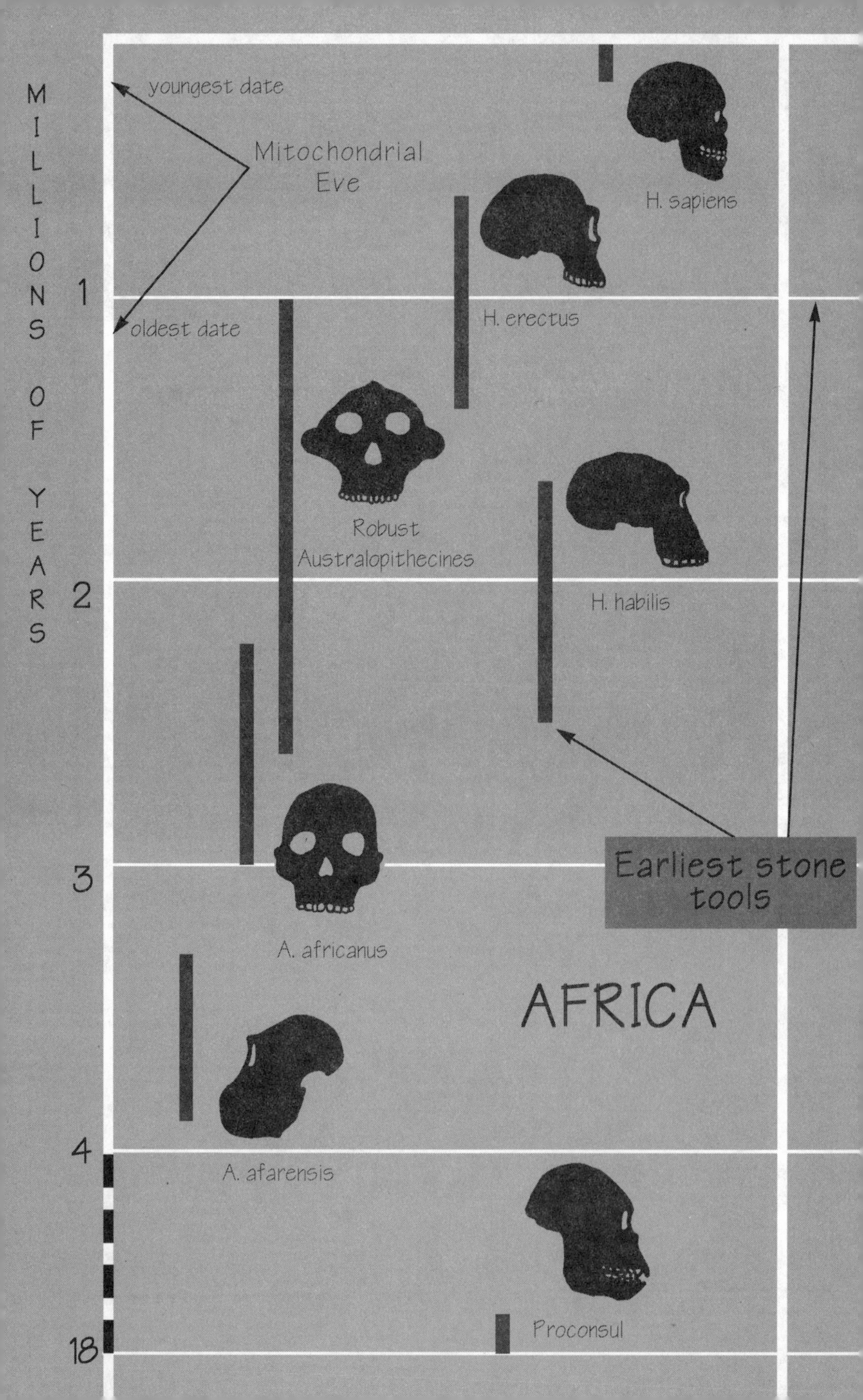
MILLIONS OF YEARS
youngest date
Mitochondrial Eve
oldest date
H. sapiens
H. erectus
Robust Australopithecines
H. habilis
A. africanus
Earliest stone tools
AFRICA
A. afarensis
Proconsul
1
2
3
4
18

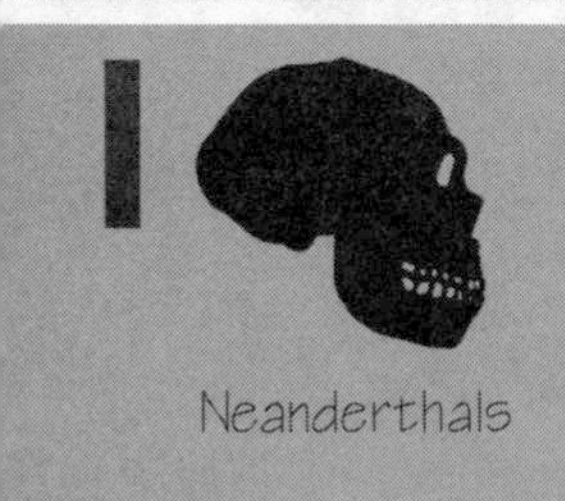

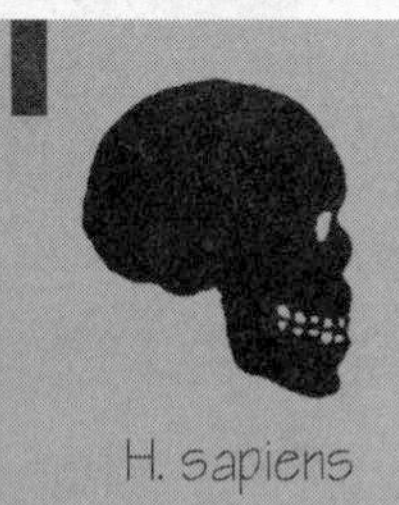

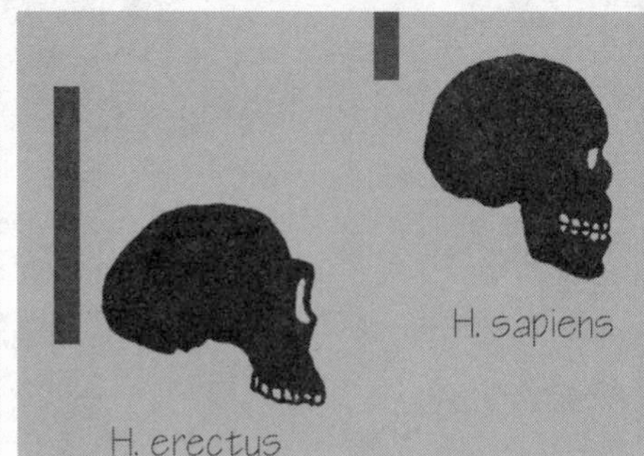

First Occupation of Europe- Perhaps by H. erectus ?

ASIA

EUROPE
&
MIDDLE EAST

APPROXIMATE DATES
OF
HOMINID FINDS

PREFACE

The brain-culture relationship was not confined to one special moment in time. Long-continued increase in size and complexity of the brain was paralleled for probably a couple of million years by long-continued elaboration and "complexification" (to use de Chardin's word) of the culture. The feedback relationship between the two sets of events is as indubitable as it was prolonged in time.

—Phillip Tobias, *The Brain in Hominid Evolution* (1971)

Just to the west of the sprawling suburbs of Johannesburg, in rolling cattle country burned dry by the recent drought, are two gentle hills, their crests separated by a little over a kilometer. Although most of the slopes of the hills are made up of parched soil dotted here and there with reddish rock outcrops, there is an occasional tree. Most of these trees are *Celtis africana*, rejoicing in the common name of white stinkwood and growing out of almost invisible clefts in the rock where deeper soil has collected.

The clefts themselves mark entrances, now almost filled in, to a great system of caverns that ranges for kilometers under the hills. These caverns were gradually carved out by groundwater seepage through the limestone and have served as catchments for detritus from above for at least the last three or four million years. It is likely that during much of that time, leopards dragged their prey into the branches of the stinkwood trees' distant ancestors. The heads of their prey would sometimes become detached and bounce in grisly fashion into the depths below. Owls nested in the trees and coughed up pellets of undigested fur and bone that occasionally fell into the crevices, carrying their cargoes of tiny rodent bones. And sometimes torrential rains would wash in other animal remains. Gradually this accumulation of material filled in the more accessible parts of the caverns.

This hodgepodge of detritus slowly built up in the depths of the damp limestone caves. A ceaseless drip of lime-laden water from above gradually cemented it into great hard concretions called breccia. Here, amid a jumble of antelope, pig, baboon, and monkey bones, rested the occasional remains of some of our most remote ancestors, hapless victims of the predators of the time (see plate 1 following p. 230).

As roof collapses and weathering opened up new entrances to the caverns, material began to build up in cones below them. In time, before the remoter reaches of the caverns below had been completely filled, those openings became blocked. A new cycle was initiated by further roof collapses that opened up additional entrances. The eventual result was a jumble of deposits that has taken a great deal of patient detective work to disentangle.

Over time the composition of the material filling in the caverns changed. Stone tools, first primitive and then more advanced, began to appear. And the users of these tools were different as well, substantially closer to modern humans. It seems that the peoples of this more recent time were not simply the victims of predators. They were actually living in or near the rock shelters at the mouths of some of the caverns, hunting and butchering a variety of animals, and using tools to dig up tubers and other plant material. Occasionally, their tools tumbled into the black recesses of the caves below. And, still less often, the remains of one of these people fell in as well—though how and why will always be a mystery.

The paleoanthropologist Ron Clarke from Witwatersrand University is currently working on the Sterkfontein deposits, where he is patiently excavating and cataloguing every scrap of bone, both human and animal, that he can find. He has been trying to understand the complex geological events that led to this remarkable collection of *objets trouvés*. The accessible finds, which sometimes have to be blasted out of the hard concretions filling the caves, range over perhaps the last three million years. Below them there may be other treasures that would, exposed, cast light on an otherwise completely inaccessible period of prehuman history.

Clarke could actually show me where these treasures might be. Like the temptations of Tantalus, they lie just beyond his reach. He led me down a rickety wooden staircase into the depths of a cavern below the excavations. At the bottom we found ourselves in a low, wide chamber with scars on its walls and ceiling, showing where miners early in this century had removed a huge boss of limestone that had filled the chamber. We felt the weight of endless history pressing on us, for the lowest part of the paleontologists' excavations was some 10 meters above our heads. The ceiling of the chamber had been formed from the very earliest deposits that had sifted down from far above to rest on the now-vanished limestone boss. We had the disorienting experience of looking *up* at the bottom of the deposits rather than looking down at them from the top.

Clarke shone his light on the ceiling, which was filled with bones and fragments of bone. Parts of a hunting hyena jaw, dating from over three million years ago, had been gingerly chiseled out of this deposit. And the anthropologist Phillip Tobias later told me that he had glimpsed, in an inaccessible corner of the cave, what might have been a bit of a skullcap of one of our ancestors.

There is every chance that this suspended deposit conceals fossils of enormous importance, perhaps from a time earlier than any fossils of human ancestors that have been found anywhere else in South or East Africa. If so, they may help us to understand a central event in our prehistory of which we have essentially no record. We know that our ancestors were able to stand fully upright 3.7 million years ago, but we know nothing of how they achieved this remarkable feat. It is likely that a short time earlier their gait had been far more apelike, for molecular evidence suggests that the common ancestor of humans and chimpanzees—a creature that certainly did not walk upright—could have lived as little as four million years ago. Perhaps, somewhere in this dark mass of rock over our heads, lay the remains of some of our ancestors who were in the process of attaining upright posture. If such remains are ever found, we might be able to begin to understand some of the first steps in the long evolutionary journey that has led to ourselves—and that will continue after we are gone.

The ceiling loomed enigmatically above our heads, hiding its secrets (although Clarke tells me that they have begun to excavate it since my visit, turning up masses of animal bones). We clambered up the staircase toward the brilliant African sky with the feeling that Sterkfontein still has many things to tell us about our beginnings.

Sterkfontein is not the beginning of our evolutionary journey, which actually extends over billions rather than millions of years. But the last few million years have, unsurprisingly, received the most attention from scientists and the public, for it is during that span of time that we have parted company from the other animals in such a startling fashion.

Indeed, there is no more fascinating question in science, religion, or philosophy than the question of our origins. Charles Darwin shied away from it in *The Origin of Species,* but the eager people who bought and discussed his book were not so cautious. Immediately, the canard was born that we are descended from the apes.

Of course, we are not. We and the apes are descended from a common ancestor that, although certainly more apelike than human, was different from any human or ape living today. This ancestor has left no trace in the fossil record, although it is possible to venture some guesses about what it might have been like. But how different from us was it really—not only in its appearance and its behavior but in its genes? And how did we humans—as Shakespeare says, in action how like an angel! in apprehension how like a god!—come to have evolved from that ancestor? What forces drove us, and are still driving us, to occupy our unique position among all the living creatures on the planet?

Scientists have now learned far more about the process of evolution than Darwin knew, though it has been difficult for them to communicate this complex mass of recent knowledge to the general public. My goal is to try to close the gap between the scientist's perception of how evolution works and the public's; in essence, to demystify the process of evolution. To do this I will concentrate primarily on our own evolution, though there will be times when I stray quite a bit, in order to contemplate organisms very different from ourselves. There is good reason for this—evolution is universal among living organisms, and we, though we smugly imagine that we are no longer a part of the natural world, are just as subject to evolutionary laws as are microbes and mice.

And yet, we are unique! All students of human evolution keep coming back to that astonishing fact, a fact that has bewildered even the greatest thinkers. Have we somehow, unlike any other organism, escaped from those evolutionary laws? We tend to think we have, because we seldom if ever take an evolutionary view of our own species. Each of us is, after all, confined to a brief life spanning a few decades, and at the present time we are—at least in the First World—luxuriating in an existence of unparalleled ease and safety. As a result, we can easily be misled into thinking that we have somehow fooled Mother Nature. But we have not—we are like the soldier in Ambrose Bierce's *Occurrence at Owl Creek Bridge,* who dreams that he has escaped his hanging and been able to flee home—until the rope jerks him up short.

Here I will attempt to put us squarely back in the evolutionary framework. The first section of the book presents a series of dilemmas about human evolution, and the following three sections try to answer them by drawing on our growing knowledge of paleontology, genetics, and the human brain.

The first is the question of just where our species first appeared. The problem is really twofold, because it requires us both to figure out when our ancestors first became human and to discover where they were living at the time. As we will see, these are very tough questions indeed! There are no simple answers, and it is unfortunate that a very distorted version of recent scientific findings about our origins has filtered out to the public. Most of us now have it firmly in our minds that we arose from a single Eve, presumably with the help of a single Adam, who lived a long time ago, probably in Africa. If you, gentle reader, harbor this impression, I will do my best to disabuse you of it. There is a kind of Eve in our past, although not the kind found in the Bible. Where she lived is still not certain, but wherever and whenever she lived, she was not alone with her Adam. She had companions, and very interesting companions they were. And she was not the beginning of the story, for our ancestors that lived on the lip of the cave of Sterkfontein, with their upright posture and their ability to make primitive tools, date back to a time long before that problematic Eve.

There is also no guarantee that the time at which Eve lived coincided with the time of our first appearance as a species. There is now no doubt that our early evolution took place in Africa, but just how human were our ancestors by the time they left their African homeland and spread to Europe and Asia? We could answer that question if we knew when they left. Was it a long time ago, or recently, or were there perhaps two or more migrations? There is evidence for all these possibilities, though the evidence has often been colored by the personal and political beliefs of the scientists involved.

To make things even more difficult, the question of the time of our origin as a species has political and social ramifications that can be—and have been—seized upon by those with racist beliefs. If our origin was long ago, before our ancestors were "fully human"—whatever that means—then those ancestors who migrated from Africa must not yet have become human. This means that our ancestors must have made the transition to full humanity more than once, in different parts of the planet. If our origin was recent, on the other hand, then the transition need have occurred only once, perhaps before we even left Africa. Remarkably, again there is evidence for both these possibilities and neither currently can be ruled out. The former is the one that poses the possibility of a racist interpretation, for if a transition to full humanity took place more than once it suggests that different human groups may have crossed that evolutionary finish line at different times. And, unpleasantly, that some human groups may have evolved further than others. I want to tackle these problems squarely in this book and show that they depend on semantic traps to which both scientists and the public have fallen victim. Racist beliefs have no place in a consideration of human origins.

The third dilemma grows out of the statements of the last paragraph. What is all this stuff about transitions to full humanity and evolutionary finish lines, as if our species were the winner of some evolutionary Olympics? Scientists know that evolution does not work that way, that it can go in any direction, and that it is entirely dictated by how the environment changes and by the opening up of new ecological niches. And yet, discomfitingly, the literature on human evolution is filled even now with references to "prehumans evolving toward full humanity" and "the appearance of early modern *Homo sapiens.*" It is almost impossible even for scientists to rid their language of this impression of directionality, even though they may be doing their best to apply modern evolutionary theory to humans. Perhaps we cannot rid our language of this appearance of directionality *because directionality is really there.*

As many authors have pointed out, the force that is likely to have driven our evolution in this apparently directional way is a new kind of force, the process of cultural evolution. As our cultures evolved in complexity, so did our brains, which then drove our bodies toward greater responsiveness and our cultures toward still greater complexity in a feedback loop. Big and clever brains led to more complex cultures and bodies better suited to take advantage of them, which in turn led to yet bigger and cleverer brains.

Evolutionary feedback loops are actually quite common. In 1915 the British statistician and geneticist R. A. Fisher described a feedback loop driven by the need to compete for mates and called it runaway sexual selection. If the sexual selection were strong enough, he thought, it might actually result in characteristics that could harm the organisms that exhibited them. A peacock with a large tail has a great advantage in finding mates—but only if its tail is not so large that it can no longer escape from predators!

Other kinds of feedback loop have been studied since, such as the coevolution of some species of flowers and the particular insect or bird species that pollinate them. The uniqueness of our own evolutionary story lies in the fact that we, alone among the millions of species on the planet, seem to have been caught up in a process of *runaway brain evolution.*

As we explore this feedback process, we will see how it helps to illuminate and perhaps solve many of the major dilemmas about our evolution. We will see that Eve and her companions must already have been well down this runaway racetrack and that even the creatures living in the cave shelters of Sterkfontein had, at that remote time, been caught up in the runaway process for millions of years. If our various ancestors really did make the transition to full humanity more than once, then it must have been runaway brain evolution that drove them in that direction. We will see that those ancestors must have had most of the genetic equipment needed to cross that imaginary evolutionary finish line. Even if some did it early and others later, the result has been remarkably similar among all members of our species.

Of course, the evolutionary finish line is an arbitrary construct that we have invented because we smugly imagine that we have crossed it. In actuality, there are no finish lines in evolution. Further, feedback loops cannot go on indefinitely. They can quickly get out of hand, as anybody who has worked with a public-address system knows. Runaway evolution can have unfortunate consequences for a species that is caught up in it, and we are not immune to those consequences. Runaway brain evolution, unlike other kinds of runaway evolution, can also have deadly consequences for all the other organisms that share the planet with us. The question we face is one that no species has faced before us: have our runaway brains made us smart enough to think our way out of this ultimate evolutionary dilemma?

Anyone who has the chutzpah to try to write a book that spans the last two billion years of our history needs a lot of help. This book has benefited immensely from discussions with or help from Giuseppi Attardi, Elizabeth Bates, Ursula Bellugi, Wesley Brown, Ted Case, Howard Chertkow, Ron Clarke, Jim Crow, Hilary Deacon, Marian Diamond, Niles Eldredge, Frances Gillin, Jeffrey Hall, Alison Jolly, Bill Kimbel, Ed Klima, Robert Livingstone, Philip Morin, Carol Padden, David Pilbeam, Trevor Price, Vilayanur Ramachandran, Paul Renne, Sam Ridgway, Vincent Sarich, Himla Soodyall, Ian Tattersall, Francis Thackeray, Phillip Tobias, Erik Trinkaus, Rose Tyson, and Alan Walker.

Elizabeth Bates, Ursula Bellugi, Wesley Brown, Ron Clarke, Jim Crow, Jeffrey Hall, Ed Klima, Carol Padden, and Vilayanur Ramachandran read parts of the manuscript at various stages and offered many corrections and insights. My editor, Susan Rabiner, kept the project on track. Ted Case, Philip Tobias, and Trevor Price cheerfully read the whole thing and helped greatly to pull it together. I shoulder the blame, of course, for errors that remain.

A timeline of major events in our evolutionary history and a map of important fossil discoveries are provided at the beginning of the book. Those who lose their way among the proliferating genes and fossils may find succor in the glossary.

THE RUNAWAY BRAIN

Introduction

The Himalayan, Moscow and Angora rabbits . . . are only a little larger in body and have skulls only a little longer, than the wild animal, and we see that the actual capacity of their skulls is less than in the wild animal, and considerably less. . . , according to the difference in the length of their skulls. The narrowness of the brain case in these three rabbits could be plainly seen and proved by external measurement.

—Charles Darwin, *Variation of Animals and Plants Under Domestication* (1868)

The Prisoners of Chillon

In 1963, when I was a graduate student in genetics at Berkeley, I was taken by a friend to the basement of Tolman Hall. There, the psychologists David Krech and Michael Rosenzweig were performing experiments on a genetically uniform strain of rats.

The rats had been divided into two groups shortly after weaning. The first group was kept in bare cages in a soundproofed, darkened room, illuminated only occasionally by a dim red light (rats can hardly perceive light at that wavelength). We tiptoed in to change their food and water. The light was too dim to see the rats, but we heard them rustling tentatively in their cages. Our task done, we tiptoed out again and closed the thick door behind us.

The second group of rats was in a brightly lit animal room, full of bustle as workers came and went. They were kept in cages crowded with playthings—balls, exercise wheels, and so on. These rats were far more active than the ones in the darkened room. They surged around inside their cages and interacted continually. My friend went to another much larger cage that had been set up to one side, containing a maze with movable partitions. She rearranged the partitions to close off some passages and open up others. Then she deposited some food in the middle of the maze.

We carried the rats' cages over to the big cage with the maze, put the smaller cages inside, and opened the lids. Immediately the rats poured out, a cheerful throng, investigating every corner of the big cage and beginning to dash up and down the passages of the maze. It did not take them long to find the food, for exploring the maze was a daily ritual. After an hour or two in this new and exciting environment, they were persuaded back into their own cages.

The contrast between the life-styles of these two groups of rats could not have been more stark. Krech and Rosenzweig had designed their experiments to determine what influence, if any, the environment might have on the development of the rats' brains. They had been inspired by earlier observations that the learning ability of rats was genetically malleable, and by other experiments that showed it was actually possible to detect biochemical differences in the brains of rats that had different learning abilities.

Investigations of the learning function of the brain had a long intellectual history on the Berkeley campus. Back in 1922, the psychologist Edward Tolman (after whom Tolman Hall is named) had set out to see what the genetic basis for intelligence might be. He started with a genetically heterogeneous strain of rats and tried to select from it rats that could easily solve a maze and others that could not. His experiments made little progress, and he turned the project over to a young student, Robert Tryon.

Tryon used more sophisticated genetic techniques and an elaborate automated maze. Within three rat generations he had managed to produce strains that were distinctly different both from each other and from the rats with which he had started. He named these selected strains maze-bright and maze-dull. During five subsequent generations of selection, the strains continued to diverge in their abilities. Tryon continued his selection for another twelve generations but obtained no further response.

The descendants of Tryon's maze-bright and maze-dull rats were maintained over the succeeding decades in the Berkeley rat colony, where they were often used as demonstration material in psychology labs or sent to other investigators around the world. In all the years that followed, the maze-bright and maze-dull lines retained their differences.

Meanwhile, the study of the chemistry of the brain had begun. At the same time that Tryon was selecting his rats, the German chemist Otto Loewi was discovering that the small molecule acetylcholine was important in the transmission of impulses from one nerve cell to another. Acetylcholine was the first and is still the best known of the more than fifty different neurotransmitter molecules that have been discovered. The Berkeley biochemist Melvin Calvin, who was then in the middle of his Nobel Prize–winning work on photosynthesis, knew of Loewi's work. He suggested to Krech and Rosenzweig that there might be differences in acetylcholine levels between Tryon's strains.

Using the primitive biochemistry of the 1950s, Krech and Rosenzweig ground up brains from both maze-bright and maze-dull rats and found that the maze-bright rats did indeed have more acetylcholine. It seemed that an abundance of this neurotransmitter was a marker for highly active brain cells.

As psychologists rather than geneticists or biochemists, however, they were more interested in the effects of the environment on the brain than in any genetic differences. This was why they had set out to discover whether changes in the environment could produce similar chemical changes in the brain. And the result was the two colonies that I visited in the basement of Tolman Hall, made up of rats that had been chosen randomly to live either the grim existence of the prisoner of Chillon or a life more rich and varied than even the most yuppified rodent parent might have been able to bestow.

The first measurements Krech and Rosenzweig made on these rats were the obvious ones. The rats' brains were tested to determine their acetylcholine levels, and indeed those from the enriched environment were found to make unusually large amounts of acetylcholine, just as the maze-bright rats had done.

The differences in the brains of the enriched and deprived rats were far more profound than simple differences in the levels of a neurotransmitter, however. Marian Diamond, the psychologist who has since taken over the project, knew that acetylcholine levels were a very crude measure of something as complex as brain development. The early results of Krech and Rosenzweig suggested that the brain might improve with exercise—like any other muscle, as the old joke had it. But did these differences in acetylcholine levels really indicate differences in brain function? One way to find out would be to look for other differences, perhaps alterations in the brain's very structure.

As Diamond and her students probed deeper into the brains of the deprived and enriched rats, they did indeed find many more differences. The thickness of the cerebral cortex, particularly the outermost layer, was noticeably greater in the rats from the enriched environment. As they examined thin sections of the brains under the microscope, they found even more fundamental differences. The enriched rats had more connections among their neurons, and indeed even the kinds of connections involved were subtly different. Some of these differences could be detected only at the electron-microscope level.

Diamond's group grew better at probing the rats' brains as the years went by and more techniques became available. A remarkable pattern soon emerged: the more detailed and sophisticated their investigation, the greater were the relative differences that they found between the stimulated and the unstimulated brains. For example, although the thicknesses of the cortical layers of the brains of the enriched and deprived rats differed by only a few percent, the enriched rats had far more connections among their neurons than the deprived ones did—on the order of 50 percent more. Had Diamond and her co-workers confined their measurements to the earliest crude and simple ones, they would not have realized how different the enriched and deprived brains really were.

They repeated the experiment in a slightly different way. This time, isolated rats were kept, not in dark dungeons, but in cages—still in solitary confinement—with opaque sides. The brain differences turned out to be just as great. It seemed to be the lack of company more than anything else that was producing the neural deficit in the isolated rats.

Next, Diamond wondered whether environmental enrichment could sometimes be too much of a good thing. Perhaps overcrowding and the resulting stress could produce negative rather than positive changes in the brain. Some years earlier, John B. Calhoun of the National Institutes of Mental Health had done some well-publicized experiments to investigate the effects of overcrowding on mouse populations. The crowded mice showed a variety of unusual reactions—females were unable to finish building their nests and often ate their young, and males showed a great increase in homosexual behavior. Both sexes spent far more time fighting than did the animals in the uncrowded cages. Might their brains also have shown the effects of this stress? Diamond tried to replicate Calhoun's conditions with her rats, but she could not crowd the rats enough (and still obey the new legal requirements for animal care) to have any effects on their brains. The crowded rats seemed to thrive. The complexity of the rats' environment appeared to be the key to their brains' development.

Poco a Poco Accelerando

My experience in the basement of Tolman Hall came back to me vividly years later, in 1990, in the course of a lunch with Bob Livingstone, a professor emeritus in the neurosciences department of the University of California at San Diego. Bob and I were talking about a lecture I was to give in his course on the nature of humanity, a vast topic to which I was to contribute a modest something on human genetics. Our conversation turned to the evolution of the brain, and he made a very interesting arithmetical point.

Our nearest living relative, the chimpanzee, is born after a gestation period of seven and a half months. The chimp essentially reaches adulthood at about age nine. Its physical skills at first develop quickly compared with those of a human baby, but then, as the skills to be acquired become more complex, the rate slows. A baby chimp can extend its arms and hold its head steadily erect at two weeks after birth; a human baby cannot do so until twenty weeks have passed. The chimp can walk on all fours at twenty weeks, a skill it takes a human baby forty weeks to acquire. But bipedal standing and walking require over forty weeks of growth and development in the chimpanzee, almost as long as the fifty-four to fifty-eight weeks required of a human baby. Once this dramatic developmental milestone has been reached the human baby spurts ahead, acquiring new motor and verbal skills that are quite beyond the chimpanzee's capabilities.

At the time of birth, the brain of a chimp is well developed and occupies a volume of about 350 cubic centimeters, about the same volume as a can of soda pop. During growth to adulthood, the chimpanzee's brain increases only slightly in size, to about 450 cubic centimeters. Indeed, there is little room for it to grow, for the fontanels of the skull are small and have essentially closed at birth. Our own brain, however, follows a very different course.

Just as with the chimp, our brain at birth is about 350 cubic centimeters in size. If it were much larger, then the birth process, limited as it is by the size of the mother's pelvic opening and the size of the baby's head, would be intolerably difficult and dangerous.

The three bones making up each half of the pelvis are called the ischium, the pubis, and the ilium. At birth they are still distinctly separate bones in both human and chimpanzee babies. But in a year-old chimpanzee they have fused together completely. Although the bones making up the chimpanzee's pelvis continue to grow in size, its shape is essentially determined at that point. In a human child, in contrast, it takes seven to fourteen years for complete fusion to occur. This long period of separation allows the bones not only to grow but also to change in their relative position. During our childhood development, the ischium and pubis move up and forward, in the process altering the whole shape of the lower abdomen and the angle at which femur and pelvis join.

Even after all this movement, which is essential to our upright posture, the opening through which the baby's head must pass has not grown any smaller in a human female's pelvis than in that of a chimpanzee. It has, however, become less circular in cross section. It is this that makes our birth process more difficult. If the pelvic opening were even slightly smaller or less circular, or if the baby's brain were even slightly larger, most human births would be impossible. The evolutionary trends of brain and pelvis are on a collision course, and the result is that our brains at birth are constrained to be no larger than those of chimpanzees. Runaway brain evolution may actually have dictated this collision, for the impetus toward developing ever larger brains seems to have been so powerful that it has for the first time produced organisms—ourselves—whose process of birth is prolonged and dangerous. Like Fisher's runaway sexual selection, runaway brain selection can have negative consequences. We have survived these consequences because through our acquisition of culture we have been able to ameliorate them.

After birth, the development of the human brain takes a dramatically different turn from that of the chimpanzee. Once it has been released from the constriction imposed by the size of the birth canal, the human brain explodes in size. By age four it has tripled in size. Although it develops more slowly thereafter, it eventually reaches about 1400 cubic centimeters, four times its size at birth.

Livingstone pointed out to me something that is quite obvious in retrospect. The human brain does most of its developing outside the womb, but the chimpanzee brain develops primarily within it. The consequences are profound. During most of the development of a human's brain, the infant is bombarded with stimuli from the outside world, stimuli that are largely denied to the developing brain of the chimpanzee. Indeed, they are denied to an even greater extent to the brains of our more distant relatives such as the loris and the potto. The brains of these lower primates have reached 80 percent of their adult size by the time of birth.

The effect of this long period of postpartum development on the human brain must be immense. One cannot help but think that once this process had begun, once the brains of our ancestors started to spend a greater proportion of their developmental time outside the womb, this fact in itself must have opened up many new opportunities for further evolution of the brain.

Extensive development of the brain after birth, however, is not the whole reason why our brains are unique. Dolphins tend to be larger and heavier than humans, but even a small dolphin with a body the same size as a human's has a brain about half as large as ours, and larger dolphins can have brains much larger than our own. Their brains typically have more convolutions on their surfaces than those of humans and a much greater area of cortex. Yet, insofar as their intelligence can be measured, they seem to be about as bright as dogs.

Dolphins, having lost their hind limbs, have tiny vestigial pelvic bones that do not interfere with the birth process. Freed from the tyranny of the pelvis, they can carry their young for a longer period than most land animals—twelve months is a normal gestation. When the young are born, they are relatively mature animals that can swim expertly almost from the moment of birth. A dolphin baby can be a meter long and weigh 12 kilograms. The brain at birth is already 50 percent of its adult size, and it swiftly reaches 80 percent within a year and a half, compared with four to five years for humans.

Nonetheless, a good deal of growth of the dolphin's brain does occur after birth, which should provide the same kind of stimulus to further brain evolution that the brains of our own ancestors must have received. Why has this apparently not happened?

It seems that an increase in the sheer size of the brain is not necessarily accompanied by increased complexity of function. The dolphin cortex, although extensive, is a much thinner layer of tissue than our own. It is made up of neurons that have much simpler structure than the neurons in our own cortex. Seals, with a very similar and perhaps even more extensive range of behaviors and abilities, have much smaller brains. It may be that the dolphin's brain has, in part, simply expanded to fill the space available for it, and indeed the swiftness with which it reaches its adult size suggests that this may be so. To evolve a human-like brain, it seems, not only must there be a longer exposure to the environment during development, but the environment must be a highly complex one.

Of course, were dolphins able to begin creating such a complex environment, there would be nothing to prevent them from entering a gene-culture feedback loop like our own.

In a way, the experiments of Krech and Rosenzweig reproduced some of the factors that must have led to the different developmental paths taken by the brains of humans and other animals. Although the brains of their rats had largely completed their development before the experiment was begun, enough plasticity remained for environmental stimulation to have a great effect. Almost certainly, the differences between the stimulated and unstimulated brains of the rats would have been even greater if it had been possible to begin the experimental regimes of stimulation or deprivation at the moment of birth rather than at the time of weaning.

In our prehuman ancestors, the combination of constricted fetal brain development and increasing adult brain size meant that more and more development had to take place outside the womb. As a result the fetus was flung ever earlier, progressively more naked and unprepared, into the world. As these babies grew up, intense selective pressures must have acted on the genes controlling the development of their brains, far more intense than would ever have been brought to bear if they were still isolated in the womb. Such selective pressures would have been very different from those that shaped our teeth, or our pelvis, for they were in the form of *intellectual* rather than physical challenges.

We know that at the same time as our ancestors' brains were growing larger, their posture was becoming more upright, fine motor skills were developing, and vocal signals were graduating into speech. As a result, interactions among members of the group were becoming progressively more complicated. And the very inventions that a few of our ancestors made—fire, clothing, shelter, new words and concepts, new kinds of tools—had the result of making the environment of every member of the group far richer and more varied. The members of the group or tribe best able to take advantage of this increased environmental complexity were the ones most likely to leave their genes—including genes for accelerated and altered development of the brain—to the next generation. And a small fraction of these survivors in their turn, by their social and technological inventions, made the environment even richer and more challenging.

Because of tool use, language, and developing social interactions, the environments into which our ancestors were born soon became far more complex than those of other animals. Indeed, as the pace of social and cultural evolution accelerated, the environment of each generation became progressively more and more different from the environment of the generation preceding it—to the point where now in industrialized societies the environment of one generation is profoundly, startlingly, different from that of the previous one. Now, indeed, most of us live much of our lives completely out of touch with the natural world.

The rapidity of this change, and the quality of the change itself, are unique. There has never been anything like it in the history of life on our planet. Of course, environmental change is one of the essential requirements of evolution. Yet for the most part such change has come slowly. As we look at the whole history of life, we see that each successive layer of deposits in the geological record is filled with the remains of more and more species of increasing variety. This is because the opportunity for complex interactions between species has slowly increased over spans of tens or hundreds of millions of years, and has given rise to new ecological niches that support yet more species. In our present world this has resulted in ecosystems of incredible intricacy, such as tropical rain forests and coral reefs. Yet all this change took place over enormous spans of time.

Of course there have often been periods of very rapid environmental change, followed by rapid bursts of evolution. One of the most dramatic of these changes was the sudden extinction of the dinosaurs almost exactly sixty-five million years ago. Evidence is now very strong that this was the result of the impact of a 10-kilometer-wide asteroid on what is now the Yucatán peninsula. That catastrophic event seems to have killed off most land animals more than a few kilograms in weight, along with much of the ocean's plankton. It abruptly and dramatically simplified the complex ecosystems of the time, causing many previously occupied ecological niches to be vacated and even opening up a number of brand-new niches. The subsequent adaptive radiation of the mammals to fill these empty niches, eventually resulting in the appearance of animals as different as whales and bats, was relatively swift in evolutionary terms. Still, it took place over millions of years. And during that time, once the planet itself had recovered from the trauma of the collision, the environment remained relatively constant. Ecological niches that had been left vacant by the demise of the dinosaurs simply awaited the mammals that would fill them. Mammals did not create them, although undoubtedly they did modify them somewhat.

It is quite safe to say that never before the appearance of humans, even during the traumatic periods that followed mass-extinction events, has there been the kind of generation-by-generation acceleration of environmental change that our recent ancestors imposed on their environment. So it is not surprising that our evolution has a unique pattern, in its swiftness and consequences unlike anything else that has taken place in the three-and-a-half-billion-year history of life on our planet. If this feedback loop is real, it does much to explain why our own evolution has been so clearly different from that of other organisms, why it has been driven with such apparent single-mindedness toward what seems to us to be the goal of *Homo sapiens sapiens.*

Of course, the ideas I have just recounted here are not new ones, although I intend to examine them in what I hope are fresh ways. Many evolutionists and geneticists have suggested a feedback between the environment and the development of our brains. Some have made brief but cogent encapsulations of the idea. For example, in their *People of the Lake* (1979), Richard Leakey and Roger Lewin paint a very similar picture:

> As the direct descendant of *Homo habilis, Homo erectus* had built up an evolutionary momentum that was to propel it inexorably toward, first, primitive *Homo sapiens* and ultimately to modern man. The emergence of the basic grade of *Homo sapiens* was probably around half a million years ago, perhaps first in Africa, or in Eurasia, perhaps in many different places at about the same time. The complex interaction of physical and intellectual capabilities within a self-created framework of culture probably operated on many populations of *Homo erectus,* urging them to the *sapiens* state. Not a divine guiding hand, but a biological inevitability.

The most lengthy exploration of this viewpoint was by Charles Lumsden and Edward O. Wilson, who detailed the possible consequences of a brain-culture feedback in their book *Genes, Mind and Culture* (1981). But they erred by making the feedback too direct and too specific. They began by postulating features of prehuman or human culture that they called *culturgens,* features that could be as simple as particular tools or as complex as grammatical forms of language. Then they supposed that the appearance of these features would have a direct influence on the frequencies of particular genes in the population, selecting for some of these genes and against others. They used the mathematics of population genetics to show that, provided there was such a direct connection, there could be a rapid change in the genetic composition of the population. And they pointed out that once this had happened, the genetic changes resulting in more complex and powerful brains could lead to the invention of yet more culturgens.

Lumsden and Wilson were criticized unmercifully for two reasons. First, they were castigated by many social scientists and some biologists for even suggesting that there could be a genetic basis for cultural behavior. Indeed, that point of view had already landed Wilson in hot water when he explored it in his book *Sociobiology* (1975). His critics had immediately pointed out that there was very little evidence for a genetic component to behavior or intelligence in humans—something that happened to be quite true at the time. Further, most of the small amount of evidence then available was highly suspect.

The critics of the Lumsden and Wilson book were quite right to complain that the authors had built their argument about the evolution of human behavior on shaky evidence, but the critics themselves ignored all the evidence from other organisms that showed genetic components to behavior, evidence that Wilson had painstakingly amassed in his earlier book. And the critics were also ignoring one central fact: if there really are no genetic differences that contribute to differences in behavior and intelligence among individual humans, then it is difficult to see how the human brain could have evolved in the first place!

The second point on which Lumsden and Wilson were criticized was their suggestion that there could be a direct connection between culturgens and genes. Of course, taken to its extreme, such a direct link can sound pretty ridiculous. Did Lumsden and Wilson mean to suggest that there is a gene for being able to program a VCR, and if you don't have it you are doomed? Of course not—they tried to be more sophisticated than that. They pointed out repeatedly that any relation between culturgens and genes was likely to be a complex one. But when it came to examples, they quickly became prisoners of their mathematical models. Again and again in the course of their book they backed up their arguments with simple models in which a particular culturgen influenced the genetic makeup of their hypothetical populations. This was done through postulating direct selection for and against a few *alleles*—alleles are different forms of a particular gene—at a single genetic location or *locus* on a chromosome. Selection would make these alleles rise or fall in frequency in the population, and as a result the gene pool of our species would be changed. Although they briefly flirted with the idea of models in which many genes could interact with a culturgen, they quickly retreated to more mathematically tractable examples.

In another book published at the same time (*Cultural Transmission and Evolution: A Quantitative Approach*), Luigi Luca Cavalli-Sforza and Mark Feldman of Stanford University also treated the evolution of culture in a mathematically sophisticated way. They built complex models that again drew from population genetics to show how cultural transformation can actually spread faster than genetic change. But they did not explore the possibility of a feedback loop between genes and culture, and indeed they specifically ruled out the idea that culture could have much of an influence on the gene pool. They succeeded in sidestepping the political problems that faced Lumsden and Wilson, but their caution prevented them from exploring many of the consequences of cultural evolution.

For many reasons, a few of which we have already explored, I find myself rather sympathetic to the Lumsden and Wilson viewpoint—but only after considerable modification. Throughout this book, I will return repeatedly to the theme of a genotype-culture feedback loop, because I think that it explains so much of the apparent uniqueness of human evolution. But I will try to avoid falling into the "one-gene–one-culturgen" trap, with all its potential for ridicule, that gaped for Lumsden and Wilson. And I also want to explore, and if possible avoid, another trap that lurked in waiting for them. That trap is less obvious.

The difficulty arises when we consider the centrally important matter of the nature of the human gene pool, and the very different views that different evolutionists have of it. In generating their mathematical models, Lumsden and Wilson blithely drew on our gene pool as if it were the Bank of Commerce and Credit International (BCCI). That bank, as you know from newspaper accounts, was able to make huge loans to its favored customers by drawing on largely imaginary assets. The bank was able to fool outside auditors into thinking that its assets were much larger than they were. This deception was carried out by rapidly transferring into any country in which an audit was to take place whatever slender real assets BCCI possessed in other countries.

Lumsden and Wilson began by bedecking the environment with culturgens. Then they dipped into the gene pool and lo! they were able to find just the genes they wanted. These genes, they supposed, could then be selected through their interactions with the culturgens. How convenient. Did these genes arise by recent mutations? Unlikely, since most new mutations tend to be harmful. All our genes have been shaped and honed by a long history of natural selection, so most random alterations are likely to make the genes' products work more poorly than before. Occasionally, of course, new mutations will alter the function of such genes in a way that benefits the organism, but new mutations cannot produce *entirely new genes*. The evolution of a new gene takes very long periods of time, and needs the contribution of many different mutational events and much selection.

It does not help the Lumsden and Wilson model to suppose that the genes interacting with these new culturgens were actually older than the culturgens, that they already had a long evolutionary history. After all, why should those genes have been there even before the appearance of the culturgen that could select for them?

If we are to assume that the gene-culture feedback loop exists in some form, then we must examine the nature of the human gene pool that helps feed it. Does the gene pool somehow, like BCCI, produce just the genetic assets that are required as our culture-driven feedback loop accelerates? Or is it, like a more stable and cautious bank, provided with assets that are sufficiently rich and diversified that they can power our current feedback loop and perhaps to spare? If so, then where did those genetic assets come from? Did they originate well before the days of VCRs or even of stone tools? What were they doing during all that time before the emergence of culture? Bear these questions in mind as we confront the singular problem of human evolution.

I

THE DILEMMAS

1

Eve's Companions

And Cain went out from the presence of the Lord, and dwelt in the land of Nod, on the east of Eden. And Cain knew his wife; and she conceived, and bare Enoch.

—Genesis 4:16–17

CLARENCE DARROW: Mr. Bryan, do you believe that the first woman was Eve?
WILLIAM JENNINGS BRYAN: I accept the Bible absolutely.
DARROW: Do you believe she was literally made out of Adam's rib?
BRYAN: Yes.
DARROW: Did you ever discover where Cain got his wife?
BRYAN: No, sir, I leave the agnostics to hunt for her.

—Transcript of Scopes trial (1925)

In this section I want to introduce some of the major controversies about human evolution. Some of these are new and some are old, but all of them—as yet—are still unsettled. I will begin with one in which scientists are wrestling with hard genetic data and end with one that trespasses into the domains of philosophy and religion.

In recent years, no discovery about human evolution has captured the public's imagination more than the story of the mitochondrial Eve. This excitement made it all the more disillusioning when the news appeared that the story might not be quite right. To understand what the scientists have been saying, and where they may have gone wrong, we have to go back to an era long before the time when the mitochondrial Eve herself lived.

Approximately two billion years ago, in the middle of that immense span of time known as the Precambrian era, an event occurred that would transform the planet and most of the creatures living on it.

During most of the Precambrian, which lasted from three and a half billion down to some seven hundred million years ago, all living creatures were single-celled, simple in both their structure and their living habits. Although they often gathered in sheets and layers on rocks or at the bottoms of tide pools, these layers were nothing more than collections of single cells that had congregated in areas rich in food or light. Such simple cells did not—could not—give rise to the complex, multicellular, multitissued organisms that make up so much of life today.

Then this evolutionary barrier was broken by an infection. A race of parasitic bacteria, distant ancestors of the green and purple sulfur bacteria that now color the hot springs of Yellowstone, invaded the cells of another group of simple, single-celled organisms. After they had come together, these newly combined organisms were able to enter a new universe of possibilities. The bacteria themselves managed to retain much of their own identity, even after they had entered their host cells. They eventually evolved into mitochondria, tiny sausage-shaped structures that are found within most living cells today, and which are their biochemical powerhouses. Once the invasion had taken place, the simple organisms that harbored these bacteria were able to use the power that the bacteria provided. As a result they could evolve into new kinds of organisms that were made up of many cells. Some of their remote descendants eventually became ourselves.

The scenario for this and other *symbiotic* events in the early evolution of life has been worked out in great detail by Lynn Margulis of Boston University. The possibility that the ancestors of mitochondria might have invaded their host cells in the distant past had actually been suggested in the last century, indeed almost as soon as mitochondria were first seen under the microscope. This speculation was at first based on the simple observation that mitochondria *looked* like bacteria! But much genetic and biochemical evidence showing that this invasion really had taken place has been amassed since.

If you could have observed this event at the time that it happened, you would probably not have been impressed with its evolutionary potential. It is unlikely that the bacteria and their hosts came together naturally like ham and eggs. Instead, this parasitism that eventually turned into symbiosis—the term means living together with mutual benefit—probably began as a disease. Even today, it is not unusual for a harmful bacterium to live inside the cells of its host. The rickettsias that cause typhus and Rocky Mountain spotted fever first attach to the outer membranes of the cells of their hosts and then penetrate the cells' interiors. There they live off the host's energy supply and, shielded from the outside environment, multiply in safety. The difference between a rickettsial infection of today and the ancient infection that eventually led to the evolution of mitochondria is that the ancient parasite brought with it a new energy-generating capability. In this regard it was so superior to its prehistoric host that the benefit soon outweighed any damage that the parasite might have inflicted.

This ancient parasitic bacterium was very different from its host. It was surrounded by a very special membrane that had various enzymes embedded in it and that had a strong difference in charge between its inner and outer surfaces. This charged membrane and its enzymes were capable of precise surgery on molecules. Single molecules of the two compounds adenosine diphosphate (ADP) and phosphoric acid could be gathered up by the membrane and oriented in such a way as to force them together to form the energy-rich molecule adenosine triphosphate (ATP). In the process, two pieces of a water molecule were extracted from the newly forming compound, then flung away in opposite directions along the gradient of charge across the membrane. This process took a lot of energy, supplied by a strong electron flow among the molecules embedded in the membrane and driven by the great pulling power of free oxygen. In effect, the parasites could breathe oxygen—their hosts could not.

Because the parasites could breathe, they could produce ATP more easily and in far larger amounts than their hosts. It was this huge new source of ATP, a molecule essential for most of the processes that go on in a living cell, that opened up wonderful opportunities for the host-parasite combination. In their remote descendants, the ATP-producing membrane has become the inner membrane of the mitochondrion (the singular form of the word—the plural is mitochondria). Of course, even as long as two billion years ago the ability to breathe oxygen and produce massive quantities of ATP was not a new evolutionary trick. Both the bacteria and their hosts had probably evolved separately for a billion years before they came together.

We can only speculate about what the hosts themselves might have been like, but there are a few hints. Frances Gillin, at the University of California, San Diego, medical school, has intensively studied tiny single-celled intestinal parasites called *Giardia.* These parasites, which are able to live in a number of mammalian hosts, have now contaminated many of the formerly pristine streams of California's Sierra Nevada, and they can give violent cases of diarrhea to hikers incautious enough to drink untreated water. Remarkably, even though *Giardia* have many of the properties of the cells of higher organisms, they and a number of other intestinal parasites lack any sign of mitochondria.

Did they have mitochondria once and then lose them, or did they never have them? Gillin thinks that they never had them, though there is yet no clear-cut molecular evidence to back up her hunch. Other scientists disagree. But if *Giardia* do turn out to be the remote descendants of cells that predate the arrival of mitochondria, then those cells must already have had many of the features that our own cells exhibit—a nucleus with chromosomes, flagella for swimming, and various ways to move proteins around in the cell that bacteria do not have. So there is a good chance that the host cells also brought important abilities to this emerging symbiosis. Lynn Margulis has marshaled many arguments suggesting that, even by the time the ancestors of mitochondria arrived, the host cells were already the result of many earlier symbiotic unions of simpler organisms. She thinks that much of the early evolution of cells can be explained by those dramatic events.

Every multicellular plant and animal on the planet today has mitochondria, and all of them can be traced back to those original host cells. And, just as animals and plants have undergone great changes in the course of their descent from that early host, much has happened to the descendants of the original bacterial parasites. For billions of generations they have been safely ensconced in the cytoplasm of the host cells, where they provide energy for growth and for the manufacture of the proteins and fats that are essential to the host cell's survival. Over this great span of time they have become small, simplified, and utterly unable to live by themselves. Now they are so thin-skinned and delicate that, unless special precautions are taken, they explode when they are removed from their host cell, just as expensive caviar explodes on the tongue. And their genetic information has become simplified as well.

When they first came to live in the host cells, each of the ancestral bacteria carried a large and complex chromosome. This chromosome was in the form of a long double helix of deoxyribonucleic acid (DNA) that was joined at its ends to form a huge ring and that carried the thousands of genes the bacterium needed to carry out a free-living existence. Some of these genes have since been lost. Others have been transferred away from the mitochondrial chromosome, out of the cytoplasm entirely, and into the nucleus of the cell where they have been inserted into the chromosomes of the host. They still continue to do their jobs from this more remote location, by making the proteins and other molecules from which the mitochondria are constructed. All that is left inside the mitochondrion is a far smaller ring of DNA, less than a hundredth as large as the original parasite's chromosome.

Each of the two strands of the double helix of a DNA molecule is made up of a string of *bases*, joined together by sugars and phosphate molecules. These bases are of four types: adenine, thymine, guanine and cytosine, abbreviated A, T, G, and C. The sequence of the bases along a strand determines the information that the DNA carries. Obviously, long strands of DNA can carry more information than shorter strands. Our chromosomes are made up of very long strands of DNA indeed. Some of our chromosomes are as much as two hundred million bases long and carry the information for thousands of genes. The chromosome carried by our mitochondria is tiny by comparison, a little over sixteen thousand bases long (although its ends are still joined together to form a ring, just like its remote ancestor's chromosome). In most animals, it has precisely thirty-seven genes, far too few to carry all the information for constructing the mitochondrion itself. All the others have been transferred to the chromosomes of the host.

Since most of the original genes of the mitochondrial chromosome have been transferred, why did this process come to a halt when thirty-seven genes still remained? Indeed, why does the mitochondrion still have a chromosome at all? We simply have no idea why this truncated chromosome, a kind of genetic vermiform appendix, has persisted for so many billions of generations. But, for whatever reason, this vestigial little chromosome of the mitochondrion lives on, sequestered from the rest of the genes in the cell. It has existed apart for so long that, like groups of people living in the remote valleys of the Ozarks, it has even acquired its own dialect. The way the genetic information is coded into its DNA is now slightly different from that of its host.

These little chromosomes are passed from one generation of their hosts to the next, but by very different rules from the ones that govern the hosts' chromosomes. Still carrying a tiny collection of genes that are, though they are much changed, lineal descendants of the genes of their remote Precambrian ancestors, they traverse the generations largely intact.

It is this fact that has provided scientists with a tool for probing the genetic history of the human species. And at the same time, these innocent mitochondria have been instrumental in reopening an angry controversy about the origin of humans, a controversy that has pitted scientists against each other and that has—at its ugliest—raised the specter of the possibility that there might be a genetic basis for racial inequality.

The Mitochondrial Family Tree

The great majority of our cells are filled with mitochondria, a thousand or more sometimes crowded into a single cell. Thus, when we come to the end of our lives and the ten trillion cells that make up our bodies die, we actually die a collective death. Our enormous multitude of mitochondria, the remote descendants of those Precambrian bacterial parasites, all die at the same time as we do.

Not all our genes, of course, necessarily die along with us. We can achieve some measure of genetic immortality by bequeathing some of the chromosomes in the nuclei of our cells, along with their genes, to our children. Each child receives half the chromosomes of its father and half the chromosomes of its mother. Thus, both father and mother contribute roughly equal amounts of genetic information to the next generation.

A few of the mitochondria, too, can escape the doom awaiting the great majority of their companions by being passed on to the next generation, carrying with them their little loops of DNA. The mitochondria, however, are confined to the cytoplasm of the cells, not the nuclei. Only if a little of this cytoplasm is passed on to the next generation can a few of the mitochondria be passed on as well.

Here is where the unusual nature of mitochondrial inheritance becomes obvious. The egg passes cytoplasm filled with mitochondria to the next generation, while the sperm does not.

Sperm start out as normal cells, but they quickly lose most of their cytoplasm and shrink down to become little more than a package of chromosomes with a frantically waving tail attached. Most of the mitochondria are lost as well, and the few that are left are located at the base of the sperm's tail where they supply the ATP needed to drive the sperm in its single-minded quest for the egg.

Eggs are huge balloonlike cells, with a voluminous cytoplasm richly endowed with all the materials needed to start the embryo's development after fertilization. When sperm and egg fuse, the ordinary nuclear chromosomes carried by the sperm are passed safely to the egg. But the shreds of mitochondria in the sperm are usually left outside the egg, along with their tiny mitochondrial chromosomes. Almost always, only the plentiful mitochondria of the egg, carrying their little chromosomes with them, are passed on to the child.

If the child is a girl and has children of her own, her mitochondrial chromosomes will be passed on—at least to the following generation. If the child is a boy, then whether or not he has children, his mitochondrial chromosomes will be lost.

As a result, the inheritance of mitochondrial chromosomes is like the inheritance of surnames in western Europe and America, except that they are passed down the female line instead of the male. Like surnames, mitochondrial chromosomes can be lost or can increase in number in the population. And, like surnames, they can change. Surnames can change as a result of societal pressure—for instance, England's ruling house of Hanover changed its name to Windsor during World War I. Names can change even by mistake—my own surname can be traced back to eighteenth-century Scotland, where it had been written *Will*, but somehow the terminal *s* was added at around that time. (The change happened only in my family, so I must have plenty of more remote relatives named Will.) Once such a change has happened, it may persist for many generations before the name is changed again.

The changes in mitochondrial genes are the result of mutations of their DNA, and they also can persist for many generations. However, both surnames and nuclear genes have avenues of change open to them that are closed to mitochondrial genes. For instance, surnames can combine with each other. It is rather out of fashion now, but double-barreled surnames drawn from both sides of the family used to be quite common among the upper classes. These names were constructed in an attempt to keep relatives on both sides happy. Genes in the nuclei of our cells can change in a somewhat different way—they can *recombine* with each other by the process of genetic *crossing over.* Our mitochondrial chromosomes, however, seem not to have either of these options. Of course, because they come only from the mother, there is ordinarily never a time when two very different kinds of mitochondria share the same cell. But even when two very different kinds of mitochondrial chromosomes are forced to coexist in the same cell, they seem never to recombine. Giuseppi Attardi of the California Institute of Technology has conducted an extensive search for such recombinants and has not found them.

The slow accumulation of mutations without subsequent recombination means that over time the mitochondrial chromosomes in various human families gradually diverge and become more and more distinct with time. If two mitochondrial chromosomes are passed separately down through the generations for millions of years, they gradually become more and more different, until eventually they become as different as they can be and still retain their function.

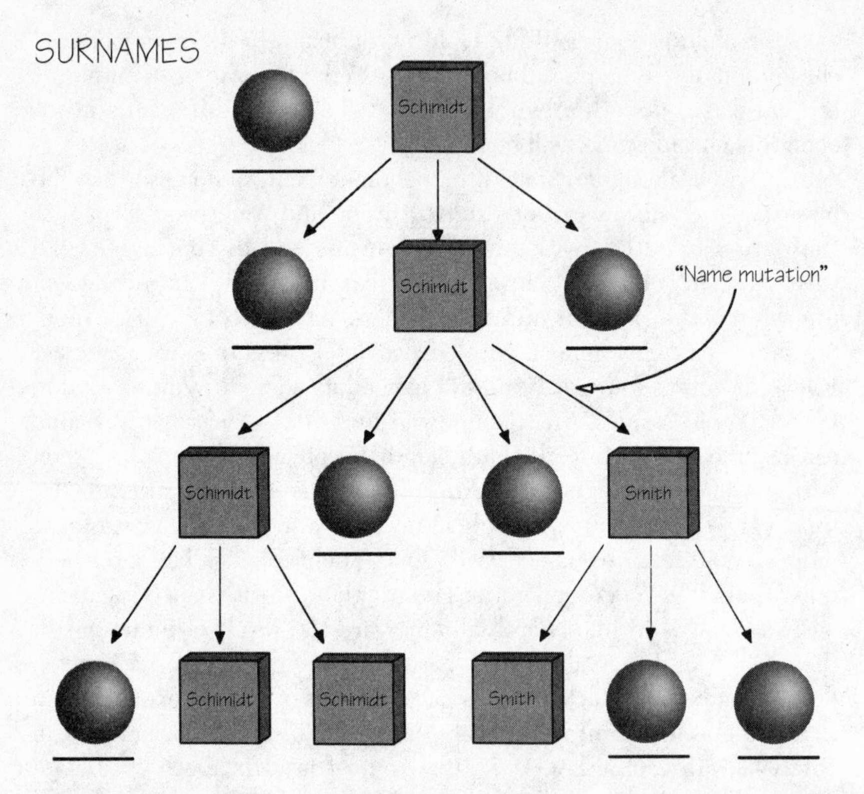

FIGURE 1.1. Three kinds of inheritance. As is traditional in human genetics, males are represented by squares and females by circles. Surnames are inherited down the male line because a woman who marries and changes her name does not pass her original surname on (her inability is represented here by a heavy line beneath her symbol). "Mutations" can happen to surnames for a variety of reasons. Further, all representatives of a surname can be lost if in a given generation none of the males carrying it produce male offspring. On the other hand, quite by chance, a surname can become more frequent in the population over time.

Mitochondrial chromosomes are inherited in a similar fashion, but are passed down the female rather than the male line. Here, mutations that occur in the DNA are passed down through the female descendants, but each mutation is passed only to the direct descendants of the woman in whom the mutation occurred. Males have mitochondrial DNA as well, but, as indicated by the heavy lines, they are unable to pass their copies on to their descendants.

Nuclear genes can have a more complicated inheritance, some examples of which are shown in the third part of the figure. A_1–A_5 are different *alleles* of a particular gene. Here, both parents can pass alleles on to their offspring. Like mitochondrial DNA, alleles of nuclear genes can mutate, but they can also recombine, as is shown here. Alleles A_3 and A_4 have recombined to give rise to allele A_5, which is different from either of them. Such events make it impossible to trace our nuclear genes back to a single nuclear gene ancestor.

Mitochondrial DNA

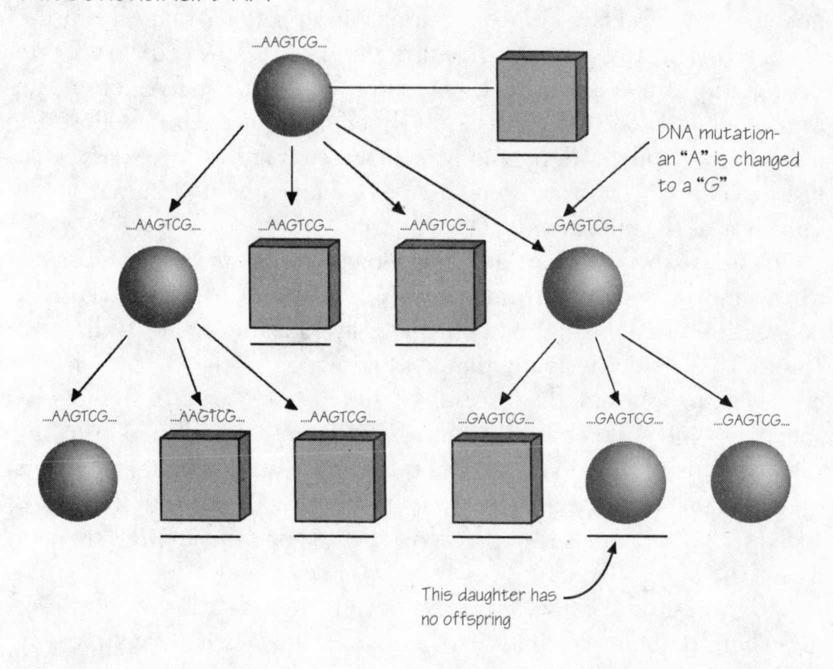

Nuclear Gene

A_1A_2 A_2A_3

Recombination between A_3 and A_4 produces a new allele, A_5

A_1A_3 A_1A_2 A_1A_2 A_1A_3 A_3A_4

A_1A_1 A_3A_1 A_3A_2 A_1A_3 A_3A_5

Just like mutations in the genes carried by the chromosomes of the nucleus, the effects of mitochondrial mutations can range from the severe to the invisible. For example, Leber's hereditary optic neuropathy is marked by a degeneration of the central region of the retina and its associated nerves. And although it can occur in both sexes, it is inherited only from the mother. An affected father never passes the disease on to his offspring. The disease, then, is inherited in exactly the same way as the mitochondrial chromosomes.

Of course, this inheritance pattern alone does not mean that a change in the mitochondrial chromosome causes the disease, but a direct connection was actually made in the mid-1980s by Doug Wallace of Emory University. He found that, on the mitochondrial DNA molecule carried by people with the disease, a G has been changed to an A at one specific point. This change is enough to damage the function of the mitochondria of these patients, but the damage is very slight and affects only cells that are especially active metabolically. The cells in the center of the retina are among the most active in the body and are thus the first to feel the effects.

The potential for mutation somewhere among the multitude of mitochondrial chromosomes carried by our species is very high. The total number of these chromosomes in the bodies of the more than five billion people on the planet is astounding, represented by the number 5 followed by twenty-six zeros! In each generation a small fraction of them mutate, but most of these new mutants die when the person carrying them dies. Only a tiny minority of these mutated chromosomes is passed on to the next generation. We feel sure that some of these mutations must be even more severe than the mutation that causes Leber's neuropathy, but they are never found because anyone unlucky enough to carry mitochondria with such a harmful mutation will die of its effects.

Not all the bases in the mitochondrial DNA are so critical, however. At least half of them can change into other bases without any noticeable effect at all. Wesley Brown, a postdoctoral fellow working in the laboratory of Howard Goodman at the medical school at the University of California, San Francisco, at the end of the 1970s, was the first to find traces of these invisible mutants. He began looking for them because of his clear understanding of how mitochondrial DNA was inherited. He reasoned that this unusual mode of inheritance should make the mitochondrial chromosome an ideal tracer with which to follow the course of human evolution.

Because mitochondrial DNAs do not recombine with each other, each human's mitochondrial chromosome retains a record of that person's evolutionary history. Two closely related people have a good chance of carrying the same mitochondrial chromosome, just as they have a good chance of having the same surname. But more distantly related people carry chromosomes that differ by the number of mutations that have accumulated during the generations that have separated them. This is the equivalent of somebody named Smith in America having a distant relative named Schmidt back in Germany. Both are descended from ancestors named Schmidt, but the American name has undergone a mutational change.

Surnames, of course, have many independent origins, most of them dating from medieval times and created by the need of governments to keep track of their citizens for taxation and impressment into the military. They may come from occupations—Smith or Taille-fer, Skinner or le Peletier. Or they may have become attached to people because of where they lived—people residing near the village green were often called Green. Brown's own name probably derived from the fact that some of his ancestors had dark complexions or brown hair.

Brown realized that, unlike surnames, all the mitochondria of the human species must have only one origin. This meant that, if he had a time machine, he could trace all the mitochondria of the human species back and back through time, discovering on the way various ancestral types that knit together the millions of different separate lineages into fewer and fewer, until he finally reached the ancestral mitochondrial chromosome from which all the present-day mitochondrial chromosomes are descended. In the process, he might reach so far back in time that the creature carrying that ancestral chromosome was not even human.

Brown wanted to employ present-day mitochondria as a kind of surrogate for a time machine and use them to build a speculative but roughly accurate family tree in which they would form the tips of the branches. Because he could not be sure of what had happened in the past, the details of his tree might be wrong, but he hoped to be able to determine its general shape—the ways in which twigs could be traced back to branches and eventually the branches back to the trunk. To make a good tree, he had to get information about a lot of present-day mitochondria taken from many different people from around the world.

In his search he was lucky on a number of counts. Working in a state-of-the-art lab, he shared the same bench with people on the forefront of DNA research. And he was working with a molecule that was beautifully suited to analysis by the techniques of the day. Proper analysis of any

DNA molecule requires that it be available in fairly large quantities and that it be free of contamination by any of the many other types of DNA present in the cell. These days this is done by cloning the gene or by magnifying it by the remarkable technique known as the polymerase chain reaction (PCR). But at that time, even though gene cloning was still in its infancy, the mitochondrial DNA had already effectively been cloned for him because of the sheer numbers of mitochondria in the cells of the body. And it could easily be purified by separating it from the far more abundant but also far more complex DNA of the nucleus.

The DNA molecules of the chromosomes in our cell nuclei are *linear*, enormously long strands that stretch from one end of the chromosome to the other. The DNA of the mitochondrion, however, forms a little circle, and the circle has been modified even further. You can visualize this by picking up a rubber band and twirling it between your fingers. As you put extra twists in the circular rubber band, it knots up into a tangle. In a similar way, certain enzymes put extra twists in the DNA of the mitochondrion, knotting it up into what DNA chemists call a *supercoil*.

Supercoils can be separated from the rest of the DNA in the same way that you might separate short and bulky rotelli pasta from long and slender vermicelli by allowing a mixture of these two kinds of pasta to settle in water. The rotelli, even though its density is the same as that of the vermicelli, reaches the bottom of the container first. Using what was in essence this simple technique, but aided by a very powerful centrifuge, Brown was able to separate quantities of mitochondrial supercoils from contaminating nuclear DNA.

At just about this time, Fred Sanger and his colleagues at Cambridge were working out the last details of the sequence of the entire human mitochondrial DNA molecule, all 16,569 bases of it. This biochemical tour de force would help Sanger win his second Nobel Prize (see my *Exons, Introns, and Talking Genes* for the details of this remarkable story). Brown, of course, was not interested in the properties of a single mitochondrial DNA molecule, but rather in how many differences there were between one human mitochondrial DNA molecule and another. It was quite beyond his resources to sequence the whole length of the mitochondrial molecules from many different humans—he knew that Sanger's very large group had put in dozens of person years of backbreaking effort into the sequencing of a single one. So he compromised, by extracting a little bit of information from each of a large number of different molecules.

It was lucky for Brown that *restriction enzymes* had recently become available. He could use these enzymes to attack the mitochondrial DNA and find out something about its sequence. Many enzymes are known that degrade DNA, reducing it to the molecular equivalent of a heap of rubble. Restriction enzymes, however, are far more specific. Each restriction enzyme has the remarkable ability to recognize a certain sequence of bases in the DNA and then snip the DNA precisely at that point. A great many of these enzymes, manufactured by bacteria, are now known. Brown used some of them to cut his isolated mitochondrial DNA. Then he sorted the resulting fragments by electrophoresis, through which an electric current separates out different sizes of molecules by forcing them to travel the length of a Jell-O-like gel. The smaller fragments tend to travel more quickly through the gel.

If he used an enzyme called *HpaI,* made by a bacterium named *Hemophilus parainfluenzae* that can cause inflammation of the pericardium, he knew that it would recognize a particular sequence of six bases of the double helical DNA molecule, and only that sequence:

⇓
. . . GTTAAC . . .
. . . CAATTG . . .
⇑

(G is always paired with C, and A with T
across the two strands of the double helix.)

The little arrows show where the enzyme cuts. This exact sequence turns up in only a few places along the length of the mitochondrial chromosome. When the average mitochondrial chromosome is treated with this enzyme, it falls into approximately four pieces, with sizes ranging from 2,400 to 5,700 bases. Brown quickly found that, if he cut the mitochondrial DNA from many different humans with this enzyme, most would fall into four fragments but some fell into more or fewer.

In these unusual chromosomes a fragment would be broken into two, or two fragments would be joined together into one. This meant that the enzymes were revealing occasional DNA sequence differences among the chromosomes, differences that had been caused by mutational changes.

How could one fragment break into two? This could happen if, somewhere in the fragment, a sequence of six bases like GTTATC, differing only by one base from the sequence recognized by HpaI, had changed by mutation into GTTAAC. This mutation would create a new site for the enzyme to find. And two fragments could be joined into one by the reverse of this process—a GTTAAC sequence that the enzyme could recognize might change into ATTAAC, a change that would then make this region of the DNA invisible to the enzyme.

Each enzyme, of course, allowed Brown to examine only a small fraction of the entire mitochondrial chromosome. Even though he could not be sure exactly what the differences were among the various chromosomes, at least he could get an approximate idea of their number. When he used more enzymes, he got a better estimate of the number of mutational changes that had taken place. Soon, he had enough information to draw some tentative conclusions, although not yet enough to construct a convincing and detailed family tree.

The first thing he noticed was that there was not a great deal of variation from one human to another. Using the information from his enzymes, he could calculate the average number of differences between any two human mitochondrial chromosomes picked at random. This average was, he found, only about one base in every three hundred.

This close resemblance among human mitochondrial chromosomes meant that they all must have had a fairly recent common ancestor. Sometime in the relatively recent past of our species lived a woman, perhaps a true human or very close to a true human, who carried mitochondria from which all the immense numbers of mitochondria possessed by humans today were descended. The question was: When did she live? And, if he could build a detailed enough family tree, could the shape of that tree provide a clue to where she had lived?

To build this tree he had to gather more data, from other species as well as from humans. And he had to go through calculations that were fraught with possible errors, errors that would soon come back to haunt other workers in the field of human mitochondrial evolution.

He began with a very important supposition. This was that changes in our mitochondrial chromosomes have tended to accumulate over thousands or even millions of years at a fairly constant rate—or at least, if there was any variation in the rate, it was capricious and without pattern. If it was not—if for example variation accumulated more quickly in people living in one part of the world than another—then this could greatly distort his tree by making some branches much longer than others. This distortion would throw into question any estimate he might make of the time back to the mitochondrial ancestor, and might even make the real shape of the tree impossible to determine. Nothing in the data he had accumulated, however, suggested such a bias—for example, he found that there seemed to be roughly the same amount of mitochondrial variation

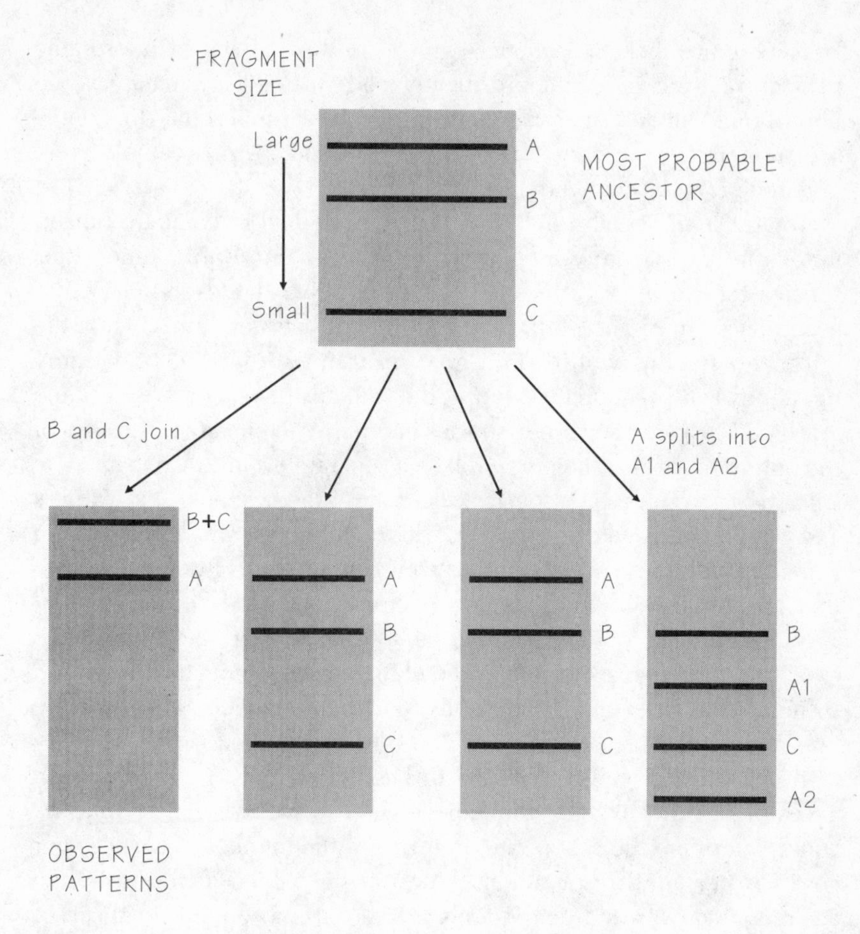

FIGURE 1.2. A simple mitochondrial family tree. Four different mitochondrial chromosomes have been digested with the same restriction enzyme, and they fall into fragments that are separated by electrophoresis. The fragments at the top of each pattern are the largest, those at the bottom are the smallest. Two of the patterns are the same, and the other two patterns can easily be derived from them by two different mutational events. In one, fragments B and C are joined to make a larger fragment B + C. In the other, fragment A is split into fragments A_1 and A_2. We can guess that the most probable ancestor of these four chromosomes is the same as the pattern that is seen twice, because any other tree would require that two mutations happened in the same lineage. But this guess could be wrong—the real ancestor is not necessarily the most probable ancestor!

among the members of each racial group. So Brown felt safe in assuming that the rate was indeed approximately constant. Still, his data gave no hint about how fast that rate was. Had a new mutational difference appeared among the ancestors of the chromosomes in his sample every hundred years? Every hundred thousand?

To attach an actual number to this rate, he had to compare human mitochondrial chromosomes with those of an *outgroup,* some other species that forms a different but still closely related branch of the evolutionary tree. The number of differences between humans and the outgroup species would then give him an estimate of how many mutations had accumulated during the millions of years since the time at which the ancestors of our species had parted company with those of the outgroup and had begun to follow a separate evolutionary path. The trick was to decide which species would make the best outgroup. Horses and whales, it turned out, would not do. Even though we share common ancestors with these animals, those common ancestors lived so long ago that by now the mutable regions of horse and whale mitochondrial chromosomes no longer resemble ours at all. Brown decided that the ideal outgroup species would be the chimpanzee. Chimpanzees are still so near to us in evolutionary terms that their mitochondria resemble ours quite closely.

Brown also, of course, needed to know how long ago the common ancestor of humans and chimpanzees had lived. If he knew that, he could determine the actual rate, in terms of the number of changes per year (or per century or per millennium), at which differences had accumulated between our mitochondrial chromosomes and those of chimpanzees. These differences would have accumulated independently down both the human and the chimpanzee branches of this evolutionary tree. He hoped devoutly that the rates of accumulation of these changes had been the same down both branches, for otherwise his calculations could again be thrown badly off.

After his postdoctoral stint at San Francisco and before establishing his own laboratory at the University of Michigan, Brown spent two years as a research associate in the laboratory of Allan Wilson at Berkeley, where he taught Wilson and his students how to isolate mitochondrial DNA and analyze it. Wilson, a lanky New Zealander who died a tragically early death from complications of leukemia in 1991, was one of the founders of the field of molecular evolution. Drawing on his pathbreaking comparative studies of the proteins of humans, chimpanzees, and other mammals, Wilson was able to supply the piece of information that Brown needed, the most recent time at which a common ancestor of humans and chimpanzees had occupied the planet.

That date was a mere five to seven million years ago. When Wilson and his colleague Vincent Sarich first published their estimate in the mid-1960s, they had sparked intense controversy. Up to that time it had been widely believed that the ancestors of humans and chimpanzees had been separate for a long period, perhaps as many as thirty million years. This span of time was almost half as long as the entire Age of Mammals, which started sixty-five million years ago with the demise of the dinosaurs. Thirty million years of separate evolution, it was generally felt, should give sufficient time to have accumulated the very clear and obvious differences between ourselves and our great ape relatives. And it gave enough evolutionary room for us to have evolved our enlarged brains and highly complex behavior.

Wilson and Sarich knew from the fossil record that most of the major groups of mammals had established themselves soon after that great Cretaceous-Tertiary extinction event of sixty-five million years ago. The actual cause of that event, a collision of the earth with a comet or an asteroid, had not yet been surmised at the time they did their work, but the existence of the event itself and the time at which it took place were well established from the radioactive dating of volcanic rocks. All the fossil evidence showed that, soon after the dinosaurs had relinquished their hold on the planet, there had been an explosion of evolution among the mammals. Wilson and Sarich found that the proteins of each of the major mammal groups were all about equally different from one another. This was exactly what would be expected if most mammal groups had been separated for about the same period of time and had been accumulating changes in their proteins independently of each other. And this in turn was exactly what the fossil evidence showed.

Humans, chimpanzees, and gorillas, however, showed only about a tenth as many differences among their proteins as more distantly related mammals. This meant that they must—if the molecular clock had indeed ticked steadily all that time—have separated at a point, not halfway through the Age of Mammals, but nine-tenths of the way through. Rather than a leisurely span of thirty million years, which gave plenty of time for the evolution of upright walking, fine motor skills, language, and the many other things that distinguish us from the chimpanzees, all these changes had somehow to be crammed into a fifth of that time.

Wilson and Sarich were beginning to win their battle against the anthropologists by the time Brown began to use their estimate in his own calculations. As more and more data were added from other molecular studies, the figure of five to seven million years was growing firmer all the time.

However, all the data on which Wilson and Sarich based their numbers were derived from proteins that they or other biochemists could isolate in quantities from blood or other tissues, proteins like hemoglobin and gamma globulin. It had already been found that the genes for different proteins evolve at different rates, and indeed that the gene for each kind of protein has its own characteristic rate—though luckily for molecular evolutionists these rates do not seem to vary much within a particular class of proteins. Further, these proteins are all coded by genes on the chromosomes of the cell's nucleus, not by genes on the mitochondrial chromosome. So the rate of evolution of these various nuclear genes still said nothing about the rate of evolution of the mitochondrial chromosomes—the mitochondrial rate might be much faster or much slower, and different parts of the mitochondrial chromosome might also evolve at different rates.

An Infinite Regression of Eves

When Brown began to look at chimpanzee mitochondria to see how different they are from ours, he immediately found that their chromosomes told a far more complicated story than those of humans. To begin with, there are at least two, possibly three, and possibly even more species of chimpanzee. Pygmy chimpanzees from the Congo are unable to produce offspring when mated with common chimpanzees, and there is growing evidence that common chimpanzees themselves fall into at least two and perhaps three distinct groups that might also be separate species. All this was reflected, Brown found, in their mitochondrial DNA. The DNA of common and pygmy chimpanzee mitochondrial chromosomes differed by one base in twenty-seven on the average, ten times as much as in humans, and the DNA from the two major groups of common chimpanzees differed by one base in fifty. And even within each of these two groups there was three times as much variation as in humans, an average of at least one base in every hundred.

Brown was struck by the fact that chimpanzees seemed to have a very different kind of genetic history from humans. It appeared at first sight that there had been far more evolutionary activity in the form of speciation among the chimpanzees than among humans, even though we have undergone far more change in our appearance and behavior than chimpanzees have done since the time of our common ancestor. Had our ancestors also split into many species during that time, and were we the only survivors? There was no way of telling, for all the mitochondria of those other near-human species would have gone extinct along with the species themselves.

Still, by averaging the various bits of chimpanzee data, Brown was able to use the Wilson and Sarich estimate of the time of the split to determine how quickly the mitochondrial chromosomes of humans and chimpanzees had accumulated differences. The rate was roughly 1 percent per million years. Because the mitochondrial chromosomes of different human groups differed from each other, on average, by only between a quarter and a third of a percent, his calculations gave him the remarkable result that at the base of the human mitochondrial tree there was a female who lived perhaps a quarter to a third of a million years ago, only a tiny fraction of the way back to the common ancestor of humans and chimps.

But did this mean that she was the only female living at that time, that she was truly an Eve, perhaps married—or at least mated—to a single Adam? Had our ancestors come so close to extinction? No, for there is good evidence from nonmitochondrial genes that the human species has never undergone such an extreme reduction in numbers. Many of our genes come in those numerous different forms called alleles. While no individual can carry more than two different alleles of a given gene (one from each parent), there may be many different alleles in the population as a whole. We share a good many of these alleles with the chimpanzees and our other relatives—too many for them all to have survived passage through a single Adam and Eve. They might, however, have managed to survive if that ancestral population had briefly been reduced in numbers to a few dozen individuals.

Because a single Adam and Eve can be ruled out, there are two other possibilities for what the population of our ancestors was like when it occupied the base of Brown's mitochondrial tree. Brown chose one of them. It is possible, he said in 1980 when he first published his calculations, that "present-day humans evolved from a small mitochondrially monomorphic population" that lived between 180,000 and 360,000 years ago. By *monomorphic*, he meant that all the mitochondrial chromosomes of that ancestral population were the same, without even the limited diversity that is found in present-day humans. How could this have happened? It could have happened if the human population size then had in fact been very much smaller than it is now—not as small as a single Adam and Eve, but perhaps only a few hundred or even a few dozen. Such a temporary restriction in population size is known to geneticists as a *size bottleneck,* because the population as it moves through time is effectively squeezed through this restriction in numbers like liquid through the neck of a bottle. During the time of this restriction, which might have lasted as little as one generation but might have been very much longer than that, it would have been easy to lose some of the different types of mitochondrial chromosome carried by Eve's ancestors entirely by chance. In the same way, it is easier for you to end up with all quarters or all pennies in your change if you habitually carry a small number of coins than if you carry a large number.

Or, a mitochondrial type might have arisen that had a great advantage, prehaps protecting against the ravages of some disease. Eventually, it would have swept away all the other kinds of mitochondria. Perhaps, Brown thought, the time of the mitochondrial ancestor marked a dramatic event in the life of our species, an event characterized not only by a severe reduction in numbers but also by strong selection and much genetic reshuffling. This was speculation pure and simple, for the only sign that such events might have taken place was what had apparently happened to the mitochondrial chromosomes, and these tiny chromosomes carry only three one-millionths of the total genetic information in a human cell. Still, as we will see in a later chapter, population size bottlenecks are often accompanied by precisely such a reshuffling of the majority of the genes that are carried in the nucleus. Perhaps Brown really had glimpsed a hint of a dramatic genetic event in our past, an event that took place a quarter to a third of a million years ago.

He soon ran into criticism. John Avise of the University of Georgia pointed out that there was no need to suppose that there was anything unusual about the size or genetic condition of the human species at the time of the mitochondrial ancestor. Indeed, the population at that time might have been quite large, with an abundance of different mitochondrial types. Brown, after all, was only able to look at the twigs that had survived down to our own time. There were many other branches of the tree that were invisible to him because, through a whole series of accidents, none of the descendants of those chromosomes had survived until the present. The females carrying them had died without issue or had produced only male children or perhaps had given birth to girls who in turn had failed to have female children. Over the span of many generations, *even if those lost chromosomes had initially been very common,* they had eventually all disappeared. Avise suggested that Eve's companions might have been numerous, but that all their mitochondrial chromosomes had been lost during the tens of thousands of generations that followed. They would, however, certainly have passed on to us copies of a good many of their nuclear genes.

To find out who was right, Brown or Avise, we must give Brown that time machine he wanted. Let us provide him with one capacious enough to carry all his enzymes and electrophoresis equipment back through the ages. We will make his first stop some fifty thousand years before the present, only a small part of the distance back to the mitochondrial Eve. Tests of the people who lived then would show that they, too, carried a diversity of mitochondrial chromosomes.

If Avise were right, then Brown would find that the mitochondrial chromosomes of those people, too, could all be traced back to a mitochondrial Eve—but that their Eve would be different from ours! The root of the mitochondrial tree would appear to have moved back in time, so that their Eve lived earlier than our Eve. And there would be roughly as much mitochondrial diversity among the people of fifty thousand years ago as there is today.

The explanation of this phenomenon is straightforward. There would be some mitochondrial types descended from that earlier Eve that still existed in the population during his time machine's first stop, but that had been lost between the time of his visit and today. The presence of these mitochondrial chromosomes would allow him to trace the mitochondrial tree back to that earlier Eve. If he were to start up the machine again and go further and further back in time, he would find the same situation. He would be faced with an infinite regression of Eves, always retreating alluringly before him and each as unattainable as a mirage. And at each stop there would be approximately as much diversity among the mitochondrial chromosomes of our species as there is at the present time.

If Brown himself were right, however, his journey by time machine would be quite different. At each stop as he approached the base of our mitochondrial family tree he would find less and less mitochondrial diversity. When he reached the base he would find little or none. He would find Eve herself, or at least a small group of people many of whom had the same mitochondrial type. If, in order to get through the size bottleneck, he arbitrarily set the dials of his time machine to an even earlier time, he would discover that this earlier population had its own mitochondrial diversity but that only one of those diverse chromosomes would survive the bottleneck. And perhaps, as he moved further back in time, he would find still-earlier size bottlenecks, possibly marking other important events in the history of our ancestors.

The length of each leap back to such a bottleneck would depend on the amount of mitochondrial diversity he found immediately prior to the subsequent bottleneck. If the diversity were limited, he would not have to go back very far to find the previous common ancestor. If it were extensive, he might have to go back a long way. By making leaps of various lengths, he would be able to repeat this process indefinitely. Provided that none of our various ancestors slaughtered and ate him along the way, he would sample populations of *Homo erectus,* and then perhaps leapfrog back through *Homo habilis,* through *Australopithecus afarensis,* through whatever the ancestors of those creatures were, through the apes of the Miocene epoch, back to the insectivore progenitors of the primates, and beyond. If the batteries of his time machine held out, he would be able to follow this trail all the way back to the time when the free-living ancestors of the mitochondria first came to live in the single cells of our own inconceivably remote forebears. At every step of the way, he would find traces of many lineages of mitochondria that have, perhaps quite by chance, been lost between then and now.

I have drawn a little sketch of these two views in figure 1.3. In the Avise view, the human population has remained a more or less constant size for a long time (except for its recent dramatic expansion). If Brown were to take a sample from such a population at any point in time, he would find roughly the same number of mitochondrial lineages because new chromosomal types have been continually arising by mutation and others going extinct by chance at about the same rate. (Actually, due to chance events, the number of lineages can vary by a factor of four or so in such a population, but the amount of variation found in any sample will give a very accurate idea of when the common ancestor lived—if there is lots of variation, she lived a long time ago; if there is very little, she lived recently.)

In the Brown view, on the other hand, there was a severe size bottleneck in the past, but after that the population was perfectly free to expand to any size. If Brown is right, then all the mitochondrial variation at the present time has arisen since that bottleneck.

Which view is correct? Some years after all this I began to think about the problem and came up with an argument that favored Brown's view rather than Avise's. My argument, like those of most people working in the field of mitochondrial evolution, depends on the assumption that most of the mutations that are seen in mitochondrial DNA do not affect their carriers in any noticeable way.

In examining the Avise view, I asked a simple question: How small must the Avise type of population have been over those long stretches of

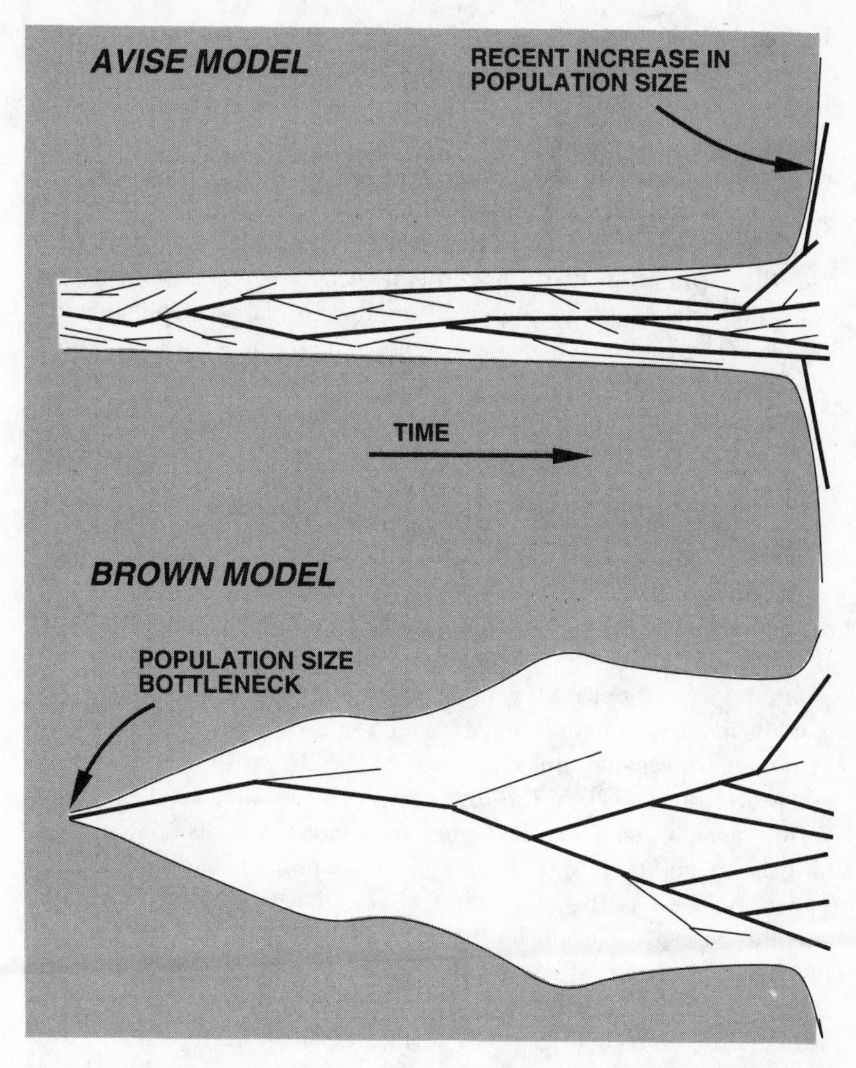

FIGURE 1.3. A simple sketch of the Avise and Brown models of our mitochondrial history. In the diagram, time goes from left to right. Mitochondrial chromosome types that have persisted down to the present day are shown as dark lines; those that have been lost are shown as light lines. In the Avise model, the population has remained roughly constant in size for a long time, and during all this time mitochondrial types have been gained and lost at approximately equal rates. In the Brown model, the population has gone through a size bottleneck, most of the variation has been lost, and since that time the population has gradually gained new mitochondrial types.

the past in order to produce the limited mitochondrial diversity that is currently seen in the human species? The answer, it turns out, is that it must have been remarkably, dangerously small. A simple calculation shows that, if our common mitochondrial ancestor lived 250,000 years ago, the population size of our species must have been approximately six thousand individuals over almost all the span of at least ten thousand generations from that time to this. If the ancestor lived 500,000 years ago, the population could have been twice as large, about twelve thousand individuals. Although it is true that this larger population was less likely to go extinct in any given generation, pushing back the time of Eve means that the population must have survived at that size not for ten thousand but for twenty thousand generations.

The Avise view therefore requires that before our recent expansion in numbers the size of the human population must have been very small for a very long time. Such a small population would be at great risk of extinction. On the other hand, aside from that one dangerous moment of the size bottleneck, a Brown-type population is under no such restriction—once through the bottleneck, it could have been of any size.

When I wrote a note elaborating on the consequences of these two different views of our past, another scientist who reviewed it suggested that it is impossible to decide between these two views because they are both equally probable. But they are not. The Brown view, with its brief but severe bottleneck in the distant past, is the equivalent of a tipsy person on an ocean liner who teeters for a moment on its bow, recovers his balance, and then walks the length of the promenade deck in safety. The Avise view is the equivalent of somebody equally drunk who attempts to cross Niagara Falls on a tightrope. Who is more likely to survive? The fact that we are here at all provides in itself a strong argument that the Brown view is more likely to be correct.

Recently, more evidence has surfaced that there has been a size bottleneck or some strong selective event affecting the mitochondria in our past. I favor the possibility of a bottleneck—such bottlenecks seem to have happened among our close relatives. Recently, the isolated populations of mountain gorillas in Central Africa have been found to have almost no mitochondrial variation, while lowland gorilla populations are filled with it. Probably at just about the time of our mitochondrial ancestor, the numbers of our species diminished dangerously—not to two, as the Garden of Eden myth suggests, but perhaps to a few hundred or even briefly to a few dozen over the space of a few generations. During this time, one particular mitochondrial type came to predominate in the population, the way one surname might become common in an isolated village. Then, later, perhaps as the population was recovering from the bottleneck, the other types that remained died out, so that when we look back from the present we can trace all our mitochondria back to a single type.

Chimpanzees show much more diversity among their mitochondrial chromosomes than we do. This means either that their most recent mitochondrial Eve may have lived three or four times as long ago as ours or that the rates of change in the two branches of the evolutionary tree leading to humans and chimpanzees are very different. It is likely from other evidence, however, that the rates of human and chimpanzee evolution do not differ much. If their Eve, too, marked some important evolutionary event, then the latest example of this type of event occurred much longer ago in chimpanzee history than in our own.

Might a larger number of such dramatic events have occurred in our ancestry than in the ancestry of chimpanzees? Given the time machine, of course, we could find out. We could go back to the time just before our mitochondrial Eve and measure the mitochondrial diversity among the humans (or prehumans) living then. Would it be very great, which would imply a long period of large population sizes and a relaxed and gradual accumulation of different chromosome types? Or would those ancestors' diversity, too, be restricted, suggesting that another severe bottleneck had taken place in their own fairly recent past? My guess, and it is of course only a guess, is that we would indeed find evidence for such a recent previous bottleneck. It would be wonderful to know whether our genetic history, already so different from that of chimpanzees, has been marked by a series of severe bottlenecks. But alas, the mitochondrial slate has been wiped clean by the most recent bottleneck, and in the absence of a time machine we can go back no further in time than our most recent common mitochondrial ancestor. We can, however, use other lines of evidence to try to penetrate beyond that barrier.

Eve Herself

Wesley Brown's mitochondrial data were not very extensive. It was difficult to get enough human tissue to isolate the needed amounts of mitochondrial DNA. Placentas, or afterbirths, were ideal sources of tissue, but they were hard to come by. The family tree that he could build from this limited amount of data was a very sparse one, and rather confusing. There was a hint that the origin of the tree—the ancestral mitochondrial type, from which all the branches of the tree were descended—might have been in Africa, but it could also have been in Asia. Doug Wallace, the scientist who later discovered the first harmful mitochondrial mutation, was collecting similar data and was convinced that the origin of the tree was in Asia. None of this work attracted a great deal of notice, beyond the narrow world of genetics.

Then two hardworking students in Allan Wilson's lab, Rebecca Cann and Mark Stoneking, were inspired by Brown to get involved in the problem of building mitochondrial trees. Stoneking had access to some rather exotic material, placentas from people living in New Guinea and Australia. Also, an improved technique for separating out the fragments of DNA from restriction enzyme digests had recently become available. The result was that many more enzymes could be used, so that more mutational changes could be detected and more information could be gathered from each precious DNA sample.

Cann and Stoneking collected DNA from as many very different humans as they could. In addition to the Australian aborigine and New Guinea samples, they obtained placental material from people in the racially diverse San Francisco Bay area whose ancestry was Asian, European or African, giving a total of 147 individuals. Because they could use many new enzymes, they were able to collect much more data than Brown had been able to get. Even with all this additional data, however, they still found as Brown had that the human mitochondrial population was remarkably uniform, with chromosomes differing on average by one base in every three hundred. But when they first scanned their earliest sets of data, calculating trees by hand, it looked as if some of the African chromosomes were very different from the others. Perhaps they could show that the deepest branch lay in Africa.

Their growing mass of data was soon impossible to analyze by hand, so they turned to the computer, in particular to a powerful program called PAUP (Phylogenetic Analysis Using Parsimony) written by David Swofford of the University of Illinois.

PAUP is really a way to get an approximate answer to a very famous conundrum in mathematics called the traveling salesman problem. Like so many mathematical problems, it is trivial to state, easy to solve for simple cases, and effectively impossible to solve for more difficult cases. At the moment, there is no mathematical way to deal with such difficult cases, although ways have been found to obtain approximate solutions.

Imagine you are a traveling salesman, charged by your home office with visiting all the cities in your territory while minimizing your travel expenses. If you need visit only three or four cities, the solution is fairly easy—find the cost of travel between each pair of cities and then play around with the numbers until you find a way to visit each one at least once for the smallest total amount of money. In order to keep your cost under control, you want to find the most parsimonious solution.

Even with only a few cities to visit, however, the number of possible choices of route is quite large. Unless you patiently go through all of them, you will come away with the nagging feeling that there might have been a better solution to the problem that you missed. And when you try to add more cities to your itinerary, the number of possible choices goes up exponentially.

Cann and Stoneking were faced with essentially this problem. There were an immense number of possible evolutionary trees that could be drawn to join together all their 147 individuals. However, some of these pairs of mitochondrial chromosomes were more closely related than others were, so that the cost of evolutionary travel between them, in terms of the numbers of DNA changes that must have taken place, was less than it was between more distantly related pairs. They knew that a good way to draw a tree based on their data was one that reduced the evolutionary cost to the lowest possible level. This tree might not be—indeed, it probably was not—the real tree, but it would be the best approximation to that true tree that they could extract from their data. This maximum-parsimony tree lurked somewhere in that mass of data—the trick was to find it.

Unfortunately, as the number of possible trees goes up the chance of stumbling on the most parsimonious one goes down. PAUP, or any other program designed to wrestle with such a problem, can easily get trapped in what is called a local minimum. The program may find a quite parsimonious tree, one with what seems to be a relatively small number of relatively short branches, then begin to explore many other related trees and discover that they are all worse. So it concentrates on the parsimonious tree, rearranging it and refining it until it can do no better. Then it stops. But on the other side of the surrounding hills of less parsimonious trees there may be a valley with an even more parsimonious tree at its bottom, a valley that the program completely missed because it never stumbled on that kind of tree and began to explore variations of it. Indeed, there may be an immense number of other valleys, many of them deeper than the one the program has trapped itself in. If the number of possibilities is very large, as it was with the Cann and Stoneking data, there is currently no way of ensuring that the program has explored all these valleys.

Cann and Stoneking, urged on by Allan Wilson, began to build trees using PAUP. The program required a starting point, and they gave it what seemed to be the obvious one suggested by their earliest calculations, in which some of the African chromosomes were grouped together. PAUP began to explore trees with this configuration.

And it reached a local minimum.

The tree that emerged, and that Cann, Stoneking, and Wilson published in 1987 in the influential journal *Nature,* is shown in figure 1.4. For the most part it is an enormously confusing welter of branches, twigs, and twiglets, in which people from all parts of the world are mixed together. But there is one unequivocal feature, a very clear separate branch consisting entirely of Africans that springs from the deepest part of the tree. And it was on the basis of that branch that Cann, Stoneking, and Wilson claimed that a woman carrying the ancestor of all human mitochondria lived somewhere in Africa sometime between 140,000 and 290,000 years ago (they were able to refine Brown's original numbers a little). While the paper did not mention the *E* word, Jim Wainscoat of Oxford in an accompanying comment in *Nature* did the obvious and christened her the mitochondrial Eve.

Immense publicity accompanied their paper, with worldwide news stories and an article in *Newsweek* with a cover picture that showed an imaginative reconstruction of a rather hirsute Eve. Although Wilson's group, both bemused and delighted by the publicity monster they had created, kept doing their best to explain the difference between a biblical Eve and a mitochondrial Eve, the majority of the public were persuaded that scientists had uncovered evidence for a single African female ancestor in our distant past. The fine point that Wilson's group kept emphasizing, that this mitochondrial Eve had an unknown number of companions of both sexes, many of whom had bequeathed copies of their nuclear genes to us, went essentially unnoticed. Indeed, as recently as May 1992, an article by Phillip Ross in the usually accurate *Scientific American* repeated the error that Eve was alone: "Most important, the study of paleo-DNA promises to test the theory of Wilson and his colleagues that all living humans are descended from a single woman, dubbed Eve, who, they concluded, lived in Africa." So did an article on Neanderthals by James Shreeve in the *Smithsonian Magazine* in December 1991: "In 1987, some biochemists at the University of California at Berkeley announced that their study of a particular kind of DNA revealed that all living human beings shared a common ancestry with a single female who lived in Africa some 200,000 years ago." Had Shreeve said that they shared a common ancestry *for*

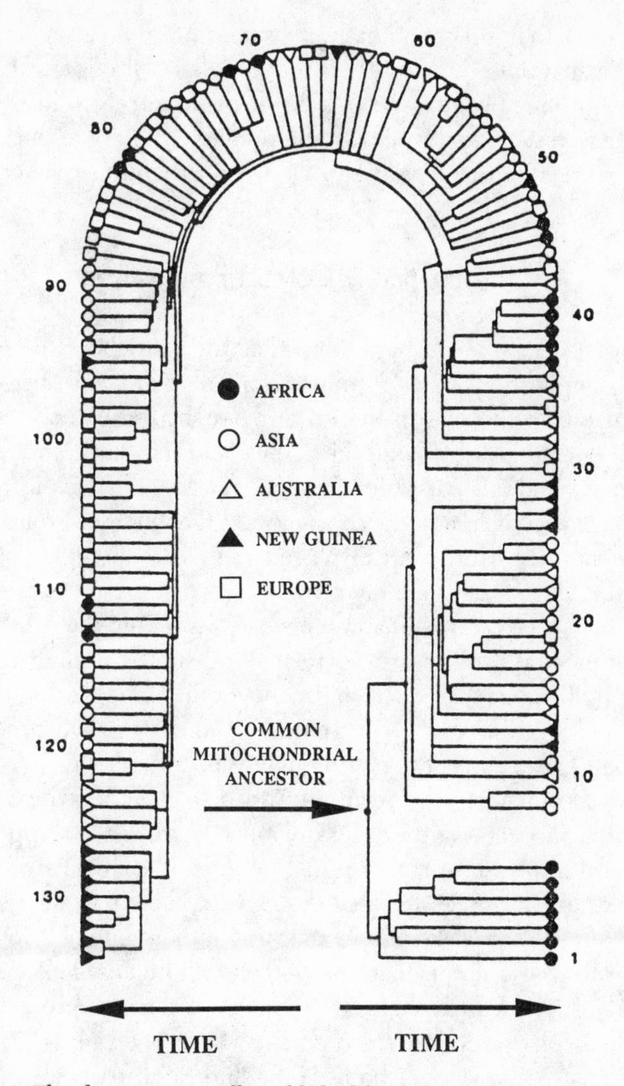

FIGURE 1.4. The diagram originally published by Rebecca Cann, Mark Stoneking, and Allan Wilson in their famous 1987 *Nature* paper. As drawn, the tree has a clear root.

that particular kind of DNA, he would have come a little closer to the complex reality of the situation.

Rebecca Cann went on to the University of Hawaii, where she is currently using mitochondrial DNA to try to trace the migrations of Pacific peoples. Mark Stoneking joined a biotechnology company, where he began to explore the newly invented PCR technique to investigate nuclear rather than mitochondrial genes, and more recently has moved to Pennsylvania State University. It fell to Allan Wilson, a brilliant and articulate public speaker whose halo of prematurely white hair and aquiline features squared well with the public's image of a spokesperson for science, to carry the news of this discovery around the planet.

The Politically Correct Family Tree

Wilson realized from the beginning that the mitochondrial tree had enormous implications for the study of human origins. An African origin for humans at some time in the distant past had of course long been accepted, but the paleontological evidence showed that some groups of early humans had left Africa for Europe and Asia as much as a million years ago. The mitochondrial tree suggested that another group had left Africa far more recently and displaced the early migrants. What had happened to those early migrants?

A stunning revision of human prehistory was required to explain their disappearance. Wilson was used to making such stunning revisions. After all, he had already convincingly shown that humans and the great apes actually had a very recent common ancestor, five times more recent than had previously been thought. He turned with relish to the new task of convincing the scientific world that modern humans had, again much more recently than anyone had imagined, appeared in Africa and fanned out through the Old World. Further, the mitochondrial data suggested that it was very likely that they had, in a tide of conquest, swept the people they found there into oblivion.

Who were those unfortunate prehumans who had fallen prey to the onslaught of true human beings? Remains of peoples predating Eve, now known collectively as *Homo erectus,* have been found in eastern and northern Africa and in both China and Southeast Asia. Many of these finds had been dated to well before the 290,000 years ago that Cann, Stoneking, and Wilson had suggested must be the earliest time for the origin of modern humans. The most famous representative of *Homo erectus,* Peking man, had lived in northern China from perhaps 700,000 to 200,000 years ago. Some of the European finds that might have been *Homo erectus* had been dated to as long as half a million years ago. In order to reconcile this paleontological evidence with the mitochondrial data, it was necessary to assume that these peoples were no more than doomed offshoots of the human family tree, driven to extinction as surely as European settlers had destroyed the Tasmanian aboriginals in the nineteenth century. And, because of the way the mitochondria are inherited, it was possible to go a step further and state that no gene flow had taken place between conquerors and conquered. The conquered peoples had been driven to genetic as well as physical extinction.

The normal pattern of conquest includes rape and forced marriage—women of the conquered peoples are forced to bear the children of the conquerors. Because women pass their mitochondrial genes on and men do not, an assiduous researcher examining mitochondrial chromosomes from the peoples of Europe and Asia should surely have found genetic traces of both the conquerors and the conquered.

Yet that had not happened. The deepest branch of our mitochondrial tree could be traced to Africa, and this branch seemed to have arisen long after the arrival of *Homo erectus* in Europe and Asia. This meant that all those early arrivals died out, or at least did not leave any mitochondrial chromosomes to posterity.

Flights of fancy were soon embarked on to explain this. One of them was that these conquered peoples were so brutal and primitive that their conquerors could summon up no sexual interest in them. There have, of course, been similar attempts to reconstruct the sexual mores of peoples of the past. The Neanderthals, a much more recent group of peoples than *Homo erectus*, appear to have been replaced in Europe by true humans some forty to sixty thousand years ago. Jared Diamond in *The Third Chimpanzee* (1992) has suggested that true humans were probably reluctant to mate with the Neanderthals, so that there was no genetic exchange between Neanderthals and modern humans. I wish I could be as sanguine as Diamond about the delicate sensibilities of your average raper and pillager.

Wilson himself went the farthest out on this particular limb. Perhaps, he suggested, humans emerging from Africa had an overwhelming advantage over *Homo erectus* because they could speak and *Homo erectus* could not. And perhaps this was due to some gene or genes on their mitochondrial chromosome! Because they could not communicate with the dumb, brutish females of the conquered species, the conquerors kept their gene pool aloof and unsullied.

Of course, one can apply the same counter argument as I applied to Diamond's scenario. Conquerors tend not to be picky about who or what they rape and might look on their victims' inability to talk as an asset. Further, it is very unlikely that the few genes that are still found on the mitochondrial chromosome have much to do with a highly complex character such as the ability to talk—mitochondria are busying themselves with far more fundamental processes in the cell. It seems much more probable that, if there really was no genetic exchange, it was because there could not be. Perhaps *Homo sapiens* and *Homo erectus* had, like the common and pygmy chimpanzees, diverged too far from each other to produce viable or fertile young.

This revisionist view of human prehistory was not the only thing that was suggested by the Cann and Stoneking data. If their version of the mitochondrial tree was right, and the genes of European and Asian *Homo erectus* never did enter the human family lineage, then other ideas about human evolution were also cast into doubt. An elaborate web of inference about the origin of human races would come crashing down.

Wilson and his co-workers supposed that true humans exploded out of Africa and displaced other near-human groups everywhere else in the Old World. This supposition rested on an unstated assumption, which forms a kind of politically correct version of the origin of human races. The transition to true humans, Wilson thought, must have happened quite suddenly at about the time of the mitochondrial Eve. As they fanned out through Europe and the Middle East and then into the farthest reaches of Asia, these peoples were fully human in their intellectual capacity, their ability to use language, and their ability to invent culture. The various human races must have evolved subsequent to that time, and it was reasonable to suppose that during that brief period of further evolution they all retained much the same genetic makeup. Even though our freshly human ancestors behaved in a rather beastly fashion as they swept through the rest of the planet and killed the pitiful primitive prehuman creatures they found there, at least they were all genetically roughly equivalent.

The mitochondrial data seemed to reinforce what William Howells had called, ten years earlier, the Noah's Ark model (biblical imagery seems unavoidable even when scientists deal with these matters!). Like Noah's family, true humans had radiated out rapidly from a particular point and populated the earth. And, like Noah's three sons, Shem, Ham, and Japheth, they had established different races in the process.

This was in vivid contrast to another, very different view of human origins, called the regional-continuity model by Howells and the multiple-origins model by students of the late anthropologist Carleton Coon. Coon and his followers, notably Milford Wolpoff of the University of Michigan, had gathered together much fossil evidence that in their opinion showed that various human races must have arisen at a time before the fully human condition had been reached. These different races, they suggested, had each made the transition, perhaps independently, to full humanity. Coon, as we will see in the next chapter, was castigated as a racist for these views because the multiple-origins view implies that different human races need not be genetically equivalent. Some races, according to this model, might be more fully human than others!

Wolpoff has attempted to soften the possibilities for racism built into this view of human evolution by suggesting that the genes that caused the transition to full humanity have been able to spread over long distances as a result of migration. This spreading would enable all these different human groups to advance in a kind of genetic lockstep. To make the scheme work, however, he has to assume that all this genetic migration was somehow accomplished without causing racial differences to average out and disappear. Genes for a certain kind of skin color or hair shape must be so advantageous or disadvantageous in different parts of the planet that races can continue to be clearly differentiated from each other—even though at the same time these races are exchanging alleles of other genes that aid them in their evolution toward full humanity. Mathematical models can be built to make this work, but exactly how realistic they are is very much a matter of opinion.

Just as in his earlier battle over when the split between humans and the great apes had taken place, Wilson was confident that he would triumph once more over these genetically naive anthropologists. He had an ace up his sleeve, for most evolutionists agreed that it was very unlikely that the human species could have had multiple origins. Evolution does not repeat itself in that way—new species become established through unique *speciation events*, and the mitochondrial story certainly provided ample evidence for precisely such a single speciation event in our past.

And then the mitochondrial story began to unravel.

Climbing Out of the Local Minimum

In their *Nature* paper, Cann, Stoneking, and Wilson had included all the data on which their tree was based. As a result, many of us played around with the data and gave it to our students as an exercise. When we did so, we quickly discovered that the published tree was not the most parsimonious tree, and that other trees could easily be constructed that had a smaller number of steps linking the various mitochondrial chromosomes. The computer programs could find these trees when they were supplied at the outset with randomly generated trees to begin their search, trees that did not have all their African chromosomes clustered together. Thousands of these shorter trees could be found, in numbers limited only by the available computer resources. And while all of these trees had that precious, unequivocal, deeply rooted branch with Africans in it, on which so many of Cann and Stoneking's conclusions about human evolution depended, in almost none of them did the branch consist only of Africans. In various trees the branch was found to have Asians in it, and sometimes Europeans and even Australian aborigines.

This fact was given wide publicity recently in papers in the journals *Science* and *Systematic Zoology* by Alan Templeton of Washington University and David R. Maddison of Harvard's Museum of Comparative Zoology. In a reply to Templeton's note, published in the same issue of *Science,* Mark Stoneking joined several other authors in admitting that, when they had reanalyzed the data from their original mitochondrial tree, and even when they looked at some newer and very extensive data, none of it unequivocally supported the African origin of modern humans.

Again, the media picked up this story. The news was spread to a puzzled public that the scientists were wrong and that Eve was not African after all. After being garbled through several filters, of course, nothing much was left of the original account. The public had first been led to believe that a solitary African Eve was the mother of the entire human race and then was informed that she did not live in Africa at all. As a result, the entire world presumably turned to the sports pages in irritation.

What had really happened? Something quite complicated and interesting, and something that by no means disproved the African origin of Eve and her companions. PAUP, and other computer programs that build maximum parsimony trees, can use only the information they are given. And even the best genetic information has errors built into it, errors that are quite unavoidable.

Because mutations in the mitochondrial chromosome happen at random, more of them might by chance happen in one mitochondrial lineage than in another. In a line leading down through the tens of thousands of generations from the mitochondrial Eve to somebody in Asia, for example, ten different mutations might accumulate. In another line leading to a New Guinea tribesman, twenty different mutations might accumulate. This introduces a great deal of uncertainty into the tree, notably in the matter of where to put the tree's origin.

It is as if our hypothetical traveling salesman did not have a map of the cities he was to visit but was asked to draw a map based on the various airline fares between them. He would certainly be able to draw such a map, but it might have only a vague resemblance to the real map because airline fares are based on the amount of traffic as well as the distance. As I recently discovered, it is cheaper for me to fly from San Diego to San Francisco, a distance of some 600 miles, than it is to fly from San Diego to Ontario Airport east of Los Angeles, a mere 70 miles away. If I were to draw a map of California based on airfares alone, it would be highly distorted. The computer program PAUP, of course, has only the genetic equivalent of these fares to deal with and knows nothing about the true underlying tree.

There is another difficulty, as I pointed out in a paper that was in press at the time Templeton's and Maddison's papers appeared. The way that trees are often presented by the computer, or by the draftsman who draws a picture of them, can make the tree look better than it really is. This can be seen when you look again at figure 1.4. There are two distortions of the data that have been built into the picture. The first is that all the lines leading from the common ancestor down to the people living at the present time have been made to appear the same length. This, of course, makes all Eve's descendants line up in a neat and pleasing row. Looked at from an evolutionary perspective, this is the right way to draw the tree, because the same number of years separate the mitochondrial Eve from all her descendants who live at the present time. But it ignores the unfortunate fact that different numbers of mutations have taken place during those various divergent lineages. To make all the lines the same length conceals just how uncertain the data really are.

The second distortion is one that at first sight appears to be almost unavoidable. You can see that some of the descendants of Eve are clustered into great complex masses of branches in the family tree. In order to make the tree look neat and keep all the lineages parallel, the draftsman (or the computer) must insert long lines at right angles to these clusters of branches. This keeps the clusters of branches separate. You can see in the figure that one particularly long line has been inserted to separate the African branch from the rest of the tree. Other lines, some of them even longer, have been inserted in order to separate the other branches of the tree. These lines are solely an artistic convention—nothing in the data themselves dictates where these lines should be put or how long they should be. And they can have the unfortunate effect of imparting an appearance of organization to the tree, an organization that it might not really have.

When these distortions are removed, you get a tree such as the one in figure 1.5. I have drawn this tree using exactly the same data that were used to construct figure 1.4, although the tree itself happens to be one of the many more-parsimonious trees that can be found when a more extensive computer search is carried out. Because I did not want to insert any of those fictitious branch-separating lines, I had to draw the tree in the form of a starburst. (I wanted to call it a "Star Trek" tree, because it reminded me of Captain Kirk hurtling into hyperspace, but the editors objected.) I also made the lengths of the various branches reflect the actual numbers of mutations that were detected, so that some of the branches are much longer than others.

This tree is an unrooted tree—no attempt has been made to specify where the mitochondrial Eve might be, and indeed you can see that it would be very difficult to decide where to put her. Somewhere in the middle of the starburst, undoubtedly, but it is not obvious where. She could have been African, Asian, or even European. Indeed, the famous African branch—you can still see it at twelve o'clock on the tree—now has some Asians in it. And the messiness of the data is apparent in another way as well—individuals are found at the ends of branches that are long, short, and in between, so that they are scattered all the way from the center of the starburst to its periphery. It is this inherent messiness of the data, a messiness that was concealed by the way the tree was originally presented, that makes it so difficult to build a consistent and understandable tree.

Similar starburst trees are often drawn by evolutionists who are working with smaller amounts of data, but when the tree becomes very complicated it also becomes very untidy. When this happens, the urge to present a neat-looking tree takes precedence, and distortions such as those of figure 1.4 are the unfortunate result.

So, as the dust from this latest controversy settles, what can we say about human origins? In spite of the equivocal nature of the mitochondrial tree, it is still very possible that the mitochondrial Eve lived in Africa. One indication that this may be so is that Africans carry a greater diversity of mitochondrial chromosomes than any other human group. But an African origin is certainly not proved. And the dust has not settled completely yet. Let me give you a glimpse of controversies to come.

After the publication of the Cann and Stoneking data, Linda Vigilant, working in Wilson's laboratory and now at Pennsylvania State University, collected samples from many Africans and from numerous representatives of other racial groups. She used newer techniques to

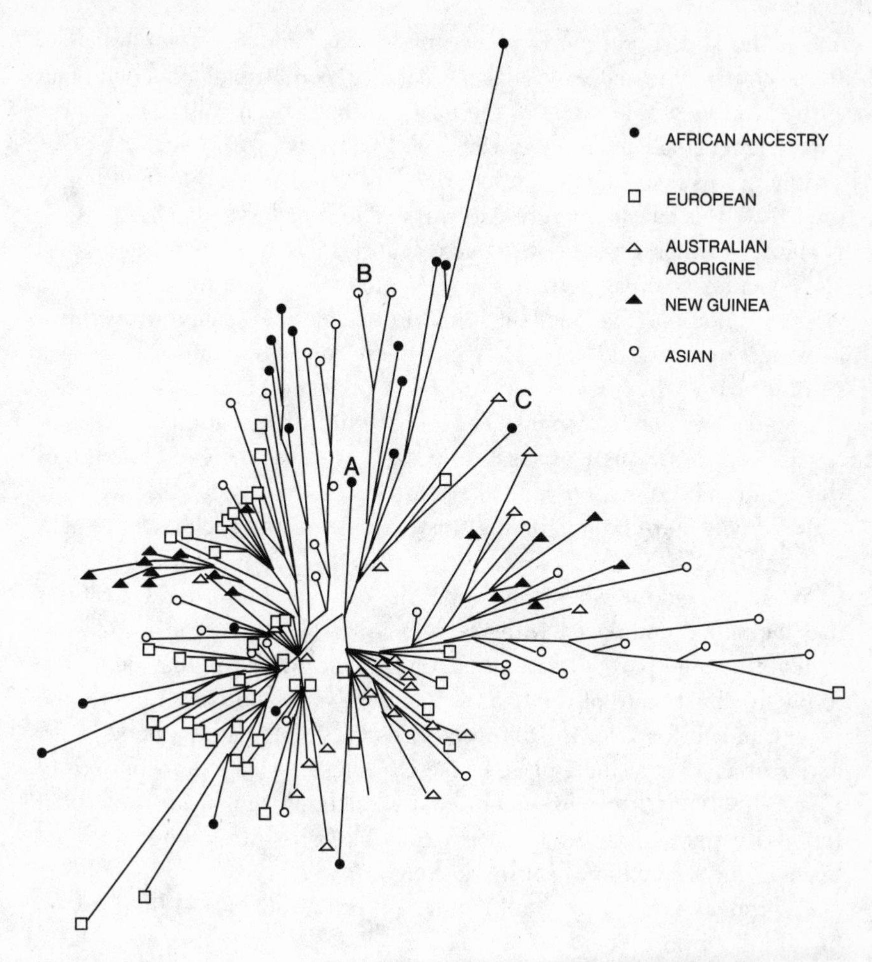

FIGURE 1.5. A starburst tree of the data from the Cann, Stoneking, and Wilson *Nature* paper. Note the great difference in length of the various branches and the consequent difficulty of determining where the root of the tree might be.

look directly at the DNA sequences themselves, so it was not possible to pool her data with those of Cann and Stoneking. Recently, she, Mark Stoneking, and others analyzed her data and decided that Eve probably lived about 130,000 years ago. Unfortunately, the tree-building method they used had enormous errors built into it, and they had to admit that Eve might have lived as little as 60,000 years ago and as long as 400,000 years ago!

Why should all these efforts at tree building have been so unsuccessful in helping us to understand where and when Eve lived? It is surely not for want of data, for the information that has been collected so far makes up one of the largest and most detailed molecular family trees yet available. In the course of reanalyzing Vigilant's data, I recently obtained a hint of an answer to this puzzle. Most of the data that have been collected, it turns out, cannot actually be used to date or to locate Eve.

When, as a result of a mutation, one base is substituted for another in the DNA, this can happen in two ways. The four bases of DNA—A, T, G and C—can be grouped into two categories based on their size. A and G molecules are bigger than T and C molecules. The most numerous class of mutations that Vigilant found in her data are called *transitions*, in which one base of the DNA has been converted into a base of a similar size (A has been changed to G or vice versa, or T has been changed to C or vice versa). The other kind of mutations are called *transversions*, in which a base of one size has been changed to a base of the other. These are far less likely to occur—for many reasons, only some of which are understood. Transversions include such changes as A to T or C to G.

What amazed me when I looked at the data was how often transitions had happened compared with transversions. The transition rates were so high in some parts of the mitochondrial chromosome that they gave too high an estimate of the overall mutation rate, and consequently too recent an age for Eve. Further, because transitions had happened over and over again at a single place in the chromosome over short periods of time, they tended to mask evolutionary relations and make trees built from the data difficult to interpret. The starburst tree, with its bewildering collection of short and long arms and mixture of peoples in every branch, is partly a result of the confusion introduced by this very high rate of transitions.

At some places on the chromosome, transitions are as much as two hundred times more likely to happen than transversions! As a result, transitions provide a much less reliable clock than transversions. Suppose one were to forget about all the transitions in the data and simply obtain the ratio of the maximum number of transversions that have accumulated between human sequences and the maximum number that have accumulated between humans and chimpanzees. The size of the ratio will tell approximately when Eve lived, as a fraction of the distance back to the parting of the ways between the ancestors of humans and chimps.

The most probable divergence time between the ancestors of humans and chimpanzees, it is now thought, lies somewhere between four million and seven and a half million years ago. When I used transversions alone to estimate the age of Eve, I found that humans had accumulated about one-seventh as many transversions as had accumulated between us and the chimpanzees. This meant that the mitochondrial Eve probably lived between six hundred thousand and a million years ago.

This result pleased me mightily, because at the outset of my investigation I had visited Phillip Tobias in Johannesburg and sketched what I was trying to do. He laid down a challenge: The earliest certain traces of human migrations out of Africa, he pointed out, can be dated to about a million years ago. Might it be possible that the mitochondrial Eve could also have lived as much as a million years ago? At the time I thought this was unlikely, but as the analysis proceeded I realized that a million years ago, older than anyone else had suggested, was indeed the outermost limit of my calculations.

The implications of this boggle the mind. If the mitochondrial Eve really lived as much as a million years ago, then the ancestors of peoples in Africa, Europe, and Asia might actually have made the transitions from *Homo erectus* to *Homo sapiens* independently of each other. The multiple-origins model, with its freight of potential racism, rather than the politically correct Noah's Ark model, would suddenly be in the ascendant. This is certainly what will happen if my analysis holds up and the mitochondrial Eve is pushed back in time.

Of course, all these are theoretical models, limited by the fact that we are trying to find out what happened in the past by using the DNA of people alive today. Is it possible to get some more direct evidence?

Attempts are currently being made to isolate DNA from the remains of one very recent extinct group of humans, the Neanderthals, who lived in Europe between roughly one hundred thousand and thirty-five thousand years ago. Ever since the first discovery of Neanderthal remains in the early nineteenth century, they have existed in a kind of taxonomic limbo. At various times they have been classed as a different species from humans, then put in with humans, and later demoted again. Currently, the consensus is that they form a subspecies slightly different from ourselves and dignified with the name *Homo sapiens neanderthalensis*.

The search for Neanderthal DNA faces two difficulties. First, museums are understandably reluctant to give up their precious Neanderthal remains, or even a small part of them, to be destroyed in the hunt for DNA. Second, no DNA from unfrozen bones tens of thousands of years old has ever been isolated. There have been two successful isolations of very old DNA. The first, from some magnolia leaves that are almost twenty million years old, was the result of an accident—the leaves had been preserved for all that time in a moist state in the absence of free oxygen. When they were exposed to the air, like the immortal girl who left the magic valley of Shangri-la, the delayed burden of the years caught up with them. The leaves turned from green to black in a few moments, and unless the DNA was immediately rescued it was quickly destroyed. The second successful isolation, from a termite embedded in amber some twenty-five to thirty million years ago, was again a case in which the DNA had been shielded from oxygen.

Neanderthal bones have been dried and exposed to the air for periods ranging from years to over a century. They are unlikely to have retained much DNA after all this time. What is really needed, of course, is a Neanderthal who fell into a glacial crevasse and has been frozen ever since. One consequence of global warming is that the last glaciers of Europe are melting. Perhaps a Neanderthal will be found thawing out of a glacier, like the 5,300-year-old Neolithic (and fully human) hunter whose corpse recently emerged from a glacier in the Italian Tyrol.

If such a find is made, it is likely that the first DNA to be isolated will be mitochondrial, because so many copies of the mitochondrial chromosome exist in every cell. Even if only a few intact mitochondrial molecules survive, they may be enough for the new PCR technique to conjure up copies of this ancient DNA.

There are two possible outcomes to such a discovery. The first is that Neanderthal DNA will turn out to be very different from that of present-day humans, so that it will form a distinct and separate branch when it is placed in the mitochondrial tree. This would strongly imply that modern humans, presumably the Cro-Magnon people who appeared in Europe perhaps 35,000 years ago at the time of the disappearance of the Neanderthals, must have driven the Neanderthals to extinction. As a result, the likelihood would be increased that our more remote ancestors could have done the same to *Homo erectus.* If we did it once, we could do it twice.

On the other hand, Neanderthal mitochondrial DNA might turn out to be very much like that of other humans. This would place Neanderthals, after a century and a half of uncertainty, firmly in our own species at last. If we and the Neanderthals, apparently so different from each other in so many ways, turn out to be closely related genetically, this would rule out a very recent Eve and increase the possibility of multiple origins for different human groups.

I cannot predict what will be found in Neanderthal DNA, but since there is a good chance that Eve lived long before the Neanderthals, my guess is that they will turn out to be members of our species. If, as a result of such discoveries, it turns out that we have to come up with some explanation for multiple origins of *Homo sapiens*, then we are going to have to learn more, much more, about Eve's companions.

2

An Obsession with Race

Racial differences imply differences in intelligence, a subject so laden with emotion that its mere mention evokes unsolicited acclaim and feverish denunciation. Even without reference to the brain or to intelligence, the simple statement that races exist drives a small coterie of vocal critics into a predictable and well-publicized frenzy. . . . I . . . formally request that no one shall quote this book as ammunition for or against any cause whatsoever.

—Carleton Coon, *The Living Races of Man* (1965)

In the fall of 1956 the anthropologist Carleton Coon and his second wife, Lisa, were invited by the U.S. Air Force to go on a round-the-world trip, using air force planes wherever possible. Their journey was to last six months and would include visits to aboriginal peoples living in eleven countries of South and East Asia and the Middle East. The trip would add to Coon's already extensive stockpile of knowledge about human variation and would give him the impetus he needed to embark on his most extensive and important work, a book about the origin of human races.

Ever since his education in anthropology at Harvard, Coon had spent a turbulent and adventurous life among primitive peoples. He was an anthropologist of a classical breed, who always took with him assorted calipers and other measuring devices with which he could determine the shapes and sizes of heads and other body parts. Such measurements were the most important tool of many noted anthropologists of the past, the most famous of whom was Johann Friedrich Blumenbach.

At the end of the eighteenth century, Blumenbach, in classifying the human species into five great racial divisions, named the one to which he belonged the Caucasian race. He had chosen the *type specimen* of this race in the form of a particularly handsome (to Blumenbach's eyes, at least) skull from the Caucasus Mountains. Coon had been trained in this hoary tradition of the search for archetypal examples of human races, but his career, unlike Blumenbach's, overlapped with the growth of genetics. He was one of the first physical anthropologists to try to introduce the new genetic way of thinking into his work.

Coon's first introduction to fieldwork was a series of expeditions, in the 1920s, to the Rif Atlas mountains of Morocco's north coast. The peoples living in these mountains, though they had been converted centuries earlier to the Muslim faith, had remained relatively untouched by the great Arab migrations across the northern coast of Africa and into southern Europe. Coon found that they were extremely variable in their appearance, with skins ranging from dark to fair. Some of them were blond-haired or blue-eyed or both, showing many traits that we think of as Caucasian.

Coon, an excellent linguist, could soon speak passable Arabic, and he picked up enough of the Rif language to communicate with these romantic and warlike tribesmen. He traveled extensively through the area, often on foot or mule back, and made careful notes of the paths that the tribespeople used to move about. He wrote down their songs and the details of their ceremonies and was struck by the fact that repeated attempts by the Spanish and French to dislodge them from their strategic position guarding the route between Morocco and Algeria to the east had all failed. These people would, he was sure, be formidable opponents in any future war. After all, this land was the country of Abd-el-Krim, the leader who had fought the Europeans to a standstill and a mysterious figure whom Coon never managed to meet.

With the outbreak of the Second World War, Coon was in a unique position. After first being rejected by the army, he was recruited into the Office of Strategic Services and sent to Morocco. Among other things, his job was to foment rebellion, particularly among the tribesmen of the Rif whom he had come to know so well.

Fomenting rebellion turned out to be quite easy. The most visible enemy presence in Morocco was that of the Vichy French. The French, although they could not conquer the tribesmen of the Rif, had always treated them with contempt. Coon smuggled his tribesmen both money and arms, and trained them in sabotage. From his base at Cap Serrat near Tunis, he was able to coordinate a secret war that was carried out with great success by the Rif tribesmen. He was also the proud coinventor of a type of mine that was disguised to look like a pile of mule droppings, so that it could be placed on roads in full view. These mines destroyed a number of enemy vehicles.

While at Cap Serrat, he was injured in the head during a dive-bomber attack. The delayed effects of the injury caught up with him a year later, when he was waiting to be infiltrated into Albania—another remote area with which he just happened to be familiar as a result of his prewar anthropological excursions. Invalided back to the United States, he was eventually awarded the Legion of Merit.

Coon's appetite for adventure and keen observer's eye had already led him to amass a great deal of information about human diversity. After the war he went on a number of expeditions to the Middle East, where he excavated a series of caves containing Paleolithic deposits in Iran and Syria and incidentally measured and observed the peoples he found living there. His interests were broad, his style was exuberant, and he was very anxious to communicate his growing store of knowledge about the diverse peoples of the world to the general public. In the 1930s he had written two novels about the people of the Rif, and after the war he wrote a book about the history of the Middle East that for years was a bible for diplomats sent to the region. And those of you who remember the Paleolithic days of television might recall a program called *What in the World?* in which Coon and weekly guests would be required to identify obscure objects taken from museum collections. The program, which never managed to attract a sponsor, persisted on CBS from 1948 to 1964 as a charming relic of the days when the public airways were sometimes used by the commercial networks for the public's benefit.

Coon could easily have remained merely a distinguished academic and quasi-public figure to the end of his days were it not for a series of collisions between his ideas and the emerging social consciousness of postwar America. These collisions began with the publication of *A Reader in General Anthropology* in 1948, in which he compared twenty very different primate and human societies, ranging all the way from gibbons to the Roman Empire, in great detail. He arranged these societies in levels of increasing complexity, undiplomatically numbering them from zero to six. This kind of thing ran counter to the statement on human rights that had just been sent to the United Nations Commission on Rights by the executive board of the American Anthropological Association, a statement that said, among other things, that "respect for differences between cultures is validated by the scientific fact that no technique of qualitatively evaluating cultures has been discovered." Coon, who thought that he could distinguish the ancient Roman culture from the less complex culture of gibbons, disagreed.

His next book ran into even more flak. Coon, trained in an older tradition of anthropology, at first knew very little about the emerging discipline of human genetics. Indeed, books on human genetics and human races that were required reading when he was a student, like C. B. Davenport's *Heredity in Relation to Eugenics* (1911) and R. Ruggles Gates's *Heredity and Eugenics* (1923), were repositories of Eurocentric racism that make pretty distressing reading today and that tended to give human genetics a bad name. Coon's book *Races: A Study of the Problems of Race Formation in Man* (1950) was coauthored by his student Stanley Garn and an expert on Australian aborigines named Joseph Birdsell. It classified the human species into thirty geographic races, although it did not attempt to arrange them into a family tree. And it tried to apply some evolutionary theory to explain why these races were different in their appearance and their physiological adaptations.

Because of its subject matter, Coon's book tended to be lumped together with all those unfortunate books on human races from the twenties and thirties. Not only was Coon taken to task for bringing up the matter of race at all, but he was also castigated by evolutionists for rediscovering a number of well-known rules that explained features of animal evolution and for applying them to humans without mentioning that they had been applied to animals first. For example, Coon in his comparison of human races had rediscovered Bergmann's rule, that animals living in colder climates tend to have smaller body sizes to minimize heat loss, and the similar Allen's rule, that appendages like ears and noses tend to be smaller in cold climates for the same reason. Although these relationships were news to many anthropologists, including Coon, they were old stuff to evolutionists. Coon began to acquire a reputation among his fellow scientists for being quaintly narrow and out of touch, and further for being obsessed by the unfashionable topics of social and racial differences.

Coon, however, felt very uncomfortable with the older anthropologists' views of human evolution. Much earlier, his mentor Franz Weidenreich had wrestled with the origin of races. Weidenreich concluded that they must have originated independently, and that these independent origins were followed by a series of separate transitions to *Homo sapiens*. Weidenreich felt that these transitions must have been brought about by some force outside the ordinary evolutionary process that had driven prehumans toward humanity. Coon knew that Weidenreich's idea was wrong, but he tended to be out of his depth in human genetics and evolutionary theory. He earnestly wanted to apply these new scientific ideas to the study of human variation and human prehistory, and he set himself to learning more about them. So the chance offered by the air force to travel around the world, visit places he had never seen, and fill in the blank spaces in his knowledge of human diversity was too good to pass up.

The Origin of Races, drawing in part on his trip, was published in 1962, and even before its publication it was seized on by racist groups in the United States as justification for their beliefs. The book demonstrated scientifically, these groups announced, that whites were more evolutionarily advanced than were blacks. A book with such an inflammatory title would probably have been grist for the racist mill even if Coon had said nothing controversial in it whatever. Unfortunately, the ideas that he had developed about race formation were the very essence of controversy.

In his new book, Coon collapsed into five the thirty races that he had originally distinguished—Australoid, Mongoloid, Caucasoid, Congoid (the majority of African blacks), and Capoid (groups from southern Africa such as the !Kung*, who are strikingly different in appearance from most blacks, and indeed from most other human groups). He then suggested, on the basis of a detailed discussion of the fossil record, that these races, which he raised to the category of subspecies, predated the transition from a prehuman to a human state.

Two particularly inflammatory statements were singled out by reviewers:

> My thesis is, in essence, that at the beginning of our record, over half a million years ago, man was a single species, *Homo erectus,* perhaps already divided into five geographic races or subspecies. *Homo erectus* then evolved into *Homo sapiens* not once but five times, as each subspecies, living in its own territory, passed a critical threshold from a more brutal to a more sapient state.
>
> [and the second:]
>
> . . . it is a fair inference that fossil men now extinct were less gifted than their descendants who have larger brains, that the subspecies which crossed the evolutionary threshold into the category of *Homo sapiens* the earliest have evolved the most, and that the obvious correlation between the length of time a subspecies has been in the sapient state and the levels of civilization attained by some of its populations may be related phenomena.

In subsequent printings, Coon modified the first of these statements but not the second. He was both wounded and puzzled by the reaction

*The exclamation mark represents the sharp click that is an important part of the wide range of sounds in the Bushman languages.

to what seemed to him an eminently reasonable view of human evolution. As a man who had spent all his life communicating with and learning about other races, he could not understand the motivations of those who turned his work to their own ugly ends.

Coon's limited understanding of the evolutionary process was brought painfully to his attention by the distinguished evolutionist Theodosius Dobzhansky in a long and well-reasoned review of *The Origin of Races* that appeared—over Coon's objections—in *Scientific American.* Dobzhansky pointed out that speciation was very unlikely to happen in the way Coon envisioned it:

> A biological species can be likened to a cable consisting of many strands; the strands—populations, tribes and races—may in the course of time subdivide, branch or fuse; some of them may fade away and others may become more vigorous and multiply. It is, however, the whole species that is eventually transformed into a new species. . . . Coon is too competent an anthropologist not to know that man is, and apparently always was, a wanderer and a colonizer, and that as people come into contact gene exchange takes place.

Here Dobzhansky states clearly the idea that much gene flow has taken place in the course of human evolution. As we saw in the last chapter, the idea of copious gene flow of certain human-generating genes throughout the species has since been added by Wolpoff and others to Coon's original scheme. Yet Coon had already explicitly ruled out gene flow in some cases:

> There are no hybrids in the Pygmy camps of the deep [West African] forests. To live like a Pygmy you have to be one. Gene flow between Pygmies and Negroes is thus a one-way stream which may have made the Negroes biologically more adaptable to forest living than were their ancestors out on the savannas and grasslands.

Elsewhere in his review, Dobzhansky actually suggested a genetic underpinning for Coon's view:

> We can agree with Coon that natural selection has favored, in human races everywhere in the world, a *sapiens*-like genotype over an *erectus*-like one. This can be stated most clearly as follows. The transformation of prehuman populations into human populations involved a feedback process between the genotype and the environment; *sapiens* genes favored the development of capacities for symbolic thinking, language and eventually for civilization. These

human adaptations in turn made human genes essential for survival and opened the way for further adaptations.

Here Dobzhansky almost, but not quite, stated a central theme of the present book—simply substitute "cultural selection" for "natural selection," and you have it in a nutshell. (The term *genotype* in the quotation refers to the entire complement of genes carried by an individual, as opposed to its *phenotype,* the individual's actual appearance and behavior that is in part a result of its genotype.)

Dobzhansky, as will be seen later, was the world authority on speciation in the fruit fly *Drosophila*. Central to his view of speciation was the idea that all natural populations, including those of humans and *Drosophila,* are full of genetic variation (his "cable of many strands") and that much of the speciation process involves the sorting out and recombining of this variation into new patterns. It is perhaps not surprising in retrospect that Dobzhansky, repelled by the racist interpretations that were being made of Coon's work, did not try to close the gap between his own genetically sophisticated views on race formation and Coon's more naive ones. The purpose of his review, after all, was to try to defuse the racist interpretation of Coon's book, although one imagines that the review's carefully reasoned tone had little effect on the demagogues.

Yet he could have closed the gap, in a way that would have given little comfort to the racists. Early in his book, Coon had given a simple diagram of his idea of multiple origins, a diagram that I have reproduced in figure 2.1. The circles represent distinct races. In the top cluster of circles, a single mutation has arisen in one of the races (the circle numbered 3), spread by migration to all of them, and caused them all to evolve into a new species more or less synchronously. In the bottom cluster of circles, two very similar mutations arose independently in races 3 and 5. Even though in this particular cluster of circles no gene flow takes place from the other races to race 5, race 5 can still evolve toward a new species along with the others because it acquired the same mutation independently.

Ironically, if you embrace Coon's genetics, then his theory of multiple origins becomes extremely unlikely—many highly similar mutations would have to arise independently in different races in order for multiple transitions from *erectus* to *sapiens* to take place. But if you embrace Dobhzhansky's view that much of the variation needed to accomplish the transition *was already present* in the gene pool of our ancestors such as *Homo erectus,* and if you couple it with strong cultural

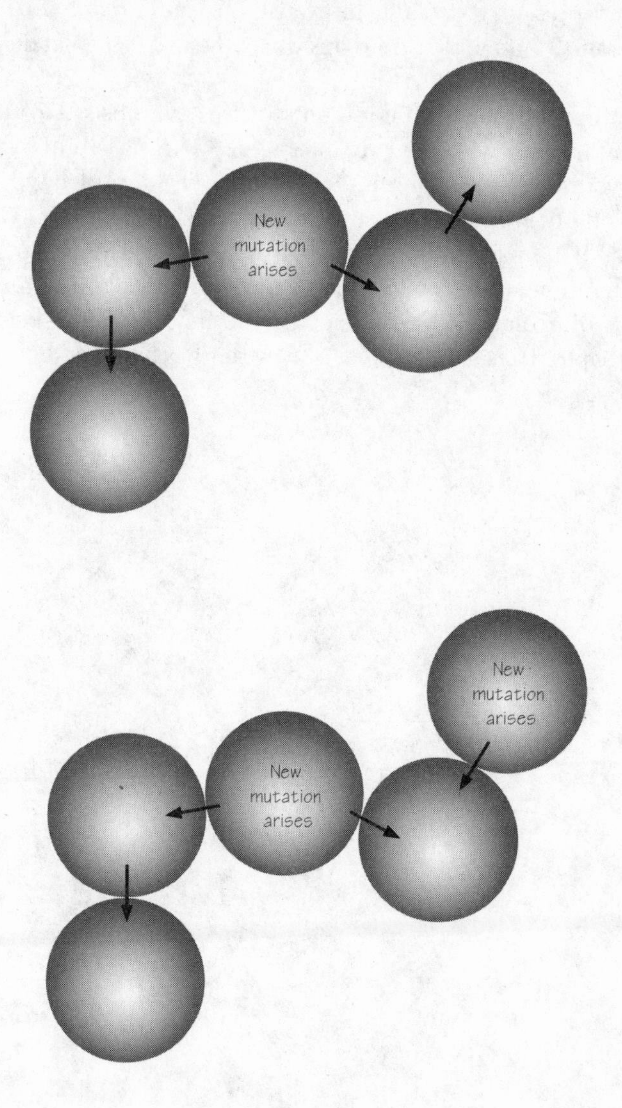

FIGURE 2.1. Carleton Coon's diagram of how a species made up of a cluster of subspecies (or races) might evolve. In the upper part of the diagram, a favorable mutant gene arises in one of the subspecies and over time spreads to the others. In the lower part, two independent occurrences of the *same* favorable mutant gene take place in two different subspecies. Although less migration is required to spread the gene through the species, this is counterbalanced by the reduced likelihood that the same mutation will arise more than once.

evolutionary pressures driving the transition from *erectus* to *sapiens,* then suddenly Coon's multiple-origins theory becomes a good deal more likely.

After setting out all the ingredients of this synthesis, Dobzhansky backs away from it with a bald statement: "The possibility that the genetic system of living men, *Homo sapiens,* could have arisen independently five times, or even twice, is vanishingly small." Why did he say this? Is human evolution really *that* unique? The feeling that it really is, that human evolution is somehow qualitatively different from other kinds of evolution, has permeated the field from the beginning. The next chapter tries to give a hint of why this has happened.

3

An Obsession with God

Evolution, which offers a passage to something that escapes total death, is the hand of God drawing us to himself.

—Pierre Teilhard de Chardin (1951)

Throw Another Bat on the Barbie

About 30 miles southwest of Beijing, China's current capital, lies the little village of Zhoukoudian. The village is on an alluvial plain, and running north and east from it is a range of low limestone hills. The limestone was laid down by the action of small marine organisms in Ordovician times, almost half a billion years ago. Most of the deposits accumulating after that time have been eroded away by the rain and by the bitter winds from central Asia that still send dust from China halfway around the planet. A few islands of this younger sedimentary rock perch here and there at the tops of the ancient hills. And there are other geological traces of the intervening years, notably coal deposits that were laid down during the Carboniferous, some two hundred million years after the Ordovician. Heavings of the earth have by now caused the limestone and the coal seams to be jumbled together.

A million years ago the hills were higher than they are now, extending further toward the site of the present-day village. They were heavily wooded, and the climate was moister and colder than it is today. About two thirds of the species of animals that roamed the hills and the nearby plains are no longer with us—among them giant marmots, short-faced hyenas, large hunting cats, and cave bears.

The bears were attracted to the region because the hills were honeycombed by an extensive series of caves, formed in the limestone by the action of seepage. Some of these caves consisted of large chambers connected to each other at different levels. There were often a number of entrances to the various cave systems, hidden by brush and trees. Inside the caves, near the entrances, a modest layer of bat guano, droppings of other animals, and bones of animals that had been dragged there by predators covered the limestone floor. In the caves' quieter depths, stalactites and thicker limestone deposits accumulated without disturbance.

All this changed with the arrival of humans.

The first signs of human activity can be dated to about seven hundred thousand years ago, with some uncertainty at Zhoukoudian and with greater certainty elsewhere in China. Although it is hard to put an exact date on these earliest arrivals, it is clear that during the subsequent half million years the detritus of human existence slowly filled up the caves and eventually blocked off the lower entrances. Ash from fires accumulated on the cave floors in uneven hummocked deposits, showing that the inhabitants raked the ashes up into heaps in much the way that campers today rake together half-burned logs to keep their campfires burning a little longer. Limestone blocks that had perhaps been used to form primitive hearths were calcined by the heat into lime. Bits of broken and burned bones accumulated, representing the remains of meals that ranged from thick-jawed deer to (perhaps a reflection of leaner times) rats and bats. The dwellers in the caves seem to have used surprisingly few tools. Some primitive flakes and scrapers, made from quartz fragments gathered from outcrops in the hills a mile or two away, eventually found their way into the deposits when they were worn out.

In spite of this long period of occupation it was a rare event when human remains were added to the slowly accumulating layers of detritus. It is impossible to tell now whether these bodies were those of people who had deliberately been buried in remote parts of the caves, had crawled away from the flickering hearths to die, or perhaps had simply been dragged away by their fellow tribesmen from areas that were occupied at the time. So rare are the bones that they must mark highly unusual events—for the most part, these people seem to have been taken elsewhere after they died.

As detritus accumulated and blocked the lower entrances, the caves' inhabitants learned to climb down through the chimneys and fissures that led into the caves from above, living, working, and reproducing on the surface of the gradually thickening layers of waste material. Millennia went by, and the climate gradually warmed. Deer remained plentiful, but the marmots and cave bears were replaced by porcupines, small straight-tusked elephants, noisy bands of large macaques, and even the occasional ostrich. Then, some two hundred thousand years ago, the people living on the slowly rising top of this primitive landfill ran out of room as even the upper chambers became choked with debris. When this happened, it seems, they left.

The span of human activity captured by these deposits is enormous by the standards we usually apply to human existence, although of course it is no more than an eye blink in geological time. Consider that five thousand years ago human civilizations in China, Mesopotamia, and Egypt were still using bronze rather than iron and only beginning to employ the wealth generated by agriculture to extend their power and influence. The first alphabets were just being introduced by the traders of the eastern Mediterranean and by the priests of Ur and Thebes. All written human history has happened since that time. Yet, the people of the caves of Zhoukoudian crouched over their smoky fires, eating their half-cooked bats, for a *hundred times* as many years as we have been recording our civilization.

Found and Lost

As the present century opened, the limestone quarries and coal mines of Zhoukoudian were worked as they had been for almost a thousand years by gangs of indentured laborers under conditions almost indistinguishable from slavery. During all this time, the laborers and the people of the village were continually finding old bones and selling them in local markets. The bones were particularly abundant in the nearby humps of limestone that the villagers called Chicken Bone Hill and Dragon Bone Hill. They were ground up and used as poultices on injuries, and suspensions of them were drunk to cure a variety of illnesses. This practice, widespread throughout China, has actually led to the discovery by scientists of many important paleontological sites, although at the same time it has probably led to the destruction of many irreplaceable pieces of information about China's distant past.

Many human and prehuman fossils, particularly those found in East Africa, are exposed by weathering and will, if they are not soon discovered, be rapidly destroyed by the very weather that exposed them in the first place. They are usually replaced by a new crop of fossil bones as the formation continues to erode away. Cave deposits, however, like those at Zhoukoudian, are very different. Once plundered or destroyed, they are gone forever. When, in 1903, the German paleontologist Max Schlosser came across a fossil human tooth that had been purchased in an apothecary shop in Peking (now Beijing), it was undoubtedly from just such a long-vanished deposit. This discovery immediately alerted paleontologists to the possibility of finding fossil humans in northern China.

Scientific excitement grew when the Jesuit Émile Licent found some chipped-stone artifacts in Gansu province near the western end of the Great Wall in 1920. Not remarkable in themselves, these finds stirred a good deal of interest in the paleontological community because of Henry Fairfield Osborn's well-publicized theory that the origin of humans must have been in central Asia. The whole region was crying out for exploration, and the logical base for mounting this effort was Peking, with its large European community.

Two prominent members of the tiny world of science in Peking were Davidson Black, a Canadian doctor at Peking Union Medical College, and the mining adviser John Gunnar Andersson. Both were enamored of Osborn's theory and planned a central Asian expedition. They suggested to a young visitor, the Austrian paleontologist Otto Zdansky, that he join them. Andersson thought that Zdansky might want to do some digging at nearby Zhoukoudian, to get some idea of what he would be up against when they ventured further west. To everyone's surprise, during the summer of 1921 Zdansky found two very old but very human-seeming teeth in the Zhoukoudian deposits. This site, a day's mule trip from Peking, was a great deal more accessible than central Asia, and Black and Andersson decided to concentrate their immediate attentions there. The land was purchased from the mining company, and large numbers of laborers were hired.

All this came in the nick of time. A railway spur from Peking had just been completed. Because this made the introduction of excavating machinery possible, it would not have been more than a decade or two before the priceless deposits of the Zhoukoudian caves were excavated, discarded, and destroyed.

The deposits in the hills above Zhoukoudian were enormous in extent. In places the walls of the ancient caves had eroded away to reveal the deposits within, like the yolk of a broken hard-boiled egg. As a consequence, it was feasible to excavate down through layer after layer in the open air rather than underground, although occasionally fissures were found that led to smaller chambers within the main caves. Many other caves nearby were excavated at the same time, some of them apparently representing different parts of what at one time had been a single large cave complex. The largest of the deposits, however, soon to be named site 1, was also the most productive.

Masses of animal bones and chips of quartz were found as the excavators slowly made their way down through the first half of the 40-meter-thick deposits. There was much evidence of human activity, but no more human fossils were found after those two tantalizing teeth.

In the winter of 1929, there were no Westerners at the site. The man in charge of the digging was Pei Wenzhong, who had received all his training at the site and was nervous about his responsibility. On the evening of December 2 a fossil-rich branch of the main cave was being worked on. When the laborers excavating this little crevice found something round sticking out of the frozen muck, they called on Pei. He dug away at the object and found that it was the top part of a skull. The excited telegram that he sent to Black in Peking read, "Found skullcap—perfect—looks like a man's."

In fact, the skullcap did not look like a man's, for there were some remarkable differences. From the top it looked not so much like a skull as a flask-shaped bottle, with a pronounced lip and a narrow neck (see plate 2). The lip of the flask was actually an enormously developed supraorbital shelf, a ridge of bone above the eye sockets that was so large that it extended well to either side of the sockets below. The shelf was continuous, rather than forming two distinct ridges above each eye as in the much more recent Neanderthal man. The dome of the skull was long and low, with a slight forehead, and was sufficiently well preserved that a good estimate of the size of the brain that once occupied it could be made. This estimate turned out to be about 1,200 cubic centimeters, about four-fifths as large as the brain of the average modern human or Neanderthal. Small as it was, it was actually well within the range of current human brain sizes. Many perfectly normal modern people of small stature have brains as small as 1,000 cubic centimeters or even smaller. Nonetheless, the shape of the skull with its low vault and enormous projecting browridge made it far more primitive in appearance than any other human skull that had been found.

Over the next seven years as the excavations continued, skulls and fragments of postcranial skeletons were found that had belonged to about forty individuals of various ages. These were primarily concentrated in the layers just below those where Pei had made his first discovery. The excavators themselves had only the vaguest idea of how old the bones and ash deposits might be, for of course absolute dating methods using radioisotopes lay far in the future. (Since the Second

World War, a variety of dating methods have been applied to the deposits. Although there are conflicting results, the bulk of the evidence seems to be that the layers where the most plentiful remains of Peking man were found date to about 450,000 years ago.)

It was possible even then, however, to get a good idea of the ages of these deposits relative to other deposits nearby. The Jesuit priest Pierre Teilhard de Chardin, sponsored by the National Museum of Paris, arrived in China in 1923 and immediately set out with Licent and others to survey the geology of the whole northern region.

Teilhard was no newcomer to fieldwork. His career in paleontology had already spanned some of the most important events of the age. While still training for the priesthood in southern England, during one of the periods when the Jesuits had been banished from France, he had met the amateur geologist Charles Dawson and aided him in his excavations of the Kentish quarry where fragments of the jaw and skull attributed to Piltdown man had been discovered. In 1913, during one of his subsequent visits, he was helping to sieve gravel from the area where the skull and jaw had been found the previous year. His sharp eye spotted a canine tooth that was later shown to fit perfectly into the jaw.

This find had boosted young Teilhard's scientific reputation but also had the unfortunate effect of adding a touch of apparent authenticity to a fossil that some paleontologists were already beginning to question. Piltdown man was finally shown to be a fraud nearly four decades later, a fraud made up of the ingenious mating of a human skull with the jaw of an orangutan, neither apparently more than about five hundred years old. The hoax was accepted for so long because it was exactly what the English scientific establishment had wanted to find—a large-brained ancestor of the English race, with an apelike jaw to show that it was primitive but with a lofty brow that made it seem so much more intellectual than the brutish Neanderthals that had been found on the Continent.

Like everyone else even remotely connected with Piltdown man, Teilhard has since been accused of being a party to the fraud. But his life of utter probity and intellectual honesty before and after the incident makes it seem as unlikely as the possibility (which has also seriously been suggested) that the Prince of Wales was Jack the Ripper. It seems probable that the tooth had become separated from the jaw while the fake was being planted and that Teilhard's discovery is simply a tribute to his keen eye.

On his arrival in China, Teilhard set to work with enthusiasm. He had an iron constitution and was able to survive with ease the rigors of the food and the primitive accommodations in that largely unexplored part of the world. Some of his most remarkable finds were on the perimeter of the Ordos plateau of Inner Mongolia, where there were huge deposits filled with the bones of extinct mammals from the Upper Pleistocene epoch, now known to be half a million or more years old. Many of these same species were also found at Zhoukoudian. Teilhard, continuing his explorations of the area for the next decade and a half, wrote paper after paper that put the deposits at Zhoukoudian in a much wider geological context.

In 1934 Davidson Black died suddenly of a heart attack, at the very desk on which the precious Peking man skulls rested. Teilhard took over briefly as acting director but was soon replaced by the German anthropologist Franz Weidenreich. By this time the excavations at Zhoukoudian, hampered by the worsening political situation, were winding down. At the close of the digging season in 1937, the major part of the work at site 1 had been completed, and Teilhard left for Europe.

Then the Sino-Japanese War swept through the area. As many as two hundred excavators had been employed at the site during the peak activity, but now with the arrival of the invading Japanese all work was halted. The laborers melted away, and some of them joined guerrilla bands of the anti-Japanese resistance.

The most important of the fossils disappeared in the confusion following Pearl Harbor, although the Japanese government apparently knew nothing about it. Two Japanese scholars who had visited the institute before Pearl Harbor had submitted a report on the excavations to the Ministry of Education. The report was forwarded to the emperor, who ordered a search for the fossils. Thus it was the Japanese who announced to the world in early 1942 that the fossils were missing. Feeble attempts were made by the Japanese to continue excavations at Zhoukoudian, but the Chinese workmen had covered the site with a layer of rubble before they abandoned it. In apparent retaliation for the lack of Chinese cooperation, in April 1942 the Japanese military ordered the remaining books and bones at the Cenozoic Laboratory destroyed.

The loss of the fossils was tragic enough, although luckily Weidenreich had managed to take casts and detailed drawings to the American Museum of Natural History in New York in 1937. Far more serious, however, was the scattering and destruction of the thousands of specimens of bones and tools and the bits of matrix in which they had been embedded and that had been painstakingly collected and organized over two decades of excavation. It was these specimens that would have given a context to the fossils. Perhaps with more sophisticated analysis they could have revealed details of how these primitive people lived.

All was not completely lost. Some of the specimens, mostly broken and without labels, were found in a storage room after the war. They were jumbled together with samples from other parts of China. After months of painstaking work, about 20 percent of the material was recovered.

The Search for Omega

It was to this world of upheaval and tragedy, both human and scientific, that Teilhard found himself returning. He had been essentially banished to China for a second time by his Jesuit superiors as a result of his attempts to link the teachings of the Church with the scientific study of evolution.

He returned to Peking from France, via America, at the end of August 1939. His time in France had been spent fruitlessly trying to persuade the generalate of the Jesuits to give him permission to publish, not his scientific writings—there had never been any problem with those—but his writings on the linkage between evolution and religion. Perhaps they might have allowed him to publish—Teilhard was persistent and persuasive and had many friends and admirers. But Pope Pius XI was dying, and the Church was in upheaval over the uncomfortably close ties that he had forged with Hitler and Mussolini. In an attempt to remove one more irritation, Teilhard's superiors sent him back to China, promising vaguely that on his return they would see about allowing him to publish—provided of course that they had final editorial say. His exile was intended to be a brief one, but China was to be Teilhard's involuntary home until the end of the war.

He found a grim situation in Peking. The imperial city, firmly under the thumb of its Japanese rulers, had lost its prewar bustle and was a cold and cheerless place. Much of its intellectual life had disappeared as well. Teilhard, who had been at the center of that life, was now forced to take up residence in a gloomy Jesuit hostel miles from his office in the Cenozoic Laboratory at the medical school. The superior at the hostel accused him of being a communist or an evolutionist—the two were indistinguishable in his eyes—and set about making things as unpleasant for Teilhard as possible.

Through all this upheaval, Teilhard was writing a book, the culmination of an anguished intellectual life that had long been split between the poles of science and religion. He wanted to reconcile those two very different views of the world and to persuade both scientists and theologians of the importance of his unifying vision:

> As a purpose in life, my science (to which I owe so much) seems to me less and less worthwhile. For a long time now, my chief interest in life has lain in some sort of effort towards a plainer disclosing of God in the world. It's a more killing task but it's my only true vocation and nothing can turn me from it.

The book he worked on during this period of exile was a synthesis of his own thinking about evolution. He called it *The Phenomenon of Man.* It was not, to say the least, what the scientific community might have expected.

Like many of his predecessors and contemporaries, Teilhard rejected utterly what he saw as the grim purposelessness of Darwinian evolution. He was convinced that evolution had a goal. But here he parted company even with his colleague Franz Weidenreich, who unlike most Darwinian evolutionists supposed that humans had somehow directed their own evolution.

Teilhard had a much more elaborate view. He thought that the goal of evolution was far larger, a synthesis of God, human consciousness, Christianity, and the ever-growing sentience of both the animate and inanimate parts of the rest of the planet. This synthesis was to take place at some time in the future that he called the Omega Point. After the Omega Point was reached, the world would have only a single mind and a single heart. To him, Christianity was an essential part, but only a part, of this goal-directed evolutionary process.

On the way to this vision he touched on such knotty problems as original sin and the Holy Trinity. The Jesuit superiors who read his manuscript with mounting horror discovered that Teilhard was supplying evolutionary reasons for these fundamental tenets of the church. Teilhard even supposed that God was not above evolving, when he "Trinitized" himself.

Teilhard was finally able to leave China and return to France in 1946, after the end of the Japanese occupation. He went to Rome the following year to try to get permission for *The Phenomenon of Man* to be published. His request was rejected, in spite of the vague promises that had been made to him before the war. Further, the manuscript helped to stimulate a highly reactionary encyclical from Pope Pius XII, *Humani Generis,* in which the whole theory of evolution was severely questioned.

The Phenomenon of Man and many other of Teilhard's philosophical writings were eventually published after his death in 1955. Despite the Church's fears they have had little effect, although they have managed to generate a small-scale cult following. In retrospect, it is difficult to understand Teilhard's puppy-doggish naïveté in assuming, despite repeated rejections, that the Church hierarchy would eventually relent and let him publish what to them was a horrifying collection of heresies. But, interestingly, it is less difficult to understand the problems that he had with evolutionary theory as it was being developed and elaborated by his scientific colleagues.

Teilhard and the Geneticists

Teilhard *knew,* with a certainty that consumed his life and tortured him with his inability to tell the world about it, that the evolution of human consciousness was distinctly different in kind from anything else that had happened in the planet's history. He thought of it as the accumulation through human evolution of reflective monads that transcended the general planetary consciousness possessed by all living and inanimate things. These monads formed a new and distinctly different kind of consciousness that he called the noosphere, which in turn was nothing more than an evolutionary way station toward that far grander goal, the Omega Point, and the establishment of the ultimate state of consciousness, the theosphere.

None of this, of course, can be found in conventional evolutionary theory, which is terribly dull and unpoetic by comparison. The currently accepted neo-Darwinian theory of evolution is firmly grounded in genetics. Its central basis is that the genetic material possessed by all the organisms of a species forms a common gene pool, which is passed on from one generation to the next. Evolutionary changes in this gene pool alter its composition over time. As the genes alter, so do the organisms that carry them. The alterations that occur in the gene pool are the result of mutations that change the genes, but this is only the first step. Over time, natural selection and chance events cause mutant alleles of various genes to spread through the gene pool, where they replace the older versions of the same genes. This occurs because organisms carrying these mutant alleles leave more offspring than the average. The result is that copies of favorable alleles tend to be passed to the next generation in larger numbers, while copies of unfavorable ones are passed on in smaller numbers. It is this process that, as we saw earlier, was diagrammed so clearly by Carleton Coon in his book on the origin of races. Sometimes the genes on the chromosomes increase in number, by processes we will examine in more detail later, and the new genes can eventually evolve to take up new tasks.

Contemporary evolutionary theory states that it is these processes that cause new species to arise. These same processes have over very long spans of time given rise to the astonishing diversity of life on the planet. It is these processes, and no others, that have been responsible for the ability of a bat to catch small flying insects in pitch darkness, for the ability of a sperm whale to dive a mile into the depths of the ocean, and for the ability of Beethoven to write the *Eroica* symphony. Evolution is not moving toward a goal, and in particular it is not moving toward the goal of human intelligence. Darwin's concept of natural selection thoroughly removed any need for a deity to mediate the evolutionary process. It has been a basic tenet of conventional evolutionary biology, ever since the neo-Darwinian synthesis linked natural selection with genetics in the 1930s, that human evolution can be explained by precisely the same forces of mutation, selection, and chance events that have shaped every other organism on the planet. No more and no less.

Yet Teilhard was not alone in rejecting conventional evolution. Confronted with the question of the evolution of human consciousness and the human brain, many respected scientists and philosophers have fallen back on theological explanations. Alfred Russel Wallace, the codiscoverer with Darwin of the concept of natural selection, argued for years with Darwin about whether, as he wrote in his autobiography, "some different agency . . . came into play in order to develop the higher intellectual and spiritual nature of man." In 1869 Darwin admonished him: "As you expected, I differ grievously from you, and I am very sorry for it. I can see no necessity for calling in an additional and proximate cause in regard to man."

Unconvinced, toward the end of his life Wallace wrote (1905), "The ultimate purpose [of evolution] is (so far as we can discern) the development of mankind for an ultimate spiritual existence. . . . Our Universe, in all its parts and during its whole existence, [is] slowly but surely marching onwards to a predestined end."

Many others were as puzzled as Wallace about how the huge and obvious differences between humans and other animals could have arisen. Robert Broom, the remarkable paleontologist who discovered the Australopithecines of Sterkfontein and Kromdraai, and whose life and work we will examine in more detail in the next section, embraced theology with only the slightest hesitation at the end of his book *The Mammal-like Reptiles of South Africa and the Origin of Mammals* (1932):

> We seem almost driven to assume that there is some controlling power which modifies the animal according to its needs, and that the changes are inherited. Apart from minor modifications evolution is finished. From which we may perhaps conclude that man is the final product; and that amid all the thousands of apparently useless types of animals that have been formed some intelligent controlling power has specially guided one line to result in man.

In his book *Creative Evolution,* the distinguished and influential philosopher Henri Bergson wrote (1944):

> Among conscious beings themselves, man comes to occupy a privileged place. Between him and the animals the difference is no longer one of degree, but of kind. . . . So that, in the last analysis, man might be considered the reason for the existence of the entire organization of life on our planet.

Much more recently, Sir John Eccles, the distinguished neurophysiologist, has wrestled with the origin of the human brain in his book *Evolution of the Brain: Creation of the Self* (1989). After a careful and usually Darwinian discussion of the processes and forces that have shaped our brains, at the end of his book he faces the necessity of trying to explain human consciousness:

> Since materialist solutions fail to account for our experienced uniqueness, I am constrained to attribute the uniqueness of the Self or Soul to a supernatural spiritual creation. . . . I submit that no other explanation is tenable: neither the genetic uniqueness with its fantastically impossible lottery, nor the environmental differentiations which do not *determine* one's uniqueness, but merely modify it. . . . For this reason I superimposed a finalistic concept on the materialistic explanations of Darwinism to which I faithfully adhered in the first nine chapters. It has to be conjectured that there was a final goal in all the vicissitudes of biological evolution.

One is tempted to imagine the hollow voice of Darwin asking, from his tomb in Westminster Abbey, "O thou of little faith, wherefore didst thou doubt?"

I could easily list hundreds of quotations like these. We have to ask why, when they were confronted with the fact of human evolution, so many distinguished thinkers abandoned Darwinism, or banished it to the world of nonhuman organisms. Why were so many of them forced to resurrect the eighteenth-century world of a ladder of creation, with humankind at the top? Or, like Teilhard, to elaborate even more rococo structures? Simply because, in spite of the goallessness explicitly demanded by contemporary evolutionary theory, it is obvious to anyone that humans are distinctly different, astonishingly different, from any other organisms on the planet. No chimpanzee, regardless of how intelligent it might be in chimpanzee terms, could write the *Eroica* symphony. And despite all the accumulated scientific information that shows beyond a shadow of a doubt that evolution is not progressive, that it can move as easily toward simplicity as toward complexity, this does not seem to have happened with humans. The paleontological history of the human species that has been unearthed by scientists over the last century and a half looks to the nonscientific rest of the world (and indeed to many scientists) like a remarkable, continuous, and accelerating progression toward a goal. That goal is undoubtedly ourselves, *Homo sapiens sapiens.* This progression of human evolution is obvious to anyone, not only to Nobel Prize–winning philosophers like Henri Bergson or to puzzled, tormented theologians like Teilhard. Teilhard knew that he and his contemporaries had advanced intellectually far beyond the people of Zhoukoudian. Those primitive people ate their half-toasted bats in the flickering light of their smoky fires, understanding their world only dimly and continually subject to formless fears. Teilhard, on the other hand, was capable of wrestling with concepts that Peking man, surely, was unable to comprehend.

What is obvious, of course, is not always true. It is exactly this collision between, on the one hand, what most people perceive as having happened in human evolution and on the other hand the neo-Darwinian explanation for evolution in general that the next sections of this book will try to address. There *is,* as I will show, a good, consistent, even conventional neo-Darwinian explanation for human evolution. This explanation can even account for why our evolution appears to have been so different from that of other organisms and why it has every appearance of being somehow directed toward a goal. There is nothing mystical, magical, or theological about this explanation, but it tells a remarkable and exciting story nonetheless. To understand this story in its full complexity, we must range across the face of our planet, across millions of years of time, and into the interior of the living cell and the depths of the human brain.

II

THE BONES

4

The Crystal Brain

It is therefore probable that Africa was formerly inhabited by extinct apes closely related to the gorilla and chimpanzee; and as these two species are now man's nearest allies, it is somewhat more probable that our early progenitors lived on the African continent than elsewhere. But it is useless to speculate on this subject, for an ape nearly as large as man . . . existed in Europe during the upper Miocene period; and since so remote a period the earth has certainly undergone many great revolutions, and there has been ample time for migration on the largest scale.

—Charles Darwin, *The Descent of Man* (1871)

In 1924, Johannesburg was still a raw frontier town founded on the incredible wealth of gold that had been discovered in South Africa's Witwatersrand. It was, however, acquiring some of the trappings of civilization. The University of the Witwatersrand had recently established a medical school, and the young Raymond Dart had just been appointed professor of anatomy.

Dart knew only a little about fossils, although he was struck by how fossil-rich this part of South Africa seemed to be. He had received excellent training in both human and comparative anatomy at University College, London, some of it from the remarkable anatomist Grafton Elliot Smith. This part of his intellectual pedigree was to have important consequences.

Smith, an Australian, had worked his way up through the English scientific establishment at a time when it was not noticeably friendly to colonials. While at the University of Cairo, and before moving to Manchester and later to London, he had called attention to the fact that it was possible to study the brains of fossil men and animals by making casts in rubber of the inside of the skull and drawing them out through an opening. Such *endocranial casts* should, he proposed, tell a great deal about the size, shape and convolutions of the brain.

Sometimes they told too much. When he had been presented with such a cast from the Piltdown man skull, now known to be that of a modern human, he had pronounced it with immense authority to be "the most primitive and most simian human brain so far recorded." (Smith is one of the stars in the ever-shifting cast of suspects in the Piltdown case, and ends up as the accused in Ronald Millar's book on the subject—never mind that he was in Egypt at the time that Piltdown man was "discovered.")

In spite of his incautious statements about Piltdown man, of which he was a passionate advocate, Smith was an excellent anatomist. During his studies with Smith, Dart had been given a thorough grounding in the external anatomy of the brain and had studied many different endocranial casts. He had been particularly struck by Smith's emphasis on a new anatomical feature, a small crease in the rear of the brain. This little crease, called the lunate sulcus because of its resemblance to a crescent moon, was clearly visible in the brains of apes. In humans, however, because of an expansion of that region of the brain, it had been displaced and buried beneath the expanded cortex. It was not visible at all in many human specimens.

Dart was also much impressed when, shortly before leaving England, he attended a meeting of the Anatomical Society and saw a skull that had recently been unearthed at Broken Hill, a mining area just north of Lusaka in what was then Rhodesia. This skull, with heavy browridges and a sloping forehead but very human dentition, was quite the opposite of the ever-more-puzzling Piltdown man.

Dart's intellectual history bears out Pasteur's dictum that chance favors the prepared mind. He was one of only a handful of people on the planet who could have understood the fossil that was about to fall into his hands, a fossil that is certainly one of the most important relics of our prehuman ancestry.

A young student in Dart's Witwatersrand anatomy class, Josephine Salmons, had a family friend who was a director of the Northern lime works. She had noticed the skull of a fossil baboon on her friend's mantelpiece. It had come from the company's works near the village of Taung, some 550 kilometers to the southwest of Johannesburg. Dart was inclined to dismiss her identification until he saw it for himself. It was a baboon skull, all right, obviously very old because replacement of the bone with limestone was essentially complete. Dart suddenly became very interested in the Taung lime works.

The limestone cliff at Taung had been mined for decades, most recently by the application of dynamite to blast away the hard concretions and break the limestone up into pieces small enough to be baked in the ovens. Every once in a while the miners would blast their way through the filled remnants of an old cave, made up of hard breccia in which fossil bones were sometimes embedded. One of the miners, a man named de Bruyn, had been collecting the more interesting of these fossils for some time as a hobby, and some of these were sent to Dart by rail.

This paleontology by post paid off in the most breathtaking way. The first two boxes from Taung arrived just as Dart was getting ready to be best man at the wedding of a friend. Dressed in his wedding finery, Dart tore the lids off the boxes. The first held nothing of obvious interest. Lying on the top of the heap of fragments in the second, however, was the endocast of a brain.

Dart recognized it immediately, even though it was a natural endocast rather than an artificial one. It had been formed from lime-laden water and mud that had filtered into the hollow of the skull and slowly filled the cavity. The skull, he could see, had lain on its left side, for the right side of the endocast was not complete—instead of a right hemisphere there was a mass of sparkling calcium carbonate crystals. The skull had lain in the deposit at Taung, perhaps for millions of years, but its brain cavity had not had a chance to fill completely with limestone. Then the miner's dynamite charge, suddenly blowing the surrounding skull away into fragments, had left the crystal brain intact (see plate 3).

The floor of the skull, the temporal and parietal bones that form its sides and roof, and the occipital bone that forms the skull's rear must all have been firmly attached to pieces of rubble from the explosion. There was no sign of them among the fragments in the box, and indeed they were never recovered. Dart held in his hand the cast of a brain different from any he had ever seen. Clearly visible on its surface were impressions of the spreading blood vessels that had nourished the membranes in which the brain was wrapped in life, and there was even some indication of the shape of the various convolutions of the cortex.

The first thing that he saw as he held the brain in his hand was that, even though it was immature, it was larger than that of an adult chimpanzee—although still far smaller than that of an adult human. The occipital lobe at the back of the brain was the best preserved part of it, and he thought he could see the lunate sulcus. It had been pushed back by the expanding cortex in front of it, pushed back much farther than in the brain of a chimpanzee. Dart immediately surmised, during the first few moments of discovery, that he held in his hand the brain of a creature clearly intermediate in its characteristics between a human and a chimpanzee.

Rummaging frantically in the box, he made an equally important discovery, a piece of concreted breccia into which the front of the brain cast fit perfectly. In the heart of this second piece of stone, quite invisible, must be the face that belonged to the brain. By itself, the stone gave little clue to what lay within it. Like the fossils of the bones forming the vault of the skull, it might easily have been discarded or flung into the ovens of the lime kiln.

Dart had no experience with this kind of thing, but he knew he must treat the fossil face with the greatest care. Over the next several months, he chipped away at the stone as gently as possible, much of the time with a knitting needle. After seventy-three days of painstaking work the front part of the rock finally split off, and he could see the skeletal face for the first time. It was the face of a child with all its milk teeth still intact and the first permanent molars in the process of erupting. It was not as forward thrusting or prognathous as the face of an adult ape, and it was also less prognathous than that of an immature ape of the same physiological age (see plate 4).

Dart immediately sent a preliminary account of his findings to the British scientific journal *Nature* and then went on painstakingly cleaning the skull of fragments of rock. For a long time the lower jaw was still attached firmly to it, so it was not possible to get a good look at the teeth. It took nearly five years of work before he finally managed to free the jaw so that the occlusal surfaces of the teeth could be examined properly. But even before this, it was obvious that the child's teeth were very unlike those of apes and much more like those of humans.

Two immediate features were apparent. First, the canines were very much smaller than those of apes, as small indeed as those of humans. Because they did not project beyond the other teeth, the child would have been quite able to move its jaw in the circular grinding pattern typical of humans, something that the apes cannot do. As a result of this movement, Dart suspected, the molars would show the typically human pattern of wear in which the cusps of the teeth are worn down evenly. When he finally freed the jaw, he found that his prediction was correct.

Second, when the jaw was viewed from above, the teeth were evenly spaced along it in a smooth parabolic curve, with the lines of molars on each side of the jaw diverging rather than running parallel as they do in apes. And one prominent feature of the dentition in the great apes, a gap in the row of teeth of the lower jaw called a diastema, was not present. Because the upper canines project so markedly in the apes, the diastema is needed to allow the jaws to close properly. But the Taung child's upper canines did not project, and there was no diastema. In short, the teeth and jaw were remarkably humanlike, indeed far more humanlike than the rest of the skull with its mild prognathism and small cranial vault.

What name to give this creature, with its apelike face and skull and yet very human teeth and jaws? The perfect name for Dart's discovery, *Pithecanthropus* or apeman, had been proposed half a century earlier by Ernst Haeckel as a hypothetical term for the missing link between apes and humans that Haeckel was sure would eventually be found. Because the physician Eugène Dubois had already used Haeckel's name when he discovered the far more humanlike Java man at the end of the nineteenth century, Dart named his find *Australopithecus africanus* in the paper he sent to *Nature. Australopithecus* simply means a southern ape, although many people have since assumed that the fossil was found in Australia! Less forgivable to the scientific establishment was the fact that the name is a hybrid with Latin and Greek roots, a dead giveaway of Dart's provincialism.

The week after Dart's preliminary paper appeared in *Nature,* the same journal published rejoinders from four distinguished anthropologists, including Grafton Elliot Smith, Dart's old teacher, and Sir Arthur Keith, who had also been involved from the beginning with the Piltdown man discovery. All of them dismissed Dart's claim that *Australopithecus* was an intermediate between apes and humans, though all agreed that the fossil was an interesting one. None of them had seen the fossil skull itself, but they were quite capable of making this sweeping dismissal on the basis of a few poorly reproduced photographs.

The English scientific establishment in general showed great resistance to Dart's discovery, even though Dart brought his precious skull to England in 1931 (where his wife nearly lost it in a taxicab). A number of experts examined it, admitted it was not quite an ape, but would go no further. The question of human origins was at that time hotly contested, with one large group of paleontologists favoring a European origin and an Asian origin being trumpeted by Henry Fairfield Osborn, the distinguished fossil hunter and director of the American Museum of Natural History. In short, it would seem, the scientific establishment would embrace anywhere but Africa as a place for human origins, and any scheme for human evolution but one that implied the dreadful possibility that our ancestors might have been black.

All these alternative views of human origins flew in the face of the careful logic that Charles Darwin had applied to the problem of human origins. In 1871, when he published *The Descent of Man,* the human fossil record essentially consisted only of a few finds of Neanderthal man in Europe. Darwin, however, pointed out that almost without exception living species are closely related to living and extinct species from the same region. The closest living relatives of humans are to be found in Africa. Darwin thought an African origin very likely, but of course he could not rule out Europe or elsewhere.

A New Broom

From the beginning, the few scientists in South Africa who saw the *Australopithecus* fossil itself were all most impressed. Perhaps the most memorable reaction came from the fossil hunter Robert Broom, who appeared suddenly in Dart's laboratory shortly after the discovery was announced, strode over to the skull, and fell on his knees before it. "I am worshipping our ancestor!" he exclaimed dramatically.

Broom was the earliest and staunchest of Dart's supporters, and indeed Dart's discovery actually changed Broom's career. Born to a poor and struggling family in Paisley just outside Glasgow in 1866, Broom had trained as a doctor. His great specialty was midwifery, which proved extremely valuable in his travels to remote parts of the world. An early sojourn in Australia had awakened his love of natural history. When he arrived in South Africa in 1897, he was dazzled by the array of fossils, particularly the early mammal-like reptiles called therapsids that were to be found in the dry hills of the Karroo.

In spite of his successes with the therapsid fossils, Broom had lately fallen on hard times. A long-running feud with the South African Museum had become more heated when Broom, chronically short of money, actually had the temerity to sell some of his fossils to the American Museum of Natural History. Barred from the South African Museum, Broom was angry and frustrated.

Broom was, from the accounts of those who watched him at work, a remarkable paleontologist. Almost always dressed in a dark suit and a wing collar, even in the field, he could chisel his way through a lump of rock and reveal the fossil inside with the speed and instinct of a Michaelangelo. Roaming everywhere on the veldt, he befriended farmers and encouraged them to bring him interesting fossils. His skill as a draughtsman allowed him to make rapid and accurate drawings of his many finds. His endless scientific curiosity led him to do things that amused his contemporaries but horrify us. He dug up old native burials to get skeletons of various "racial types." After he had performed autopsies on native men who had died in prison, he had the bodies buried in shallow graves in his garden. A few months later the bones could be dug up, studied, and often sent to eager museums abroad.

By the time of the discovery of *Australopithecus*, Broom was in his late sixties but as scientifically active as ever. Through the intervention of General Jan Christian Smuts, later South Africa's wartime prime minister, he was given a post as assistant for paleontology at the Transvaal Museum in Pretoria at the magnificent salary of 41 pounds a month. At last he was free to abandon his medical practice and devote himself full-time to hunting fossils.

A few miles to the west of Johannesburg were the caves of Sterkfontein, in an area which at that time could be reached only by a long and bumpy drive over dirt roads. As with so many other promising sites for hominid* fossils round the world, the caves were being quarried for limestone. The quarrying uncovered areas that would have been otherwise inaccessible, but at the same time paleontologists had to work quickly before irreplaceable fossils were flung into the lime kilns. Indeed, many of the beautiful stalactites that might have made the caves a tourist attraction had already been broken up and burned at the time of Broom's first visit. And Broom later voiced the suspicion that the owner of the lime works, a man named Cooper, was very well aware that fossils were being dug up and destroyed, for Cooper had written a glowing description of the caves in a little brochure that had been published to encourage local tourism. In it he said, "Come to Sterkfontein and find the missing link!"

In 1936 Broom did indeed come to the caves with some students of Dart's, and they quickly found many baboon skulls and a great variety of other animal bones. Soon Broom and his students turned up fragments of *Australopithecus* skulls as well.

As at Taung, the rock had to be blasted with dynamite, but Broom and his assistants carried out the blasting with care and examined each bit of rubble. Skulls turned up in amazing numbers. The best preserved of these skulls, which soon numbered eight, was that of an adult, probably a female. The lower jaw was missing, but the face and the dome of the skull were intact. There were moderate-sized ridges above the brows, and the face was more prognathous than that of the Taung child, although because this was an adult skull these differences could easily have been the result of the process of maturation.

On the basis of slight differences from the Taung find, Broom put

*Hominids are humans and their relatives—all but ourselves are extinct—who make up the family Hominidae. Homin*oids* are a somewhat less exclusive group, a superfamily that includes the hominids, our closest relatives, the gorillas, chimpanzees, and orangutans, and our more remote relatives the gibbons.

some of these finds into a different species of *Australopithecus* and later put them all into a different genus, *Plesianthropus* (near man). The designation of a genus, the next classification level above that of a species, is normally reserved for groups of organisms that are so different from each other that each group forms a cluster of species, clearly different from other clusters of species. Broom's few skulls hardly fit this criterion, but he was following an old fossil hunter's tradition. It is only human nature for paleontologists to emphasize the uniqueness of their own finds by assigning them a new species name or even, and more grandiosely, a new genus name. Now, however, it is generally agreed that Broom's finds were really adults of *Australopithecus africanus,* the same species as the Taung child.

Broom's luck continued to serve him well. A schoolboy who worked on the weekends as a guide to the caves had found an ape-man palate with one molar tooth still in place. When Broom visited the boy's family, he found as he had suspected that the palate and some additional teeth had not come from the Sterkfontein caves. Instead they had come from the top of a hill called Kromdraai about a kilometer away. Careful sieving of the site yielded some of the rest of the skull. When the fragments were painstakingly assembled, the result proved to be very different from Dart's or Broom's earlier discoveries. Gleefully, Broom hurled this new find into yet another newly created genus, *Paranthropus* (beside man). Because of its powerful jaw musculature, he named the species *robustus* (plate 9 shows a more recently discovered but very similar skull).

It is a fascinating sidelight on the vicissitudes of fossil naming over the decades that, although Broom's *Plesianthropus* has disappeared, *Paranthropus* is making a comeback after a period of eclipse. At first most authorities ignored Broom and lumped his new find with Dart's Australopithecines, but some scientists are now beginning to embrace Broom's separate genus.

The jaw of this fossil skull was heavy, with large but again very human-looking teeth. The molars in particular were huge, with flat and heavily worn occlusal surfaces, and the rest of the skull was obviously adapted to a powerful grinding type of chewing. As you can see from the picture, the cheekbones projected outward to allow the large jaw muscles to pass underneath, giving the skull a broad and very nonhuman appearance. You can actually feel another difference between *Paranthropus* and modern humans. Place your fingertips at the sides of your head above the temples and move them slowly upward while clenching and unclenching your jaw. You can feel a thin sheet of muscle contracting and relaxing, a sheet that thins out and disappears as your fingers move up from the side of your head toward the top. The place where this temporal muscle attaches—where it disappears under your fingers—forms a thin, almost unnoticeable marking on the surface of a human skull, the superior temporal line. The dome of the skull has expanded so much during our recent evolution that this feeble line is all that is left for the temporal muscle to attach to. As a result, the muscle is thin and relatively weak. But in the skulls of apes, of *Paranthropus,* and of many other mammals, the muscle forms a thicker sheet, attaching further up near the top of the skull. Through its contractions, it actually helps to shape the skull itself during early development. The result is a ridge of bone, the sagittal crest, that runs fore and aft along the top of the skull. The ridge forms a superb upper anchor point for the mighty sheets of the temporal muscles. Many other animals that depend for their survival on strong jaw muscles have such an arrangement—if you are sitting by the fireside reading this book and have a dog or a cat handy, you can easily feel its sagittal crest.

It is interesting that there is a great deal of variation in the position of the superior temporal line in present-day humans. In some Eskimos it is much higher than in most of us, up nearer the region where there would be a sagittal crest if modern humans had one. Until recently Eskimos spent a great deal of time chewing hides in order to make them flexible—one wonders how much of this shift in the superior temporal line is due to such continual exercise and how much to heredity.

What was *Australopithecus?*

At the little hill site of Kromdraai, Broom found the first clues to the real nature of his Australopithecines. In the course of sieving the material near where the fragments of skull had been found, his sharp eyes picked out a wrist bone, some tiny bones of the hand, and most importantly a bone of the ankle called a talus. The talus bone, in the middle of the ankle joint, is completely surrounded by other bones. As a result, it is the only bone in the body to which there are no muscles attached.

In humans the talus is in the form of a cube, which gives it its ancient name of astragalus, or die (plate 5). This regular shape allows the weight of our body to bear down straight through the ankle and onto our firmly plantigrade foot. In chimpanzees, however, the talus appears twisted. This twist is reflected in their waddling gait, which results in the body's weight being transferred unevenly through the ankle to the inner margin of the foot. Broom's Australopithecine talus, although a little twisted, was almost as cuboidal as that of a modern human. If this talus really belonged to the same species as the robust Australopithecine skull that had been found nearby, then these Australopithecines were able to walk upright—or nearly so.

In 1948 Broom began to work yet another rich site called Swartkrans, or black cliff, only about a kilometer from Sterkfontein. As at Kromdraai, most of the remains found at Swartkrans appear to be those of the robust *Paranthropus.* Startlingly, however, in some of the more recent layers of the deposits Broom found a lower jaw and some teeth of what appeared to be a far more advanced hominid. He gave this find yet another genus name, but it is now thought to be *Homo erectus,* probably very similar in appearance to Peking man. Later, some elaborate tools made of bone were also found in the upper levels, perhaps associated with the *Homo erectus* remains. More primitive bone tools—little more than digging sticks—were found scattered through all the deposits. Could these possibly have been used by *Paranthropus,* before these more primitive creatures were displaced by *Homo erectus?*

Broom published a monograph on the Australopithecine fossils in which he turned over the discussion of the fossil brain endocasts to an anatomist, G. H. W. Schepers. The best of these endocasts was the very first one that had been found by Dart, but Broom had discovered several other partial endocasts at Sterkfontein and Kromdraai. On the basis of these endocasts, Schepers drew detailed pictures of the sulci (grooves) and gyri (the ridges between the grooves) of the Australopithecines' brains. He then proceeded to make all kinds of conclusions about the Australopithecines' brainpower. His conclusions were essentially the same as those that had been reached by Dart years earlier—the occipital lobe at the back of the brain was reduced in relative size compared with that of the apes, and the parietal regions to the sides had expanded and overgrown it. This clearly put the brains somewhere between those of apes and those of modern humans, in which the processes of relative occipital reduction and parietal expansion have proceeded much further.

One region that appeared to have grown the most included a part of the brain known as Wernicke's area, which spans the junction between the temporal lobes at the side of the brain and the parietal lobes at the top. Wernicke's area is generally assumed to be involved in the *understanding* of language—patients who have had this part of their brain damaged by a stroke are no longer able to understand either spoken or written language. Although they can still articulate words properly, their sentences are devoid of meaning. If Schepers' interpretation was right, then perhaps the expansion of this area, which had the effect of pushing back and compressing the occipital lobe, indicated that the Australopithecines might have been able to communicate verbally.

Unfortunately, the photographs accompanying the monograph, and even those taken in 1925 of the Taung child shortly after it had first been discovered, showed little sign of these features. Schepers admitted as much: "No photograph . . . can ever do justice to the relief of sulcal and gyral patterns seen in these specimens." His conclusions were roundly derided in England and referred to slightingly as paleophrenology. Wilfrid Le Gros Clark, a distinguished comparative anatomist from Oxford, dismissed Schepers' conclusions and dismissed Schepers himself as "a young South African anatomist."

How much can really be told from these endocasts? We may never know, because when a fossil is removed from its protective matrix of stone it immediately begins to degrade. Acids from the hands of eager paleontologists begin to damage the polished surfaces of the stone. At various times since their discovery, the fossils have been coated with layers of celluloid solution and shellac in an attempt to preserve them, so that they are now a dingy brown color. Some of the teeth of the Taung child have been damaged by careless handling. There is no way now that we can recreate what Dart saw on that summer afternoon in 1924 when he opened the wooden crate and held a brain cast of *Australopithecus* up to the light. Argument still continues over how much can be read from those endocasts and how much is wish fulfillment. But the likelihood is great that Dart and Schepers were right and that Australopithecine brains really have some humanlike characters, for every other feature that has been found since—their human dentition and jaw, their upright posture, their humanlike hands—shows that they resemble us more closely than they do the apes.

Indeed, this was exactly the conclusion that was reached by Wilfrid Le Gros Clark in 1947 when he actually visited South Africa and examined the fossils and their sites in detail. He had prepared for his trip meticulously by taking notes on and measurements of over a hundred ape skulls before he left England. A skeptic when he arrived, he was a convert by the time he left. The talks he delivered and the papers he wrote were enormously influential in swinging scientific opinion in support of Dart and Broom.

Dart, however, was anxious to push his Australopithecines even further in the direction of humanness. For many years, discouraged by the reception of the Taung skull, he had done little fossil hunting. But in 1945 a young student of Dart's, Phillip Tobias, led Dart's science class on an expedition to the Makapansgat valley about 200 miles north of Johannesburg. Here, a series of large caves had formed a base for a native revolt against the *voortrekkers* who had occupied the area after fleeing British rule in the middle of the last century. Buried deeper in the cliffs were much older caves, one of which was rich in travertine and marble and had been quarried extensively. In the process, the quarry workers had broken through several fossil-rich layers of breccia and had hauled these unwanted chunks of material to nearby dumps.

Dart had already seen some bits of animal bone from the Makapansgat breccias. The bones were blackened, and his suspicion that they might have been burned appeared to be confirmed by crude chemical analysis (although later analyses in fact showed no trace of carbon). His co-worker Alun Hughes and others began the painstaking process of breaking up and examining the breccia blocks in the dumps, doing their best to trace the regions of the cave from which the blocks had come. The deposits in the cave showed signs of having been built up over long periods of time, but in two parts of the deposits they found signs of Australopithecines. They also found many baboon skulls, just as at Taung. And, just as at Taung, almost all the baboon skulls showed depressed fractures.

Dart assembled a collection of these skulls and showed them to R. H. Mackintosh, a forensic expert. Mackintosh agreed with him that the baboons had been killed by heavy blows with some kind of instrument, and he inferred further that at least one of the skulls from Taung had been broken open to get at the brains.

Dart suggested on the basis of the many broken bones found at the site that the Australopithecines, rather than relying on stone tools, had employed and modified a variety of animal remains. The Australopithecine culture, he suggested, was an "osteodontokeratic" one, employing bones, teeth, and horn. He assumed that the bones had been used in hunting. He thought that his hominids were actually fierce predators capable of organizing into bands that could attack and kill equally fierce baboons, views that were widely popularized by Robert Ardrey in the book *African Genesis*.

More recent and extensive excavations, particularly at Swartkrans, have indeed turned up many bone tools at every level in the deposits, but these were apparently used for gentler tasks. They show a great deal of wear on the ends as if they had been used extensively for digging.

The youngest of the Swartkrans deposits, where signs of the more advanced *Homo erectus* have been found, have also yielded evidence of more advanced tool use. Bob Brain of Pretoria's Transvaal Museum has recently suggested that a pair of awl-like tools that were found in this most recent level might have been used to make clothing or carrying bags. And, although there were no signs of fire in the older layers, hundreds of pieces of bone from the younger deposit, when examined under the microscope, showed signs of having been burned.

The finds at these various cave sites, although they seemed confusing at first, were beginning to make a coherent pattern. The first gracile Australopithecines seem to have been dragged to Sterkfontein willy-nilly by predators. Their remains accumulated in the caves over a span of perhaps two million years or more. Then, roughly two million years ago, Australopithecines began to occupy the entrances of the caves. The picture of the last two million years that has slowly emerged from the nearby Swartkrans excavation is one in which a robust Australopithecine culture that may have used simple tools was supplanted, perhaps a million years ago, by a more complex culture in which fire was used.

The oldest deposits at Sterkfontein, with their gracile Australopithecine remains that were dropped there by predators, are sometimes embedded in rock and must be blasted free. The more recent Australopithecines and the few traces of *Homo* are often (although not always) in decalcified deposits that are easier to dig and from which much more information can be obtained. Ron Clarke has been digging with exquisite care through this material and finding many new skulls and bits of skeleton. These fossils have not yet been described in detail, but a few of them resemble the more robust Australopithecines from Swartkrans—to the point where the distinction between the various Sterkfontein and Swartkrans fossils was not at all obvious to me when I was shown them. Does this mean that the distinction between the gracile and robust Australopithecines that has always been made by paleontologists is beginning to break down? Clarke disagrees, but some authorities think so.

The last robust Australopithecines disappeared from Swartkrans perhaps a million years ago. Why did they vanish? The most obvious explanation is that they were simply driven out by the more advanced *Homo habilis* and *Homo erectus* who brought a more complex culture with them. Almost nobody believes that the robust Australopithecines, with their giant teeth and immense jaws, could have contributed anything genetically or culturally to the apparently more humanlike *Homo erectus.*

Or could they? These events of more than a million years ago are even more shrouded in mystery than those surrounding the disappearance of the much more recent Neanderthals in Europe. As we will see, there are remarkable parallels between this series of events and that disappearance. Two kinds of Australopithecine, the older apparently more humanlike—or at least more gracile and with a more human dentition—occupied these regions of southern Africa between perhaps one million and four million years ago. New fossil discoveries hint that the distinction between the two may not be as great as was previously thought. Finally, about a million years ago, the younger and apparently more primitive species was replaced by more advanced hominids. Neanderthals and modern humans seem to have played out a similar story, much more compressed in time and much closer to our own day.

But we are getting ahead of our story, for the discovery of the Australopithecines was only the beginning of a series of findings about the early history of our species in Africa. The scene now shifts away from the rolling countryside and caves of Sterkfontein and Swartkrans, 3,000 kilometers to the north among the far more spectacular veldts, valleys, and volcanic cones of Kenya and Tanzania.

5

The Birth Canal of Our Species

The Phoenicians sailed from the Red Sea into the southern ocean, and every autumn put in where they were on the Libyan [African] coast, sowed a patch of ground, and waited for the next year's harvest. . . . These men made a statement which I do not myself believe, though others may, to the effect that as they sailed on a westerly course round the southern end of Libya, they had the sun on their right—to the northward of them.

—Herodotus, *The Histories* (trans. A. de Sélincourt, 1954)

The Great Rift Valley is one of the most striking geological features on the planet. Stretching some 6,000 kilometers across Africa, it begins far to the south near the coast of Mozambique and extends half the length of the continent to the Afar Triangle that separates the Gulf of Aden from the Red Sea in the north. The rift does not stop there—it splits, one half heading northeast through the length of the Gulf of Aden to join a bewildering maze of rifts on the floor of the Indian Ocean, the other half heading northwest through the Red Sea. There it splits again, and a branch veers north through Israel, where it forms the valley of the Jordan. The Dead Sea, 300 meters below sea level at its deepest point, occupies the northern part of the Great Rift Valley.

Along the length of the African part of the valley is a chain of lakes, the southern ones huge and deep, the northern ones shallow and filled with alkaline salts. The deep lakes of Malawi and Tanganyika, defining the valley in the south, separate Tanzania and Zambia. The farthest north of this chain of deep lakes used to be called Lake Albert, though it currently (and presumably temporarily) rejoices in the name of Lake Mobutu Sese Seko. Lakes Assai, Turkana, and Natron, the most prominent of a chain that extends to the northeast, separate Kenya from its neighbors. The northernmost lake of this southern chain and the southernmost lake of the northern form a symmetrical bracket that

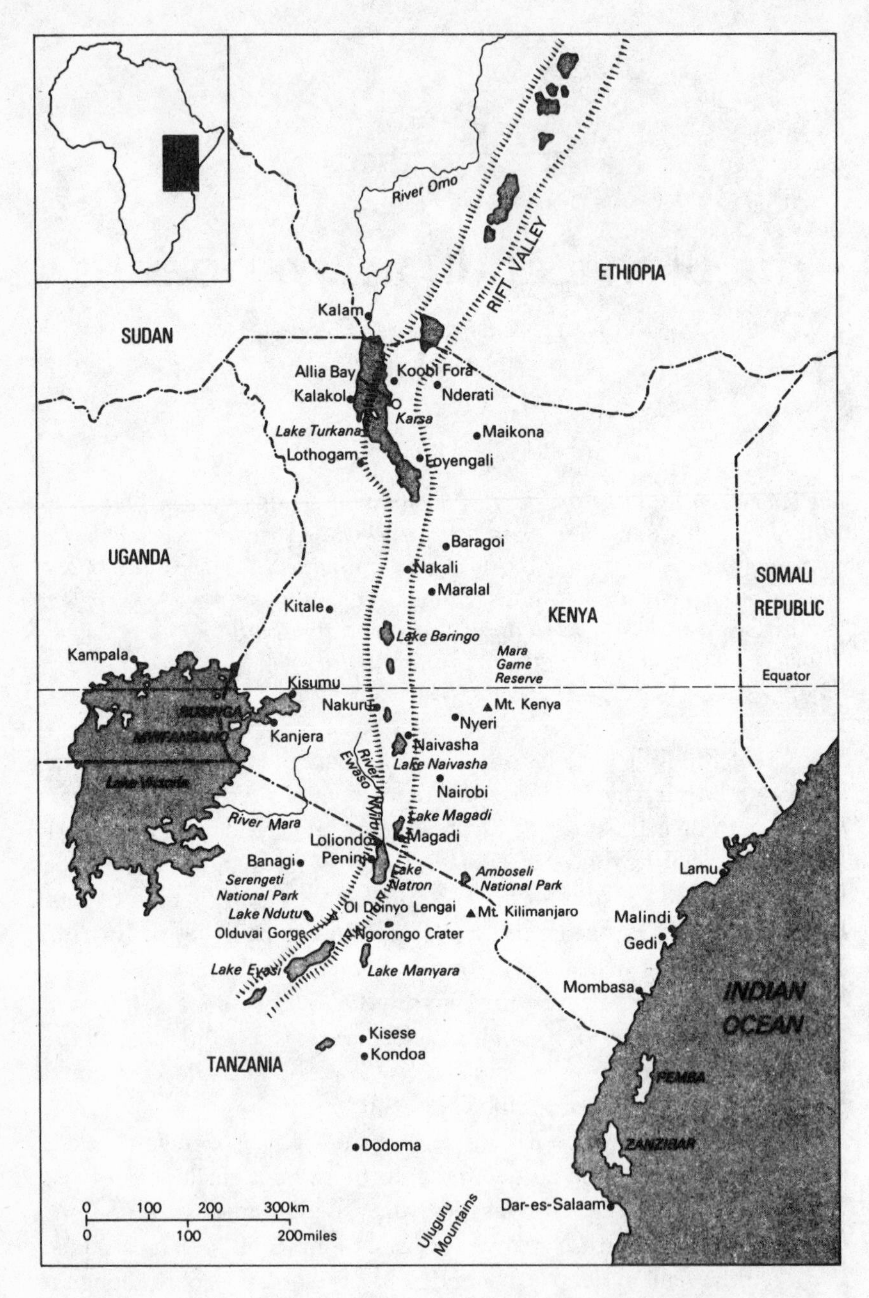

FIGURE 5.1. Map of the northern end of the Great Rift Valley, showing the location of most of the important East African sites where hominid fossils have been discovered.

encloses the basin of Lake Victoria, the source of the Nile. Lake Victoria is cradled in a region of uplift that temporarily prevents the northern and southern halves of the Rift Valley from meeting. When they do eventually meet, the sea will rush in to split Africa in two.

Although many of the rocks that it divides are very ancient, the valley itself is remarkably young, only some twenty million years old. Its youth can be traced to the fact that two major crustal plates, the African plate to the west and the Somali plate to the east, have only recently begun to move apart, driven by convective forces in the earth's mantle far below. As a result of these vast movements, many geological processes are happening simultaneously along the length of the Rift Valley. In the Afar Triangle to the north, plumes of lava come near the surface. Parts of the harsh lunar landscape formed by these plumes have been pushed above the sea's surface to form this inhospitable region of Ethiopia, in effect an above-water part of the seafloor spreading that is taking place all along the length of the Gulf of Aden. In the region where Lake Victoria separates the two arms of the Rift Valley, 1,500 kilometers to the south, the crust is thicker and has not yet split. Pressure from the mantle below has caused it to bulge upward some 700 meters and to crack in many places. Great trees of lava have forced their way through these cracks to the surface. The lava has built the numerous volcanoes that dot the area—Kenya, Kilimanjaro, Ruwenzori, and many others, some active or recently so and others worn down by erosion to nubs of their former mighty selves. Huge as they are—Kilimanjaro is by one measure the tallest mountain on earth because it lies almost on the equator and its peak actually lies farther from the earth's center than Everest's—these volcanoes are only the most obvious features of a vast uplifted region that forms the highlands of Kenya and Tanzania.

On a slightly smaller scale, much of the landscape of the valley has been shaped by climatic change. Its lakes have filled and emptied many times in the course of its short history. During the last period of very high rainfall, some ten thousand years ago, many of the lakes were 50–100 meters higher than they are at present. Now the lakes are shrinking, and human activity is accelerating the process. Particularly in the north the lakes are swiftly disappearing as a result of irrigation and the desertification of northern Africa.

The geological future of the Great Rift Valley is unclear. The split seems likely to widen and deepen with time, and eventually the sea may rush in—most probably its first invasion of the African continent will be from the Afar Triangle in the north. But the geology of the Indian Ocean is so complex that it is impossible to predict whether the Somali plate will continue to move to the east and widen this new arm of the sea or whether the new arm will remain as narrow as the Red Sea is today.

The past is much clearer. The valley has, through all its existence, provided a rich and complex environment for our ancestors. This environment, instead of being some unchanging Garden of Eden, was marked by fluctuations over periods of thousands or tens of thousands of years from wet to dry and back again. The latest dry spell is only one of dozens, and not the most severe—the northern lakes dried up almost completely between sixty and seventy thousand years ago.

It used to be considered that there was a correlation between the climate of the Rift Valley and the periods of glaciation far to the north, so that the valley's climate became colder and wetter as the glaciers advanced, hotter and drier as they retreated. Things have turned out to be much more complicated. Although the planetary pulse to which the Rift Valley's climate responds remains unclear, the result has been large changes in both weather and vegetation. During the cooler periods vast forests stretched the length of the valley, while grasslands and deserts marked the drier and hotter periods. Whether the climate was wet or dry, however, at least some of the lakes were always available to provide plentiful water for grazing animals, along with the primates, hominids, and finally the humans who lived there.

As we saw in the last chapter, however, our ancestors ranged widely through Africa and were not confined to the Rift Valley. So we cannot say that some specific part of the Rift Valley was *the* place of origin of the genus *Homo*. It is far more likely that our species was formed through the ceaseless migrations, separations, and minglings of hominids as they moved north and south along the perpetually well-watered valley itself. It is too great a simplification to assume that humans somehow emerged through some single event from the womb of East Africa. Instead, the valley acted as a kind of birth canal for our species.

Exploring the Rift

To the prehistorian of today, the whole Rift Valley tells a complex tale of our past. But none of this was apparent to the first visitors to this region from the ancient civilizations to the north, or indeed to the first explorers who came from western Europe at the beginning of the nineteenth century.

The interior of East Africa was isolated, so that very few of the tools and technologies of the modern world had penetrated there. Indeed, some skills, such as writing, never penetrated. This, of course, had an unfortunate effect on the way East African culture was perceived by the first Europeans to explore the region and write about it. The native inhabitants were, to European eyes, incredibly savage and primitive, and it went without saying that primitive peoples, lacking writing and unable to muster the resources to build large cities, could have no history. In Europe the traces of history over the last few thousand years are numerous and obvious, but to find the corresponding traces in Africa, even traces of relatively recent history, required a completely new way of looking at things.

The European travelers were greeted by what was to all appearances a cultural backwater that had borrowed its limited technology from elsewhere. Fishing communities had only begun to appear around the edges of the valley's lakes about 3,000 B.C., three thousand years later than in the Nile valley to the north. The first signs of agriculture appeared at around the same time. Almost certainly this agriculture was not indigenous, but was introduced from Egypt. Iron implements were introduced, again from the north, in the sixth century B.C., more than five hundred years later than in the Middle East, and although a primitive smelting industry soon sprang up, it remained very unsophisticated.

The first European explorers entered a world made up of a bewildering variety of tribes, both nomadic and pastoral. At least fifty different languages of five very different language families were spoken in the regions that are now Kenya and Tanzania. Yet even by this time the ancient ways of life of these tribes were being disrupted by the Arab slave trade that set one tribe against another. Because many of these tribes were forced by war or by the slave trade to move from their ancestral homelands, the delicate tracery of oral traditions that linked them to their past was often destroyed.

Eventually, a few bits of that past were uncovered, and some of them were very impressive indeed. The eight-hundred-year-old maze of ditches and earthworks at Bigo in Uganda stretches in a complex pattern over a distance of 10 kilometers. As much as 20 meters high now, the earthworks must have been far more massive when they were first built—in some cases, the ditches in front of the earthworks were actually dug 5 meters into the solid rock. Technology can also be traced back a long time. Excavations have shown that the saltworks at Uvinza in Uganda have been in operation continuously for two thousand years.

But in spite of these occasional finds, much of the continuity of history has been shattered forever. Simple questions, like the times when the present-day tribes first settled in these areas, and the characteristics of the tribes that they may have displaced, are now often unanswerable.

Yet the further one goes back in time, the more important the cultures of the Rift Valley become. Go far enough back, and you find that they actually formed the avant-garde of hominid culture. During the later Stone Age, the great advances in tool making took place in the Rift Valley at least as early as they did elsewhere in Africa, Europe, and Asia—and the further one goes back, the greater the African technological lead becomes. The oldest tools that we can attribute to hominid activity have been found in southern Ethiopia. These primitive Oldowan tools, named for Olduvai Gorge further to the south, have been dated to as early as two and a half million years ago. So primitive are they that they might have been mistaken for stones that had been accidentally fractured by natural processes, were it not for the care with which these finds were counted and their positions mapped.

Other deposits of Oldowan tools have been found in many parts of the Rift Valley. They consist of small rocks from which a few flakes have been struck, usually forming a sharp edge on only one side but sometimes forming a continuous edge all the way around the rock (figure 5.2). Because so few flakes have been struck from these tools, the edges are often very uneven. Hammer stones with signs of impact are found associated with them, and often the flakes that were struck from them are discovered nearby. Under the microscope, the sharp edges often show unusual signs of abrasion, and it is actually possible to determine whether these tools were used to cut flesh, bone, or wood.

About half a million years after the Oldowan tools first appeared in the fossil record they were augmented, rather than supplanted, by more sophisticated tools of the Acheulian type—named after the tiny village of Saint-Acheul near Amiens in France, near which some similar tools had been found in the early nineteenth century. The earliest of this type of tool, dated to 1.9 million years ago, were found recently in a rich deposit in southern Ethiopia called Konso-Gardula. Even a glance at the tools is enough to convince one that they were designed for a specific purpose. Particularly common are beautifully shaped hand axes with a rounded end for gripping, patiently sculpted by repeated blows with a hammer stone.

Sometimes, as at Konso-Gardula, the tools were left behind in remarkable profusion. In the four-hundred-thousand-year-old deposits at Olorgesailie, an ancient lake shore not far outside Nairobi that was

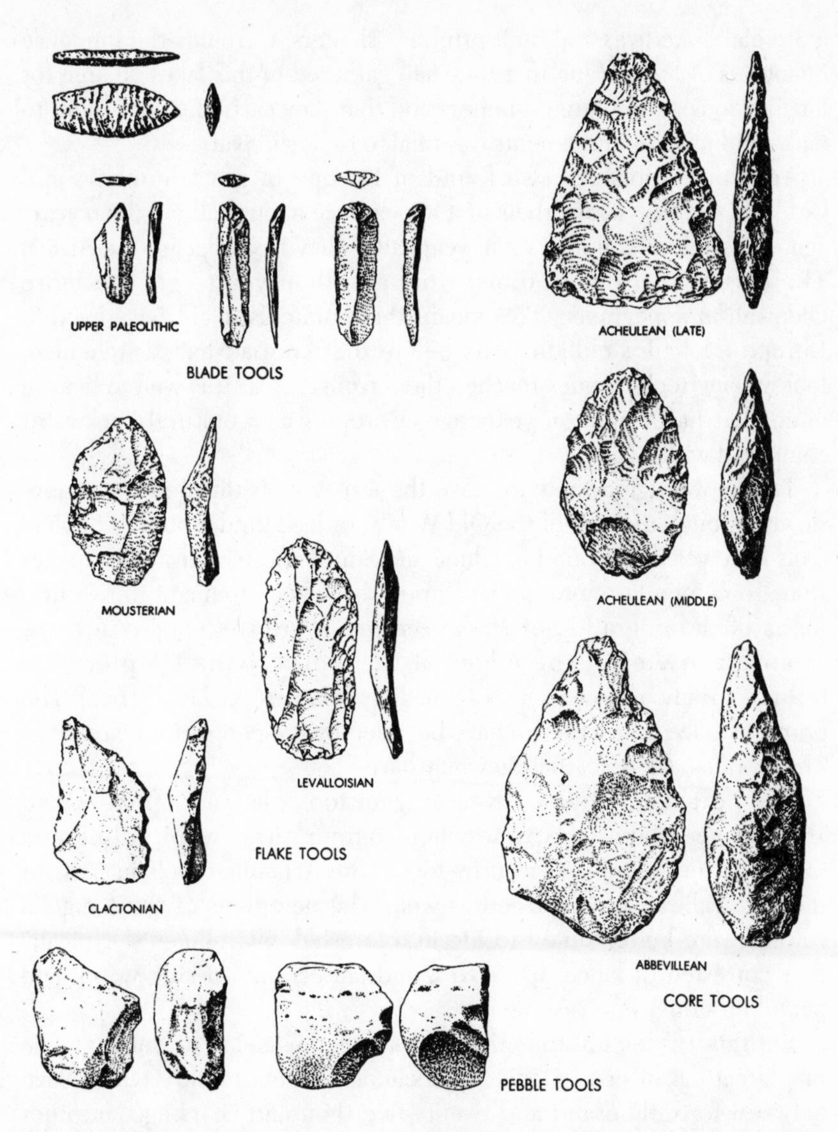

FIGURE 5.2. Some of the stone tools made by our ancestors, ranging from the primitive Oldowan pebble tools at the bottom of the picture up to the sophisticated blade and core tools made by *Homo sapiens* during the late Old Stone Age. The tools in the picture span a period of time of over two million years.

excavated by Mary and Louis Leakey during the 1940s, virtually every stone was found to be a tool of some kind. The Leakeys also thought they could see traces of living floors at various places in the site where the lake had risen and fallen. It may be that these ancient people lived quite elaborate lives and built primitive shelters. Certainly the immense quantities of tools argue that they had gathered in this lakeside area for long periods and in large numbers and that they carried stones there to make and use the implements essential to their survival.

Acheulian tools are also found in Europe in great numbers and varieties, but they make their first appearance about half a million years ago, almost a million and a half years after they first appeared in Africa. The earliest European tools, from a million years ago, are more Oldowan in appearance. This means that hominids might have lived in Europe for half a million years before the more advanced Acheulian tool-making technologies reached them from Africa. It is well to bear in mind that half a million years ago, Europe was a cultural backwater compared with Africa.

The knowledge of how to make these new tools diffused even more slowly through the rest of the Old World. Indeed, undoubted Acheulian tools have yet to be found in China and Southeast Asia. Instead, smaller stone tools, of a far more limited variety, have been found at these outer limits of the migration of *Homo erectus*. Were the people in these remote marches at the edges of the world living simpler, less technologically advanced lives than their relatives back in Africa? The grim existence that seems to have been led by the people in the caves of Zhoukoudian suggests that they may have been.

The next real cultural advance again took place in Africa. About 180,000 years ago, at a time when temperatures were falling and savannas were being replaced by forest, the Acheulian tools in parts of the Rift Valley were replaced as well. The new tools of the Sangoan culture were better suited to life in a forested environment—typically they consisted of lance tips, adzes, and chisels for working wood, and pounding stones for crushing roots.

But this was the last time that Africa took the lead in technology. The final great advances in Stone Age technology seem to have taken place between forty thousand and twenty-five thousand years ago, in other parts of the Old World as far-flung as southern Europe and Australia. A hafted ax between thirty-seven thousand and forty-five thousand years old has been found in New Guinea, the oldest such implement yet found anywhere. Some of the most advanced stone tools of the Old Stone Age, from France, are associated with the skeletal remains of the fully modern Cro-Magnon man. These tools consist of highly sophisticated spear and arrow points and composite wood and stone implements that functioned as saws and sickles. And at the most recent stage, in both Europe and Asia, the strict utilitarianism of the earlier tools was embellished by decorations. Decorative objects with no utilitarian value at all began to appear. The first such advanced and decorated artifacts began to appear in Africa about twenty thousand years ago, some thousands of years *after* the time at which similar artifacts appeared in Europe. Africa was now no longer the fountainhead of technology.

The Leakey Dynasty

The first man to really appreciate the whole sweep of East African prehistory was the remarkable Louis Leakey. Leakey, his second wife Mary, and their son Richard have done more than anyone else to shape our image of the early history of humankind and to show that the Great Rift Valley was, if not the cradle of man, at least the most likely place to search for our early ancestors. The saga of this remarkable family would make a wonderful miniseries. It has everything—sex, intergenerational rivalry, and an exotic locale. The only problem is that most of the major protagonists spent most of their time in a very unphotogenic way, crouched and scrabbling away at almost invisible objects in the dirt!

Louis Leakey was born in 1903, into a missionary family that had been sent to East Africa by the Church Missionary Society. He grew up on the isolated outpost of Kabete, only 8 miles from Nairobi but still at that time a day's journey on foot. From an early age, more African than European, he played almost exclusively with children of the Kikuyu tribe.

One remarkable feature of Kikuyu culture is the division of the tribe into clear age cohorts. As each cohort grows older it passes—in the case of the males—from uninitiated boys through junior warriors, warriors in power, junior elders, and finally senior elders. Young Leakey joined his age cohort and moved with them through some of the initiation ceremonies that marked these transitions. Kikuyu was in effect his native language, and he later turned this ability to amusing advantage. When he offered Kikuyu to satisfy part of Cambridge University's modern-language requirement, he succeeded in throwing Cambridge's staid and Eurocentric academic world into turmoil!

As a young man, Leakey became a passionate advocate of tribal causes. He angered the white community in Kenya with his book *Kenya: Contrasts and Problems* (1936), in which he impartially criticized settlers, administrators, and missionaries. He even supported—on the grounds of cultural preservation—the practice of female circumcision among the Kikuyu. At various times he mediated between the tribe and the British authorities, who through their inaction were allowing unscrupulous settlers to grab native lands. The explosive situation was eased somewhat when he managed to explain the tribal landownership system to the colonial officers. And, playing an essential role as mediator during the Mau Mau rebellion of the early 1950s, he tried to rein in the worst excesses of the independence movement while still conveying the grievances of the Kikuyu to the authorities. During the 1930s he wrote a multivolume treatise on the tribe's language and customs that was so voluminous it was published only after his death.

But all this was far in the future when, while still reading anthropology at Cambridge, Leakey began to go on fossil-hunting expeditions. From the very beginning of his scientific career, he was convinced that modern humans had originated in Africa and that Neanderthals in Europe and Java and Peking man in Asia were nothing but aberrant offshoots of the main human line. This conviction led him to the expectation that anatomically modern humans should be found at much earlier times in Africa than in Europe and Asia. Unfortunately, although there were many modern human remains in East Africa, it was very difficult to date them with precision. If recent remains had been buried in graves, for example, the burial process would often have inserted them into geologic beds that were much older. Lacking good geological information and the precise dates that later became possible with the invention of the carbon-14 and potassium-argon methods, Leakey consistently claimed that his finds were much older than they actually were.

Leakey's first real collision with controversy came with his espousal of a discovery that had become known as Olduvai man. Olduvai Gorge, a name that means in Masai "the place of the wild sisal," lies near the still-active volcano Ol Doinyo Lengai. The gorge is on the eastern edge of the great Serengeti plain that separates the Rift Valley from Lake Victoria. Legend has it that the European discovery of the gorge took place in 1911, when a German entomologist named Kattwinkel nearly fell into it while pursuing a butterfly. A later and more organized German expedition to the gorge in 1913 found that the action of the stream that had carved it, and that now flowed only during the wet season, had exposed a remarkable slice through time. Although the gorge itself, like the Grand Canyon, is quite recent, the rocks the river had carved through are much older.

The sedimentary layers forming the walls of the gorge can be divided into four major beds, which geologists on the expedition numbered from one, the oldest, to four, the most recent. In many areas the beds can clearly be distinguished because they are separated by layers of ash of a characteristic appearance. Numerous other eruptions from nearby Lengai have helped to build up the intervening sediments. The unusual carbonatite ash produced by Lengai, when it became wet, yielded a cementlike material that happens to be superb for preserving fossils.

Hans Reck, who had been a member of an earlier German expedition, took Leakey to the site where Olduvai man had been discovered, and which appeared to be in the middle of bed two. He convinced Leakey that Olduvai man must be at least half a million years old, and the two of them published a report to that effect in *Nature.*

Leakey's conviction was reinforced by a discovery he made a year later at Kanjera on the eastern shore of Lake Victoria. A few bits of the skull of an apparently modern human, accompanied by hand axes, were found associated with bones from the same species of elephant that had originally been found by Hans Reck in bed two of Olduvai. And, not far to the west of Kanjera, at a site called Kanam, an assistant made an even more exciting discovery, a clearly human jawbone associated with the extinct elephant-like *Deinotherium.* Similar animals had been found in the oldest bed at Olduvai. Leakey did not examine the site carefully before rushing into print.

The next few years were spent slowly and reluctantly eating his words. An independent expedition to Olduvai in 1934 turned up stratigraphic evidence showing clearly that Olduvai man was a recent burial. Carbon-14 dating has since shown an age of about fifteen thousand years, not the half million that Leakey had originally trumpeted. And to make matters worse, Leakey got into trouble about the Kanam jaw. Sublimely sure of himself, he invited Percy Boswell, his chief critic, to Kenya to examine the Kanam site. Boswell quickly found that the supposed photograph of the site that Leakey had published was of a different area entirely. In his severe report to the Royal Society and to *Nature,* Boswell was harshly critical of Leakey's working methods and his integrity.

This period was also a time of great upheaval in Leakey's personal life. In 1936 a very public divorce from his first wife, Frida, was necessarily embellished with the elaborate demonstrations of marital infidelity on his part that were required by the courts at the time. All this had deeply shocked his missionary parents and the strait-laced Nairobi community. It also cut him off from Frida's considerable estate, which he had repeatedly tapped to finance his expeditions.

Science, however, was the richer, for at the end of 1936 he married

Mary Nicol, a clever young artist with a strong interest in prehistory. An eighteenth-century ancestor on her mother's side, John Frere, had recognized some stone implements as being prehistoric and is often credited with being the first to do so. Trips with her artist father to the south of France had acquainted her with the remarkable cave art and the Paleolithic and Neolithic remains to be found there. She was already an enthusiastic amateur archeologist by the time she met Louis.

Mary soon fell in love with East Africa. Everything about the country interested her. Her drawing skills enabled her to preserve, in strong and simple reproductions, the rapidly fading Neolithic rock-shelter art of Kondoa, south of Kilimanjaro. These paintings are about fifteen hundred years old, and are as striking and vivid as the paintings at Lascaux. They include far more human figures than do those at Lascaux, figures that are depicted carrying out a number of activities—some of them, for example, can be seen playing musical instruments, with the music falling from their pipes in a series of dots that vividly convey the idea of cascades of notes. But art in these shelters seems to stretch much further back in time—many of the paintings were made on top of blurs of pigment that suggested the rock walls had been in use for much longer periods. Fragments of ochre and other drawing materials that have been found at the site have been dated to as long as thirty thousand years ago.

Mary Leakey also turned out to be perhaps the best writer among the Leakey clan, as can be seen in her description of the Serengeti as viewed from the slopes of Ngorongoro:

> As one comes over the shoulder of the volcanic highlands to start the steep descent, so suddenly one sees the Serengeti, the plains stretching away to the horizon like the sea, a green vastness in the rains, golden at other times of the year, fading to blue and grey. Away to the right are the Precambrian outcrops and an almost moon-like landscape. To the left, the great slopes of the extinct volcano Lemagrut dominate the scene, and in the foreground is a broken, rugged country of volcanic rocks and flat-topped acacias, falling steeply to the plains. . . . Here and there, dark rain-storms gather as the day proceeds, but everywhere else shimmers in the hot, bright sunshine.

By the time Mary joined him, Louis Leakey had learned a great deal from his careless mistakes. Nonetheless, he continued to inflate both the ages and the importance of his finds. (One exception was the astonishing collection of stone tools and the possible remains of huts at

Olorgesailie, which he dated at 125,000 years old and which later turned out to be closer to 400,000 years old.) But the new Leakey team did not confine themselves to recent and controversial finds. They were on much firmer ground when they investigated the more distant past, where a million years one way or the other did not matter so much to their critics. Some of their most remarkable discoveries date back to a time long before humans or even hominids, to the Miocene epoch.

The long Miocene geological epoch began twenty-four million years ago and lasted down to five million years ago. It saw the beginning of the Great Rift Valley and spanned three-quarters of its subsequent development. The earth was warm through much of that time. Only as the Miocene epoch drew to a close did the climate begin to cool, in a foreshadowing of the Ice Ages to come. Through much of that endless summertime the present Mediterranean was a dry valley, kept free of Atlantic waters by a land bridge across what is now the Strait of Gibraltar. The hot, deep Mediterranean valley, parts of it as much as 4,000 meters below sea level, provided almost as formidable a separation between Europe and Africa as the current Mediterranean Sea, but even so there must have been many animal migration routes wending through its less forbidding depths. We know this because, during the Miocene, the European and African faunas were very similar—several species of hippopotamus, for example, wallowed in the valley of the Thames. The Miocene was also remarkably rich in evolutionary opportunities, with a greater diversity of mammalian species than at any time before or since.

In the course of their travels, Louis and Mary stumbled on a rich treasure of Miocene fossils on the island of Rusinga in Lake Victoria. The fossils were so superbly preserved that they included many insects, which are usually among the rarest of fossil finds. Far more exciting, there were also a number of teeth and jaw fragments of apes, among the first traces of apes from that time. Some of these closely resembled a similar, earlier find made in the area by Arthur Hopwood of the British Museum. Hopwood had found an ape-jaw fragment that he had waggishly named *Proconsul africanus,* a name that suggested it was the ancestor of the trained chimp Consul of American vaudeville fame. On trip after trip the Leakeys patiently scoured this low, parched island and turned up more and more tantalizing fragments of *Proconsul.* But it was not until 1947, fifteen years after Louis had first visited the island, that Mary Leakey finally found a badly crushed, thoroughly fragmented but almost complete skull. Her brilliant reconstruction required the piecing together of hundreds of tiny fragments, some no larger than the head of a match. It still stands as one of the outstanding achievements of paleontology.

The *Proconsul* find cast light on a completely unknown period of prehuman evolution. Fragments of a dozen more *Proconsul* skeletons have since been found on Rusinga Island, and recent careful examination of the Leakeys' fossil collection has even turned up more parts of the original skull. It is now possible to gain a fairly good idea of what *Proconsul* and the world in which it lived were like. Eighteen million years ago the Rusinga area was heavily wooded. Thousands of bones and bone fragments have been found, some of them actually lodged inside the fossilized remains of hollow trees. Indeed, the trees of the Rusinga area must have been full of *Proconsul.* Judging by the shapes of their shoulder joints, they only occasionally leaped and swung from branch to branch but rather moved with some care along the branches and perhaps spent a limited amount of time on the ground.

There were at least two species of *Proconsul* living in and among those trees, one weighing an average of about 10 kilograms and the other almost 40. The dentition was apelike, with prominent canines in the upper and lower jaws. *Proconsul* had, as nearly as can be estimated, an unusually large brain compared with present-day monkeys of the same size. It also shared with humans and African apes a large sinus in the frontal bone of the skull—the same cavity that in ourselves is often responsible for sinus headaches when its drainage is blocked. Old World monkeys such as the macaques, baboons, and vervets do not have this sinus. And *Proconsul* was tailless, again closer to the apes and ourselves than the monkeys.

Proconsul was not, however, entirely an ape. Its wrists, arms, and legs had both apelike and monkeylike features, and indeed much of the postcranial skeleton showed a mixture of these characteristics. Such a situation is quite common in the fossil record—one almost never finds an ancestor that is exactly intermediate between two descendant groups. Because *Proconsul* shares the preponderance of its characters with the apes, it is now generally felt that it lies closer to the ancestral lineage of the apes and ourselves than it does to that of the Old World monkeys. Yet molecular evidence, from Allan Wilson's group and others, suggests that the split between the apes and the monkeys must have taken place at about the time that *Proconsul* lived. If *Proconsul* or its close relatives were actually alive during this important evolutionary event they might even have played a role in it. This would place them, in evolutionary terms, exactly equidistant between the Old World monkeys and ourselves.

The Leakeys' discovery of the *Proconsul* skull pushed our ancestry back almost to the time of the beginning of the Great Rift Valley. It seems that our development and that of the valley went hand in hand. Yet *Proconsul,* enormously important though it was, was too different from ourselves and lived at too remote a time to really capture the public's imagination. Mary Leakey's next major discovery was at Olduvai, 300 kilometers southeast of Rusinga, and it consolidated the Leakeys' worldwide reputation.

As we saw, Olduvai had first been visited by Louis almost twenty years earlier. Although he had been wrong about Olduvai man, the site continued to fascinate him. During the intervening two decades both Leakeys had spent much time there, and they had found so many animal fossils and stone tools, seasoned with the occasional tantalizing hominid tooth, that they were sure important hominid fossils would eventually be found in the gorge. Finally, in July 1959, Olduvai Gorge yielded its first substantial hominid fossil. It was the skull of a robust Australopithecine, again highly fragmented but with many of the fragments recoverable from the deposit near the main find. And it was squarely and without the possibility of a mistake in bed 1, the oldest bed in the gorge.

Louis immediately dignified Mary's find with a new genus name, *Zinjanthropus,* from the ancient Persian word for African. Phillip Tobias was given the job of describing the skull by the Leakeys. On the basis of its massive jaw he bestowed on it the eminently quotable nickname "Nutcracker Man," a name that quickly resounded throughout the world. And this was the discovery that gave the Leakeys the full benefit of the immense publicity machine of the National Geographic Society.

Each issue of *National Geographic* magazine, it has often been remarked, is too beautiful to throw away. As a result it forms a kind of geological stratum of its own, extènding throughout the suburbs of North America. When the magazine published lengthy and superbly illustrated articles on *Zinjanthropus* and on each of the Leakeys' subsequent discoveries at Olduvai, it quickly turned the Leakeys in the public's mind into the world's authorities on human evolution, and Olduvai Gorge into the cradle of humankind. Never mind that *Zinjanthropus* was not, in the opinion of most anthropologists, very different from the robust Australopithecines that had been found in southern Africa decades earlier. The Leakeys had discovered it, and that fact alone gave it great importance.

If one were to judge by the size of its brain, *Zinjanthropus,* like the other robust Australopithecines, was not a towering intellect. Louis immediately assumed, however, that it had been responsible for the primitive Oldowan pebble tools that had earlier been found in considerable numbers in bed 1 deposits. And once again and within a span of eighteen months, he had to eat his words. Both he and Jonathan, his older son by Mary, soon made further finds indicating that several kinds of hominid had, at various times, made their homes in the Olduvai region and that Nutcracker Man was probably not the maker of the Oldowan tools.

Jonathan's fragmentary finds were also from bed 1, and consisted of bits of two parietal bones from the sides of a skull and parts of a lower jaw. The jaw was far less massive and had more human features than that of Nutcracker Man. These fragments were all that remained of a juvenile individual with teeth very reminiscent of Dart's original *Australopithecus africanus,* and they shared with *africanus* human-sized molars and a large and delicate braincase. The find was a vivid contrast to *Zinjanthropus,* with its massive jaw and huge sagittal ridge. Louis's own find was equally important but came from a later deposit, the upper part of bed 2. Again only a fragment, it was a large dome of a skull with immense browridges that were strongly reminiscent of those of Peking man. It was the first solid indication that *Homo erectus* might have lived in East Africa.

The Leakeys' abiding faith in Olduvai had been amply repaid. Within a space of less than two years, remains of three very different kinds of hominid had been found there. All were discovered within a few miles of each other, and all (though Louis Leakey was loath to admit it) resembled major groups of hominids that had been discovered earlier in South Africa or Asia. Even if Olduvai Gorge was not the cradle of humankind, it was certainly a crossroads of human evolution.

Soon, after further very fragmentary remains resembling Jonathan's *africanus*-like creature were found, Louis Leakey cheerfully embroiled himself in controversy again by naming it *Homo habilis* (handy man). This creature, rather than the brutish *Zinjanthropus,* Leakey now decided, must have made the primitive Oldowan tools found in bed 1.

He was mercilessly criticized for bringing the genus *Homo* into the picture so early, particularly by Le Gros Clark, who could see no difference between these remains and those of *Australopithecus africanus.* Phillip Tobias, however, contended that the association of *Homo habilis* with tools, a far plainer association than anything that had been found from such an early time in South Africa, clinched the matter. It was what Leakey had always dreamed of, evidence for the extreme antiquity of the human species in East Africa.

How Old Are the Hominids?

Conditions for hunting fossils in Leakey's stamping ground were very different from those in South Africa. There, as we have seen, most of the deposits had collected in caves and were blasted out in the course of mining operations or by the paleontologists themselves. In contrast, throughout the Rift Valley the fossils have been weathered out of ancient deposits by wind and water erosion. No blasting is needed, and the deposits from which the bones come can be determined much more clearly than in South Africa. The fossils are also self-renewing—an area picked clean of fossils in one season will blossom over the next few years with a new set, exposed by the occasional torrential rains. The drawback, however, is that the fossils themselves tend to be in much worse condition than those found in caves, for they can usually be discovered only by patient walking over wide swaths of terrain. By the time the fossils are discovered they have already been partially exposed and weathered. Sometimes, as at Olduvai, they are damaged further by the hooves of the Masai cattle that roam through many areas.

In spite of all these difficulties, the great advantage these fossils have over the South African remains is that they can be dated accurately. The potassium-argon and fission-track dating methods can both be used to assign dates to the periodic volcanic upheavals, which did not happen during this period in South Africa. Both these methods act as geochemical stopwatches, which are reset to zero in the rocks that are formed from volcanic eruptions.

Potassium has been given the symbol K, from the Arabic *kali* (alkali). It is one of the commoner elements in the earth's crust and indeed in our own bodies. Much of it was born in ancient supernovas, which spewed it out into the nebulas that give birth to the next generation of stars and their planets. Like most other elements, it comes in several different isotopic forms, both stable and unstable.

The nuclei of unstable isotopes decay in a variety of ways, producing many different results. Some of the heavier ones, like the isotope of uranium ^{235}U, actually split into two lighter nuclei and release energetic neutrons that can in turn break other nuclei apart. Atoms of the radioactive isotope of potassium named ^{40}K decay more discreetly. Once in a very long while, a proton within the nucleus of a ^{40}K atom captures a particularly energetic passing electron. This converts it into a neutron. The result is that the nucleus has lost a proton and gained a neutron, and so is almost unchanged in mass. But when the proton is lost, the nucleus can no longer hang on to all its electrons. The outermost electron that gives potassium its great chemical reactivity flies away. The result is an atom of a stable isotope of the very unreactive noble gas argon, ^{40}A.

Although this newly created argon is a gas, at ordinary temperatures its atoms can diffuse only very slowly through the rock. They eventually make their way to the surface, but they escape at a very low rate. When the rock is heated and melted, however, the argon diffuses quickly through the liquid material to the surface and is lost while the remaining atoms of ^{40}K, which cannot escape, are left behind. When the rock cools and solidifies, they begin afresh their slow breakdown into trapped atoms of ^{40}A. Thus, this radioactive clock is reset at the moment the magma from a volcanic eruption solidifies.

By now, the isotope ^{40}K makes up only one in every ten thousand atoms of potassium. But because its rate of breakdown is very slow, it has not vanished completely during the billions of years since it was first produced. There was twice as much ^{40}K in the rocks of the earth's crust 1.3 billion years ago as there is now, and 2.6 billion years ago there was four times as much. At the time the earth was born, 4.6 billion years ago, about one in every thousand atoms of potassium was ^{40}K, ten times as much as at present. With each 1.3 billion years that passes, the remaining amount of this isotope is halved, but there will still be plenty around in the foreseeable future for any intelligent races that follow us to be able to carry out potassium-argon dating—perhaps in order to date the puzzling remains of our own species.

The small amounts of ^{40}K and ^{40}A in rock can be measured precisely by a technique called mass spectrometry. Mass spectrometers are small-scale particle accelerators—they take atoms or molecules of a gas and hurl them down an evacuated tube in the presence of a strong magnetic field. The farther these atoms or molecules travel down the tube before they touch its wall, the less massive they are. Thus, the different isotopes of argon from a bit of rock can be released by heating and measured almost atom by atom.

However, just as argon in the rock can diffuse out, tiny amounts of argon from the atmosphere can diffuse in. This makes it difficult to determine the ages of young volcanic rocks in which very small amounts of potassium-generated argon have accumulated. In such young rocks argon from the atmosphere can overwhelm the traces of radioactively generated argon. Potassium-argon dating is really practical only for rocks or ash that cooled and solidified more than fifty thousand years ago. And normally, to obtain the ratio of potassium to argon, the amount of potassium in the rock must be measured from a separate sample, which can give an erroneous result when the percentage of potassium happens to vary from one part of the rock to another.

An ingenious way to get around this latter problem was suggested by Craig Merrihew and John Reynolds in 1950. If a bit of the sample to be measured is first placed in an atomic pile and bombarded with energetic neutrons, a proton can be knocked out of some of the nuclei of ^{40}K, converting them into a different isotope of argon, ^{39}A. After this treatment, the sample can be placed in a mass spectrometer and the amounts of ^{39}A and ^{40}A can be measured simultaneously with great precision. The amount of ^{39}A in the sample now acts as a kind of proxy for the amount of ^{40}K. This technique also helps to circumvent the problem that arises when old rocks or ash are contaminated with younger material or have been heated subsequent to the original melting event. The experimenter, by heating the rock in a series of stages, can measure the ratio of ^{39}A to ^{40}A in the argon that is driven out at each stage. If the ratio remains the same throughout the experiment, the rock has remained undisturbed and uncontaminated. If it changes, the shape of the curve can give a clue to the rock's history. The final ratio represents the gas that was driven out of the deepest and presumably least disturbed part of the sample. Often however, the curve is a complex one, making interpretation of the results difficult.

Scientists love it when there are two independent ways of measuring something, for each acts as a check on the other. A second independent check became available when it was found that the melting of volcanic rocks resets a quite different radioactive clock. Volcanic glasses contain, along with many other elements, tiny amounts of uranium. When a uranium nucleus in such a glass splits, the two resulting fragments fly apart, bore through the glass, and leave behind an invisible track. If one end of the track happens to terminate at the surface of the piece of glass, strong chemical treatment can enlarge the track and make it visible under the microscope (see plate 6). Melting the glass, of course, destroys all the old tracks, but they begin to accumulate again after the glass cools. Simply counting the tracks gives a direct measure of the sample's age. Of course, both this technique and the potassium-argon technique can easily give erroneous results, but if the dates agree it is likely to be more than a coincidence.

These two methods have been used with brilliant success in East Africa, although they have had teething troubles that have been recounted in detail in Roger Lewin's fascinating book *Bones of Contention*. After that initial rough start, they have now begun to provide firm, unarguable dates for the East African fossils. Even the

South African material can now be dated with more certainty, as a kind of dim reflection from the East African dates. When East African animal fossils are given approximate dates by dating the volcanic material that overlays them, finds of similar or identical species in South Africa can be assumed (although with some trepidation) to be of the same date.

Recently the techniques have become increasingly sophisticated and precise. A number of laboratories around the world are providing geochronologies of unprecedented detail. One of these is at the Institute of Human Origins, which occupies a disconcertingly modern building on the north side of the University of California, Berkeley, campus, safely away from the hurly-burly of Telegraph Avenue. There I recently watched Paul Renne, the geochronology lab's director, use a precisely directed laser beam to heat single crystals teased from a bit of compressed volcanic ash called tuff. When the beam hit the tiny crystals, each only 1 or 2 millimeters long, they jumped about under the microscope as they released puffs of gas. The argon from this gas, as little as a few tens of thousands of atoms, was purified in a maze of tubes and then channeled into the mass spectrometer.

Renne told me that the technique is now so good that its precision has outstripped its accuracy! That is, it is possible to tell precisely how old a sample of rock is in comparison with a standard sample of some other age—even though the age of the standard itself cannot yet be determined precisely. Still, Renne predicts that this will soon be done. Then, a combination of argon-argon and fission-track dating with new advances in measuring the residual magnetism of rocks will enable geochronologies to be constructed in unprecedented detail. Already it is possible to separate deposits from individual volcanic eruptions that took place only a few thousand years apart, even though the eruptions may have happened tens of millions of years ago.

A New Generation of Discoveries

These dating methods became available just as a most astounding series of discoveries of hominids was being made by groups headed by Mary Leakey, her son Richard, and a young scientist trained at the University of Chicago, Donald Johanson. The intense rivalries among these groups, always illuminated by the relentless publicity spotlight provided by the National Geographic Society, are widely known and would certainly provide much of the excitement in the potential miniseries I mentioned earlier. Although there is no space here to do justice to the complexity of the events, I will try to summarize briefly the major discoveries and put them in an overall perspective.

Like his father, Richard Leakey, born in Nairobi in 1944, had an adventurous childhood, though he did not take part in the tribal rituals and tribal life that gave his father such an instinctive feel for Kenya and its problems. Determined to pursue anything but paleontology, he avoided going to college and instead founded a safari business. The organizational skills needed to mount safaris turned out to be very useful on his later scientific expeditions. And it was not long before he became involved with the Kenya Museum that had occupied so much of his father's life, and began to be drawn by the excitement of paleontological expeditions. There is no doubt that being his father's son advanced his career with unusual rapidity, but this advancement would never have happened without his capacity for hard work, deep knowledge of the country, and keen eye for fossiliferous terrain.

In 1967 he took part in an expedition to the crocodile-infested Omo River of southern Ethiopia. On a flight back from this site, the pilot was forced by a storm to detour over the eastern shore of Lake Turkana (formerly Lake Rudolf). This huge, shallow lake lies along the eastern branch of the Rift Valley, mostly in Kenya but with its northern tip just penetrating Ethiopia. As the plane flew low to avoid the storm clouds, Richard glimpsed huge areas of folded sedimentary rocks that did not show up on any geological map.

After some effort, and in spite of opposition from his father, Richard obtained support from the National Geographic Society for expeditions of his own to Lake Turkana. During the second of these, a base camp was set up on a spit of land called Koobi Fora that extended into the lake. The camp, growing more and more permanent, has remained the center of operations at Lake Turkana ever since.

It was this 1969 expedition that began to establish Richard Leakey as a brilliant fossil hunter independent of his father, and it is ironic that his growing reputation received its major boost as the result of a dating error. Soon after the expedition's arrival at Koobi Fora, the geologist Kay Behrensmeyer found flakes of stone that could have been very primitive tools. Volcanic ash that had been compressed over time into tuff was taken from the horizon immediately above these flakes and sent to John Miller and Frank Fitch in England. They used the ^{39}A-^{40}A method to determine that the tuff was 2.4 million years old, a finding that was very exciting because it meant that these signs of hominid activity were at least half a million years older than the oldest remains at Olduvai. The lowest bed at Olduvai, bed 1, had earlier been dated by similar methods at about 1.9 million years. Anything that lay below Kay Behrensmeyer's tuff of volcanic ash would have to be much older than the finds at Olduvai—and of course Richard, who had been steeped from childhood in his father's conviction that *Homo* in Africa was very old, was as anxious as his father had been to prospect for fossil *Homo* in the oldest possible beds. He did not know at the time—and for years refused to believe, in the face of mounting evidence—that this tuff was in fact no older than Olduvai bed 1.

Soon, his expeditions began to find fossil robust Australopithecines in these beds at Koobi Fora. The sheer number of the fossil discoveries was remarkable—there seemed to be about as many hominid bones as baboon bones, and baboons are still very common in the area. It seems that two million years ago the region supported a large and highly successful population of Australopithecines.

Things got more exciting in 1972, when the remains of a very different kind of hominid were found by Bernard Ngeneo. Ngeneo was one of the many Africans whom the Leakeys had trained over the years to be excellent fossil hunters and geologists, and who now make up a large part of the scientific effort in Kenya and Tanzania. The fossil, soon to become world famous, was named for various complicated reasons KNM ER 1470, known more familiarly as 1470. Although the skull was badly fragmented, when it had been pieced together it was very different in appearance from the robust Australopithecine remains that had been found earlier in the very same layers. There was no sign of a sagittal crest, the brain was far larger—200 to 300 cubic centimeters larger—and the teeth were very humanlike. There was no sign of the enormous grinding molars so typical of the robust Australopithecines. The skull shared many features with the gracile Australopithecines from South Africa such as *africanus*, but the most striking difference was its relatively gigantic brain.

It was the brain size, and not any other features of the skull, that made Richard put it into the genus *Homo*. Perhaps, he thought, it could be an important link between the small-brained gracile Australopithecines and the much larger-brained *Homo erectus*. Because the date of Kay Behrensmeyer's tuff seemed to make the skull more than half a million years older than the *Homo erectus* skullcap his father had found in bed 2 of Olduvai, this seemed very likely. The dating made it fall into just the right time frame for a missing link.

Was this creature *Homo?* Certainly Louis Leakey thought so when his son showed him the skull at the end of September 1972. Its

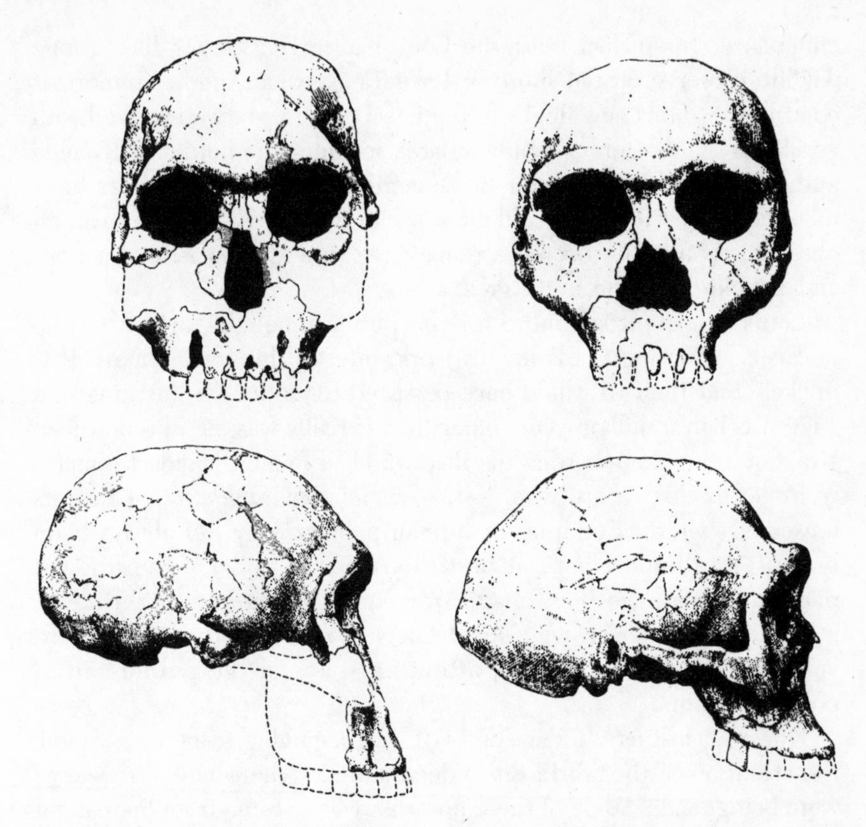

FIGURE 5.3. Richard Leakey's 1470, *Homo habilis,* on the left, compared with *Homo erectus* on the right. *Homo erectus* has a large braincase, a more "dished" face, and more prominent browridges—but the vault of the skull of *Homo habilis* is at least as lofty as that of *Homo erectus.*

surprisingly modern appearance coupled with its apparent great age reinforced his strongly held beliefs about human evolution. So it was a very happy Louis Leakey who left Kenya that same evening for one of his frantic fund-raising tours to England and America.

Two days later, in London, he was dead of a heart attack.

Louis Leakey, brilliant, charismatic, opinionated, enormously knowledgeable about every aspect of the life and history of East Africa, left a titanic legacy. Not only had he founded much of paleoanthropology in the area, but he also had stimulated the expansion of the new field of comparative ethology. He sought out and inspired a series of dedicated young workers who began to collect observations on the behavior of the great apes, our nearest living relatives in the wild. The best known of this group is Jane Goodall, who has made brilliant observations on chimpanzee troops living at a site Louis had picked for her, the Gombe wildlife preserve on the shore of Lake Tanganyika. Another important contributor to this new field was the late Dian Fossey, who worked with gorillas in the dangerous mountain area spanning the border of Rwanda and Zaire until her murder in December 1985. Many others have followed in their footsteps, and their tens of thousands of hours of careful observations have uncovered a remarkable variety of social interactions and complex behaviors in the great apes.

Louis's constant scramble for the public limelight, driven by the necessity of raising funds for his work, infected his son Richard. It is unlikely that Richard would have persisted for so long in insisting that 1470 was half a million years older than it really was, had he not been afraid of losing the record as the man who had found the oldest remains of *Homo.* If that record were lost, *National Geographic* might turn its lenses elsewhere. Funding for human paleontology has always been measured in hundreds or at best thousands of dollars, even though many millions have been made from its exploitation. A lifetime of watching his father struggle for driblets of money had made Richard very well aware of the short attention spans of the public and of potential donors.

However, just as the date of 1470 was beginning to be questioned, the attention of the world did indeed shift to where new discoveries were being made. Some of these new discoveries came from the patient and dogged work of Richard's mother, Mary.

Transit Zone

Mary and Louis Leakey had grown apart toward the end of his life, and eventually she moved full-time to Olduvai. She also began to explore a site south of Olduvai in wild country between the gorge and Lake Eyasi. The site had first been brought to her and Louis's attention back in 1935 during their first season together at Olduvai, when very little was known of the surrounding area. They were intrigued when a native visitor came by with some fossil bones he had found at a spot called Laetoli, a nearby region of eroded badlands with a tiny river flowing through it. Laetoli was a kind of avian transit zone in the middle of grim, dry country, where thousands of birds stopped for water on their north-south migration route. As Mary Leakey was to discover, it had also once been a transit zone for hominids.

The Leakeys went briefly to Laetoli for a quick look but found no tools or hominid remains. Still, the beds that had been exposed by erosion were obviously very old. The plentiful mammalian fossils suggested that they predated the lowest beds at Olduvai. What started as a cursory side trip became a much more intense investigation in 1974. Exploring the course of the Gadgingero River as it flowed through the Laetoli area, Mary penetrated a region that neither she nor Louis had yet visited and found that the river had exposed some extensive fossiliferous beds. These were overlain by thick flows of lava, samples of which she sent for dating to the geochemist Garniss Curtis at Berkeley.

At that time, Curtis was embroiled in the argument about the date of Kay Behrensmeyer's tuff at Koobi Fora. He had dated that controversial tuff at 1.9 million years, much younger than the date of 2.4–2.6 million years obtained by Fitch and Miller, the two British geochemists who had first been employed by Richard. And more distressing evidence was coming in about Koobi Fora. Although early uranium fission-track measurements had first appeared to support the Fitch and Miller date, animal fossils did not. Many of these fossils, particularly the plentiful fossil pigs, were suspiciously like those at the 1.9-million-year-old Olduvai bed 1. Then, to shatter the story completely, more careful analysis showed the early fission-track results had been in error. The revised date turned out to be the same as Curtis's 1.9-million-year date. *Homo habilis* was turning out to be much younger than Richard Leakey had thought.

Mary Leakey, unlike her son, trusted Curtis's technique from the beginning. She was very excited when he obtained a date for an ash overlay of the Laetoli beds that he was sure really was at least half a million years older than the real date for the Behrensmeyer tuff. Because the beds lay beneath the ash layer that could be dated, they must actually be older than that. Perhaps—because of their great thickness of 30–50 meters—they were a great deal older.

Intense fossil hunting in these beds, however, resulted in only a few hominid fragments, including bits of jaws, and no sign of tools. The jaws were so humanlike in appearance that she initially supposed they too, like 1470, might actually belong to the genus *Homo.* The absence of tools reinforced the idea that Laetoli had indeed been a transit zone for hominids as they made their way between more favorable regions.

But these hominid fragments were certainly old, in fact far older than any other hominids found so far in East Africa. Curtis was soon able to obtain good dates showing that they were an astonishing three and a half million years old. If they were *Homo,* they would predate *Homo habilis,* now apparently no more than 1.9 million years old, by over a million and a half years. Unfortunately, the finds were too incomplete to be sure. They might merely have been early gracile Australopithecines, because the jaws and teeth of those hominids are the most humanlike features of their skulls.

The most remarkable find at Laetoli, however, and indeed perhaps the most remarkable find in all East African paleontology, happened entirely by chance. During the 1976 season, Andrew Hill from the Nairobi Museum was one of a group visiting the site. Walking back from an excavation, the group began to amuse themselves by throwing lumps of dried elephant dung at each other—an indication, perhaps, of the paucity of recreational facilities at Laetoli. Dodging one particularly well-directed lump, Hill fell to the ground. His face was only inches away from the rutted surface of the track, and the light happened to fall at just the right angle. Suddenly, he saw what everybody else who had passed that way had missed. There were animal footprints impressed in the hard surface of the volcanic tuff itself.

These footprints, it was soon discovered, were those of a variety of animals ranging from impala to rhinoceros that had wandered through the area three and a half million years earlier. They had been preserved in deposits from the nearby volcano Sadiman, now a dormant and eroded shadow of its former self. Sadiman had blanketed the area with deposit after deposit of loose fluffy ash, which like snow had preserved the footprints of the animals walking across it. Unlike snow, its surface was preserved in turn by subsequent ash falls. Even then the footprints might still have disappeared were it not for the fact that Sadiman's ash, like many East African ash falls, was rich in the mineral carbonatite. Water from the next rainy season percolated into the ash layers, converted the carbonatite to limestone, and preserved these almost unique snapshots of animal activity.

Many fossil prints, some of which had been made by animals that could still be identified by local tribesmen, came to light over the next two years. Of course, the burning question in the minds of everyone involved in the excavations was whether there could be hominid footprints somewhere among them and what they might look like. If the hominids of three and a half million years ago walked in a very different way from the firm, plantigrade stride of present-day humans, then their footprints might be quite unrecognizable.

A few of the prints, Mary thought, looked suspiciously hominid, but they were too blurred to be certain. The first undoubtedly hominid print was found by another visitor, Paul Abell. He made his discovery while engaged in removing slabs with some handsome rhino prints for a museum exhibit in Nairobi.

More hominid footprints were quickly found nearby, extending in a line for some 10 meters. So far as could be seen, they were identical to those of modern humans, fully plantigrade, made by hominids who left a firm heel print and pushed off at each stride from the ball of the foot. The tracks of at least two hominids were preserved. One, possibly a child, had walked a little to the left of a much larger hominid. The larger tracks, perhaps made by a male, were the same size as those of a modern human. And Mary Leakey claimed that a third, perhaps a female, had followed the male and playfully placed her feet in the larger footprints. There has been much argument about this, particularly about the possibility that the apparent doubling had been accidentally produced in the course of the excavations.

Whether there were three hominids or two, one striking feature of the tracks is that the larger and smaller tracks have just the same stride length. The fact that the tracks are so close to each other makes it unlikely that they were walking side by side. Perhaps the larger hominid, in the lead, was shortening its stride a little, and the smaller one was lengthening its stride in an effort to keep up.

The hominid footprints at Laetoli constitute what a friend of mine calls an action fossil. They provide a direct link with that casual promenade of vanished hominids that took place one and a quarter billion days ago, a link that is somehow far more vivid than all the bones and stone tools now reposing in museum cases. Perhaps more than any other paleontological discovery, the footprints show vividly that by three and a half million years ago a great deal of the evolutionary change that separates us from the apes *had already taken place.* These hominids, tiny brained though they were, were already leading lives utterly different from those of the animals that surrounded them. The difference between the upright body and plantigrade stride of Leakey's hominids and the gait of any other primate was profound, because it freed the hands for manipulating and carrying objects. Even now, maddeningly, we have no indication of the stages through which our earlier ancestors passed in order to acquire this new and remarkable ability or of how long it took. But it could not have taken more than two or three million years, because the Laetoli hominids lived at a time at least halfway and perhaps more than halfway back to the common ancestor of ourselves and the chimpanzees. And all these profound and important evolutionary changes had of course taken place long before the mitochondrial Eve.

Mary Leakey's fossil finds at Laetoli were at that time the oldest substantial hominid remains that had been discovered, almost certainly half a million to a million years older than the far less precisely dated gracile Australopithecines from South Africa. The bones were too

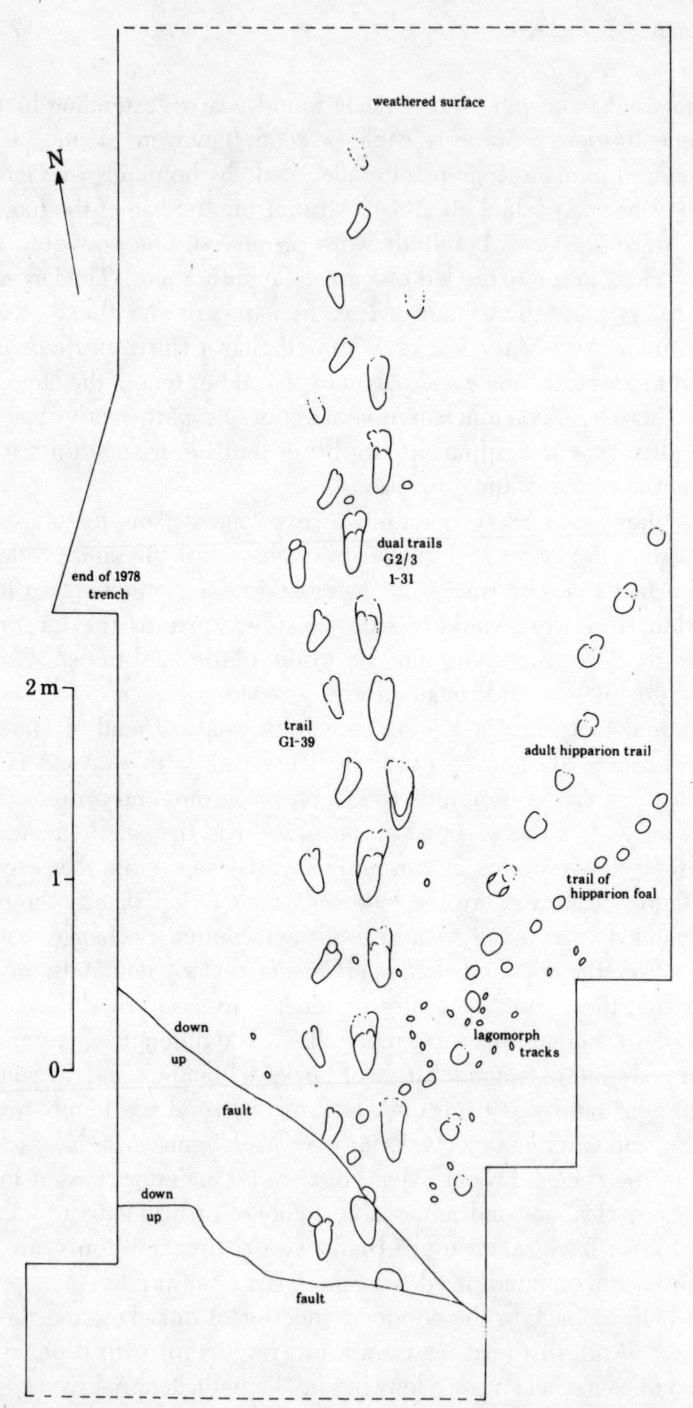

FIGURE 5.4. A view from above of the Laetoli footprints. Both hominids, in spite of the very different sizes of their feet, had the same stride length, but the prints are so close together that they were almost certainly not walking side by side.

fragmentary to be of much use, however. Had those creatures that left their bones behind also made the footprints? The answer to this question came, not from Laetoli, but from much further north.

Ethiopia

The mountains and highlands of Ethiopia are cleaved into two parts by the ax stroke of the Rift Valley, which here makes a great gash running northeast and southwest. The Omo River flows from the center of this gash down into Kenya, which has enabled a number of expeditions to use it as a gateway to explore that inhospitable region. In 1959 Clark Howell from the University of Chicago explored the area around the mouth of the river and found it to be a fossil-rich set of jumbled deposits covering at least 100 square miles. Unfortunately, his fossils were confiscated at the border and he never saw them again.

In 1966, however, Louis Leakey accomplished a political masterstroke. The Emperor Haile Selassie was on a state visit to Kenya, and Leakey showed him his fossils from Olduvai and the *Proconsul* from Rusinga Island. The emperor, struck by a pang of fossil envy, asked in irritation why Ethiopia had no such fossils. Leakey assured him that it did, and told him that they could be found in the Omo valley. The emperor immediately gave permission for expeditions to explore the rich Omo deposits.

Omo soon yielded a treasure trove of animal fossils, including a very large number of hominid remains. Some 140 species of mammals were found, ranging in age from about four million to one million years ago. But many of them, including the hominids, had been damaged, fragmented, worn down, and removed from their original deposits by the action of the floodwaters that periodically roared through the area.

One of the many young scientists who were introduced to paleontological work at Omo was Donald Johanson, a Ph.D. student of Howell's. Johanson was determined to find better fossils. Returning from Omo through Paris to finish up work on his dissertation on chimpanzee skulls, he met a young geologist at a party. The geologist, Maurice Taieb, was trying to work out the highly complex stratigraphy of the Afar Triangle at the head of the Rift Valley. This area was of particular geological interest because it was where the corners of three tectonic plates abutted. In the course of his explorations of that vast and forbidding area, which lay largely within Ethiopia, Taieb had seen many fossils.

On the spur of the moment, Johanson and Taieb made a quick camping expedition into the Afar, and Johanson saw immediately that the fossil-rich rocks there had not been subjected to the disturbance that had ruined so many of the fossils at Omo. In particular, at an ancient lake bed called Hadar, some recent gullies had been cut into strata that seemed to Johanson and Taieb to be at least three million years old.

Johanson's first full-fledged expedition to Afar, in 1973, was mounted with slender support from the National Science Foundation. It became a success and at the same time made further expeditions possible because of one remarkable find, a hominid knee joint. Nothing else of that hominid was found, but the fragments of femur and tibia fit together perfectly, and almost certainly came from the same individual. It was immediately obvious that the knee was humanlike rather than apelike.

If you run your hands down the two sides of your leg, past your knee, you can feel how the bones of the leg change direction. Above the knee the upper leg angles away from the midline of the body. Below it, the tibia and fibula continue vertically down to the ankle. This change of direction, which is not seen in the apes, could easily be seen in the ancient knee joint.

In the course of human evolution the flaring ilia of the pelvis have grown so broad in order to support the viscera that our hip joints have moved much further apart than those of a chimpanzee. Were our legs to be as straight as those of a chimpanzee, we would look like a child's drawing of a human, with columnlike legs spread far apart. Our walk would be an awkward swiveling straddle, instead of a smooth swinging stride in the course of which each foot in turn can push off from near the body's midline. Johanson and Taieb did not yet know how old their knee joint was, but the way the bones were angled showed that the hominid it belonged to walked in a very human way. Their find, which predated the discovery of the Laetoli footprints, was strong evidence for the great antiquity of human walking.

The following year was a time of worsening political turmoil in Ethiopia, for the emperor's grip on power was finally loosening. However, because of the knee-joint discovery, Johanson's next expedition was much better funded and included a variety of experts on mammalian fossils.

Almost immediately after their arrival, Alemayehu Asfaw, an Ethiopian member of the expedition, found two superbly preserved upper jaws. These jaws, and some jaw fragments, were clearly from creatures that were not quite humans and not quite apes. The molars were large, but not as huge as the great grinding molars of *Zinjanthropus* and the other robust Australopithecines. The premolars lay somewhere between the bicusped premolars of humans and the single-cusped premolars of apes, having one main cusp but with traces of another on their inner margins. The shape of the dental arcade was intermediate between the almost rectilinear shape seen in the great apes and the gently paraboloid curve of humans. And although there was a diastema where the canines of the lower jaw fitted, it was not nearly as pronounced as the large diastema needed to accommodate the huge canines of apes. These finds did not resemble the jaws of humans as closely as the knee joint discovered the year before had resembled a human knee, but they could without too much effort be supposed to have belonged to a primitive *Homo.* Richard and Mary Leakey, who visited the site soon afterward, were sure that they really were from some kind of *Homo.*

It was only a few weeks later that the fossil-rich Hadar yielded a really remarkable fossil. Johanson himself found this scattered collection of fragments of hominid bone that had recently washed out of the slope of a small gully. Only bits of the skull were present, but there were many bones from the postcranial skeleton, more such bones than had ever been found for such an old fossil. And the bones were all from a single individual—there were no duplicates. The skeleton was that of a tiny female, only a little over a meter tall but fully mature. She was probably in her late twenties when she died (see plate 7).

Wild with excitement, the camp spontaneously named her Lucy, after a Beatles song. Lucy's skeleton, after tons of gravel had been sifted to find every possible fragment, was about 40 percent complete. It included a complete lower jaw, important joints of the arms and hips, most of the pelvis, and fragments of hand and foot bones. Although her brain was apparently tiny, this was in part a consequence of her small size. The rest of her anatomy was remarkably humanlike, and her posture was certainly fully upright.

The first potassium-argon dates indicated that Lucy was about three million years old, an estimate that actually turned out to be too young. More careful stratigraphy, the discovery of old pig and *Hipparion* fossils (the latter an ancestor of the horse that they knew had to be at least three million years old) and fission-track dating eventually established that Lucy was something like 3.5 million years old. This remarkably humanlike creature had lived almost two million years earlier than Richard Leakey's *Homo habilis.* However, these revised dates, making Lucy among the oldest of the hominids as well as the most complete yet to be found, lay some years in the future. Frustratingly for Johanson, they were determined at a time when Ethiopia, like medieval Japan, was closed to the West.

Just as these fossils were being found, Hailie Selassie's ragtag empire fell and was replaced by a military dictatorship called the Dergue. Civil war soon erupted between this new central government and the province of Eritrea. Johanson's next field season, in 1975, was mounted over the opposition of the U.S. embassy in Addis Ababa, which was afraid that marines might have to be sent in to rescue the paleoanthropologists. The paleoanthropologists themselves were very nervous, as Johanson admits, but the lure of the fossils was too strong.

And they were rewarded for their bravery in full measure. Soon after their arrival, the group made a chance find of a scattering of hominid fragments on an eroding slope near the camp. Within a matter of minutes, more bones were found nearby and then still more. This led to a scene (captured on film) that must be unique in paleontology, a kind of fossil feeding frenzy in which expedition members scrambled over the slope, turning up bits of hominid wherever they looked.

When the dust, quite literally, settled and a more organized search was made, they eventually found fragments of at least thirteen individuals, including a number of jaws and enough hand and foot bones to be able to reconstruct these critical appendages in detail. It was inevitable that this astounding gathering of individuals, unique in African paleontology, should become known as the First Family.

It seems likely that many more bones of the First Family remain buried in the strata above the gully because hominid bones must have been eroding out of the slope for millennia. Even in a dry and undisturbed area like the Hadar, fossil bones last at the most hundreds of years on the surface. They are washed away by flash floods, crumbled by the action of sun and wind, and are eventually broken down into a collection of tiny unrecognizable fragments. One cannot help wondering how many bones, from how many dozens or hundreds of individuals, had already eroded out of the slope and washed away before Johanson's team found the latest crop.

Even more remarkably, unlike virtually every other East African find, there were almost no animal bones mixed in with the hominid ones, and there was no sign that the bones had been gnawed on by animals after the hominids died. Perhaps this gave a clue to the nature of the disaster that had overtaken this group. The best suggestion that Taieb and Johanson could make was that a flash flood had caught this group of hominids in a gorge, a fate that still overtakes incautious campers in desert regions today. But there was no evidence of such a flood from the strata in which the bones were embedded. Perhaps the bodies had been washed down from the flooded tributary to a calmer river or lake shore, where they sank and were covered with sediment before scavengers could find them.

The problem with this scenario is that the skeletons had been disarticulated, so they must have spent some time on the surface where the ordinary processes of decay could begin. Why, then, do the bones show no sign of animal damage? Possibly these hominids were living in an area where animals could not get to them at the time disaster overtook them. If so, they might have formed some kind of *organized* group, living in a cave or rock shelter or gully that animals could not reach. Perhaps, indeed, the First Family will eventually be renamed the First Tribe.

The bones of the First Family were a little younger than those of Lucy, eventually being dated to about three and a quarter million years ago. Whatever mischance or mischances had occurred at that time in the Hadar, this astonishing find gave a whole new dimension to the fossil story. Most fossils from those distant times tend to be single teeth, fragments of jaws, or solitary skulls. The First Family provided the first real information on variation in a hominid group. And the amount of variation found in the First Family was very great, far greater than that found in a comparable group of humans today.

The variation was in both size and shape. The jaws were especially variable, both in the range of tooth sizes and in the relative sizes of the incisors at the front of the jaw compared to the molars at the rear. Johanson thought at first that he had two species of hominid on his hands, although he was later convinced by Tim White that it was only one. White pointed out that much of the variation was spurious, simply owing to the fact that males were much larger and heavier than females. Seen from above, Lucy's lower jaw was V-shaped, with tiny incisors, but her molars were as large as those of a modern human. Most of the First Family's jaws were broader and more rounded in the front, with larger incisors relative to the molars. All the molars of the First Family, however, were smaller than the massive molars of the much later robust Australopithecines and in fact fell within the human range (see plate 8).

Eventually, White managed to convince Johanson that all this variation formed a continuum, with Lucy, separate discovery though she was, at one end. Had the extreme types of jaws in the First Family been found in different places rather than all together, they would certainly have been put into separate species. The fact that they had been found grouped along with intermediate specimens greatly increased the likelihood that they all belonged to the same species. And of course the fact that the First Family had all apparently lived and died together made another very strong argument that they consisted of a single species.

But what was the species? Tim White had spent some time with Mary Leakey at Laetoli and had uncovered some of the hominid footprints. While he was there, he had also taken the opportunity to examine the fragmentary hominid finds that her group had turned up, especially the pieces of jaw. These pieces, all from large jaws, were an excellent match to the larger jaws of the First Family. The Laetoli fragments, however, had been dated at 3.7 million years, which at that time seemed to put them much earlier than Lucy or the First Family.

Johanson and White, after much agonizing, decided to abandon the idea that these hominids belonged to the genus *Homo.* They were too old and too different from *Homo habilis.* The only logical genus in which to put them was *Australopithecus,* and so they named them *A. afarensis* after the Afar Triangle in which they had been found. *A. afarensis,* they thought, so old and so small brained, must be near the root of the hominid family tree.

Every new species must have a *type specimen,* a carefully preserved museum specimen to which other members of the species can be compared. Johanson and White committed a political error at this point: they designated as the type specimen one of Mary Leakey's Laetoli jaws instead of the more logical Lucy. They simply wanted to emphasize the similarity of the Laetoli and Hadar specimens, although they were later accused of wanting to group their specimens in with the apparently much older Laetoli fragments. (This accusation lost a good deal of force when it was found that Lucy and the Laetoli specimens were not far apart in age.) Mary Leakey was furious at the way they had trespassed on her turf, and this incident began a period of very unpleasant relations between Johanson and the Leakeys.

Now, as these controversies recede into the past, new generations of scientists are beginning to reforge cooperative relationships—though the rivalry and the desire to beat the competition to the next important discovery remains as intense as ever.

A Huge Pool of Variation

During the long interregnum imposed by the Ethiopian military and the subsequent Marxist regime, the northern end of the Rift Valley was closed to further exploration. But other finds were being made elsewhere in East Africa that formed a fascinating and deeply puzzling pattern. What began to emerge was that, between three and a half and one and a half million years ago, several different kinds of hominid made their appearance in the Rift Valley. The relationship among them, however, is totally confusing, and there is no agreement about what kind of a family tree might be drawn to relate them to each other.

Until just a few years ago, one of the biggest puzzles concerned some of the most primitive-appearing hominids, the robust Australopithecines with their large teeth and jaws and pronounced sagittal crests. There now is much sentiment that they should be put into the separate genus *Paranthropus* that Broom established for them. However, most *Paranthropus* remains have been found in strata or cave deposits dating from between one and two million years ago, making them more recent than the much more human-appearing *Australopithecus africanus,* and probably even more recent than Leakey's *Homo habilis.* Did they perhaps evolve from them? If so, it would suggest that there has not been a continuous progression in the hominid line toward a more humanlike appearance, but rather that the evolutionary story of the African hominids had been much more complicated. The hominid line would seem to have diverged recently, one lineage going on to *Homo erectus* and another veering in a very different direction toward the small-brained and large-toothed robust Australopithecines.

A discovery made in 1985 reduced the probability of this particular evolutionary scenario, although it seems to have replaced it with one equally puzzling. In that year, Alan Walker from the Johns Hopkins School of Medicine found on the west shore of Lake Turkana a heavily stained, quite complete, and extremely robust skull that could be dated unequivocally to at least 2.5 million years ago (see plate 9).

The huge sagittal ridge and massive cheekbones of this "Black Skull," as it was immediately named, are very similar to those of the other robust Australopithecines. This rather frighteningly primitive skull, to our eyes almost as grotesque as a medieval Japanese war helmet, is very different from the delicate skulls of *Australopithecus africanus* and *Homo habilis* with their dainty molars and swollen braincases.

Most hominid family trees show robust Australopithecines banished to a branch of their own, a branch that came to an end about a million years ago. How much of this banishment reflects evolutionary reality and how much reflects our own prejudices? Although the Australopithecines' diet was probably made up of a wide variety of plant materials, we do tend to equate strong jaws with cud chewing—and, whether in cows or in teenagers, with weakened intellect.

At least, Walker's discovery pushes *Paranthropus* far enough back in time that it reduces the likelihood that it arose from *Australopithecus africanus.* But it also shows that there were a number of strikingly different hominids living—and perhaps interacting—in the Rift Valley during the period of three million to one million years ago.

And, regardless of its evolutionary affinities or intellectual powers, *Paranthropus* was an extremely successful hominid indeed. Finds of *Paranthropus* are among the most numerous of any hominid. Walker's discovery adds to the evidence that it survived, apparently in considerable numbers and fairly unchanged morphologically, from at least as early as two and a half million years ago down to about one million years ago. This is as substantial a track record as can be found for any hominid species.

The history of *Homo habilis,* the presumed missing link between the early Australopithecines and *Homo erectus,* may also have been extended further back in time. In 1967 John Martyn of the University of London found a fragment of bone from the side of a skull lying on the surface of the ground near tiny Lake Baringo, far to the south of Lake Turkana. At the time, the exposed deposits in the area were thought to be of fairly recent date, so the find was simply catalogued and put aside in a museum drawer at Yale University. Later, the rocks in the area were discovered to be much older than had previously been thought. Luckily, Andrew Hill of Yale realized that enough bits of rock still adhered to the bone in its museum drawer to enable it to be dated. Laser heating of single crystals in the rock showed the fragment to be 2.4 million years old.

The bone fragment still preserves features of the brain that it once encased. They help to distinguish it from the Australopithecines and perhaps to put it into the genus *Homo.* In particular, a crest of inward-projecting bone that in life separates the temporal lobe from the cerebellum at the base of the brain has been reduced in size compared with that of most Australopithecines, suggesting that the temporal lobe of this hominid had grown very large. It is fascinating that this recent finding harks back to that moment in 1924 when Raymond Dart realized that the temporal and parietal lobes of the braincase he held in his hand were larger than those of an ape and that they had begun to crowd the occipital lobe at the rear of the brain. Dart's child probably lived a little more than 2.4 million years ago. Hill's claim has been interpreted very differently by others, but, along with other finds from Omo, it provides some indication that by 2.4 million years ago the process of encephalization, or disproportionate growth of the brain, may have gone even further in some hominids.

Further testimony to the hive of evolutionary activity in the Rift Valley during this period was also emerging. In this case, the activity seemed to be far closer to our own evolutionary lineage. In 1975 Bernard Ngeneo of Richard Leakey's group had found a very humanlike skull lying on the surface of the ground at Koobi Fora. It was badly fragmented because tree roots had grown through it, so that it had to be excavated with exquisite care. The resemblance to the Asiatic *Homo erectus* was marked, although the braincase was about 200 cubic centimeters smaller than that of Peking man. The skull had the same pronounced supraorbital ridges and the same oddly pointed occipital region that is so characteristic of the Asian fossils. Yet it was more than 1.6 million years old, which made it more than a million years older than Peking man. This fossil was not the first African *Homo erectus* to be found, but the previous fragmentary discoveries, like those made by Louis Leakey at Olduvai, were not as old or as well preserved as Ngeneo's.

A mere two or three hundred thousand years separate Richard Leakey's most striking *Homo habilis* fossil, 1470, from Ngeneo's *Homo erectus* skull. This near overlap in time of *Homo habilis* and *Homo erectus* is unnerving and makes one wonder just how valid the distinction between these species really is. Did they actually coexist at Koobi Fora? And if so, how much of a difference was there between them?

A decade later, in 1984, Kamoya Kimeu's sharp eye spotted a bit of a skull's frontal bone near the south bank of the Nariokotome River that flows into the western margin of Lake Turkana. Painstaking excavation uncovered a virtually complete skeleton of a young boy who was undoubtedly a *Homo erectus*. The find was stunning, providing the first real look at the postcranial skeleton of this central player in our evolutionary story, and the age was equally stunning. Just below the skeleton was a tuff dated at 1.65 million years ago. The next datable layer was some 46 meters above and could be dated at 1.33 million years. The skeleton, at the bottom of the layers of sediments between these two tuffs, must be about 1.6 million years old—apparently only slightly younger than the Koobi Fora skull. Yet, despite its great age, its cranial capacity is as large as that of the smallest of the Peking man skulls. The boy's postcranial skeleton is remarkably modern. Had he lived, he would have grown to over 1.8 meters in height, as tall as a tall modern human and nearly twice as tall as the tiny *afarensis* Lucy. His brain, some 900 cubic centimeters in capacity, is substantially larger than that of 1470.

All these finds suggest an incredible amount of evolutionary activity. Was there a swarm of upright hominid species living in the area—early robust Australopithecines and perhaps the last of the gracile

Australopithecines sharing the region with *Homo habilis* and the earliest *Homo erectus?* One imagines them all elbowing each other in the evolutionary race.

Or was there really only one highly variable species?

The general consensus is that there was not. The single-species hypothesis, that this swarm of creatures actually formed a single gene pool, has been espoused most vigorously by C. Loring Brace of the University of Michigan. In a physical anthropology textbook he published in 1965 with Ashley Montagu, he blithely treated all the fossil evidence from Africa as if there were only one species involved. But because the robust Australopithecines, in particular, seem so clearly different from any of the others, his idea has been dismissed.

And yet there continues to be something odd about this collection of unnervingly similar species, something that seems to prevent a clear picture of our origins from emerging, in spite of the fact that our hominid fossil lineage, incomplete though it still is, has grown to be among the best known of any group of animals.

Perhaps the most telling indication of this underlying confusion is the fact that there is almost no agreement on how to arrange the fossils in a family tree—as much confusion, indeed, as surrounds the mitochondrial family trees that we explored in chapter 1. At one time or another, almost every imaginable tree has been proposed. The only thing these trees have in common is that, because of its great age and apelike characteristics, *Australopithecus afarensis* is put at their base. But when did the robust Australopithecines branch off, and did they even arise from *afarensis?* Is there one or more than one kind of robust Australopithecine—and, if there is more than one, how different were their histories? Did the gracile Australopithecines, such as the Taung child, arise from *afarensis?* And did *Homo habilis* (if it really is something separate and not simply a way station between *A. africanus* and *H. erectus*) come from the gracile Australopithecines or again directly from *afarensis?* Finally, was there a smooth transition from *Homo habilis* to *Homo erectus,* or did *Homo erectus* appear suddenly (and from what?) less than two million years ago?

The more these fossils are studied, the more complicated the story becomes. Each group has a mixture of characteristics, some apelike and some more humanlike. In addition, each group has a set of characteristics that it shares with other hominids and another set that is uniquely its own.

Consider a structure that was centrally important to hominid evolution, the hand. The hand of a chimpanzee, unlike our own, has two distinct and very different functions. The first is knuckle-walking. When a chimpanzee wants to move rapidly, it gallops forward on all fours, pushing off from its knuckles. The second is the more humanlike function of grasping. The chimpanzee's hand is therefore a compromise and shows it. It is not as effective in walking as the forepaw of a quadruped and not as effective and flexible in gripping as the hand of a human. The palm is rigid and very muscular, and cannot be curved into a cuplike shape in the way that ours can. The chimpanzee's thumb, short and quite inflexible, looks rather as if it were stuck on as an afterthought. A chimp is able to swing its thumb across its palm to a limited extent, but it can touch its fingertips with its thumb only with difficulty and only at an angle. This is because both the rigidity of the palm and the immobility of the last joint of the thumb prevent the juxtaposition of the fleshy pad at the thumb's tip to the pads at the tips of the fingers.

What a loss of information from the environment this inability represents! Rub your thumb and a fingertip together lightly for a moment, and feel the tingle of sensations. A chimpanzee cannot do this. The differences in our tactile abilities from those of the apes are even reflected in our bones. In humans, the terminal phalanges, the last bones of the fingers, have flattened disklike swellings at their tips called apical tufts. These highly vascularized tufts of bone support the sensitive, nerve- and blood-vessel-rich tissue of the fingertips. The tufts on the fingertips of apes, in contrast, are small, rounded, and made of denser bone.

Hand bones are not common in the fossil record, but enough have been found to enable us to trace some of the complexity of hand evolution. The hand bones of the oldest hominids from the Hadar are already, at 3.7 million years, headed in a human direction. The palm was apparently more flexible than that of an ape, although not as flexible as it is in modern humans, and the thumb was longer than that of a chimpanzee. There is no sign of adaptation to knucklewalking. But the large areas of muscle attachment on the palm bones show that the palm was still relatively strong and rigid.

There are other differences as well. The proximal bones of the fingers, the ones nearest the palm, are distinctly curved, suggesting that the *afarensis* hand still retained the strong hooklike grip of its tree-dwelling ancestors. Such a hand is very useful for climbing trees and possibly even swinging from branches—although the latter is less likely because *afarensis* certainly spent most of its time on the ground. But the *afarensis* hand is not as useful for the humanlike activities of grasping and manipulating—just as in the hands of the great apes, its thumb could not be fully opposed to the fingers. We also know that the tactile links between *afarensis* and the world around it were limited, for the apical tufts of bone at the tips of its fingers were small and rounded like those of the apes.

Some hand bones that are probably from gracile Australopithecines have been found at Sterkfontein. They are very different from those of *afarensis,* for the thumb could be fully opposed to the fingers. Further, the apical tufts at the ends of the fingers are as well developed as those of modern humans. If *afarensis* was indeed ancestral to the gracile Australopithecines, then a great deal of evolution of the hand had taken place. At some time during the million years or so that separate these two hominids, a new and subtle tactile link with the environment had been forged.

A few fragments that may be from a hand of *Homo habilis* are also much more humanlike than those of *afarensis.* The shapes of the wrist bones are still apelike, and the strongly curved proximal phalanges continue to reflect that ancient adaptation, the hooked grip necessary for climbing. Yet, the tufts of bone at the tips of the fingers are again as broad and highly developed as those of humans.

All this forms a relatively consistent story, of a gradual though piecemeal evolutionary trend toward a more flexible and sensitive humanlike hand. But now we come to those brutish creatures, the robust Australopithecines. A collection of fossil hand bones that are almost certainly from *Paranthropus* found at Swartkrans, were investigated by Randall Susman of Stony Brook and Bob Brain of the Transvaal Museum in the early 1980s. These bones also turned out to be remarkably humanlike. *Paranthropus,* it appears, had an apparent ability to flex the last joint of the thumb, a strong yet flexible palm, straight rather than curved proximal phalanges, and very well-developed apical tufts. The bones of the palm actually seem to be more humanlike than the solitary palm bone that is so far known from *Homo erectus*—even though *erectus* is certainly closer to our own ancestry than are the robust Australopithecines.* If the robust Australopithecines branched off early from the lineage leading to humans, as many of the builders of family trees have assumed, then they seem to have evolved a very humanlike hand independently.

This puzzling story is not unique. The story of the pelvis is just as

*Susman's find might actually have been a hand of *Homo erectus,* a few bones of which have been found in the Swartkrans deposits. But he argues cogently that 95 percent of the other bones in the deposit are robust Australopithecines, so the chances are excellent that the hand bones are as well.

complicated and confusing. The Laetoli footprints provide incontrovertible evidence that 3.7 million years ago some creature certainly was walking upright, with an apparently fully human gait and length of stride, but of course the footprints tell us nothing about whether that creature was an *Australopithecus afarensis* like Lucy. In fact Lucy's pelvis does turn out to be slightly different from that of modern humans, with more flared and less cuplike hip bones. Yet it seems indistinguishable from that of the much later *Homo habilis,* so it was obviously a very successful structure that changed very little over a span of two million years. It is likely that this not-quite-human pelvis was quite adequate for upright walking, and that many of the subsequent modifications may have taken place in order to accommodate the increasing head size of the newborn.

Multiple Origins of *Homo*?

Although it is clear that a number of very different hominids lived in the same part of Africa at about the same time, confusion arises because they all showed a mosaic of both more and less humanlike characteristics. At the same time as the hands of the robust Australopithecines were becoming more humanlike, it appears, their dentition and skulls were becoming less so. The hands of *Homo habilis,* in contrast, seem to have lagged behind its brain. There is no tidy progression in the fossil record toward *Homo erectus,* but rather a complex set of small changes, some in a more humanlike direction, some not.

To try to make some sense out of these seemingly contradictory trends, Randall Skelton of the University of Montana and Henry McHenry of the Davis campus of the University of California made hundreds of painstaking measurements on dozens of different aspects of the skeletons of *Australopithecus afarensis,* the robust and gracile Australopithecines, and *Homo habilis.* They deemphasized the obvious adaptations related to heavy chewing that are seen in the robust skulls and concentrated on more subtle relationships among the various species. Remarkably, they discovered that the robust Australopithecines and *Homo habilis,* superficially so different, actually have numerous close resemblances. Both these groups differ markedly from the older gracile Australopithecines such as the Taung child.

Now, one likely family tree out of the myriad that have been proposed is that both the robust Australopithecines and *Homo habilis* are descendants of the gracile Australopithecines such as the Taung child. If this should turn out to be true, then the unnerving aspect of the Skelton and McHenry study is that robust Australopithecines and *Homo habilis* must have developed these dozens of different characteristics in parallel. This is a highly unusual pattern in evolution—although one or two characteristics may evolve in parallel in separate but closely related species of animals or plants, it is far less likely that dozens will.

It is here that I hope you get a feeling of *déjà vu*—if Skelton and McHenry are right, then the complex pattern of evolutionary trends resulting in early *Homo* has a lot in common with something that we first encountered in chapter 2, Carleton Coon's much-derided multiple-origins theory of human races. Suppose for a moment that the robust Australopithecines were, in genetic terms, not very different from *Homo habilis.* Then, since the robust Australopithecines shared most of their genes with *Homo habilis,* this could mean that they had the capability to exchange genes with that species and march in a kind of evolutionary lockstep with it. Then the two groups might not have had many more genetic differences than those that later were to distinguish the Neanderthals of Europe from the Cro-Magnon people who replaced them.

Could creatures so apparently different as the robust Australopithecines and *Homo habilis* really be so genetically similar? There is no objective way of knowing, for both these species lived a thousand times farther back in time from us than the ancient Egyptians. Their genes have long since been oxidized into indecipherability.

Still, we can ask whether it was *possible.* Paleontologists as a group tend to be discomfited by the notion that such different creatures might have been able to exchange genes. Yet there is no biological reason to rule it out. When biologists talk about a species, they mean a group of organisms that is genetically isolated from all other groups. To determine whether a species really fits this definition requires detailed observations of its breeding habits, in order to see whether all its members really do breed only among themselves or whether they sometimes breed with members of other "species." If they do stray across what appear to be species boundaries, then the fate of the resulting offspring must be followed in order to determine how great a barrier there might be between the species. Are the offspring sterile, or perhaps so unfit as to be unable to have offspring of their own? Or are they perfectly healthy, so that the two "species" are freely capable of interchanging their genes? Of course this is only part of the story, for even if two groups are *capable* of exchanging genes, they may or may not actually do so in a state of nature.

To make the situation more difficult for biologists, differences between species might be quite invisible even to the trained eye. There have been repeated cases of apparently indistinguishable groups of monkeys or apes being put together in the same colony in a zoo, where they are found to breed very poorly. Only then is it discovered that they are really two or more distinct species. Even groups of primates that seem quite interfertile may really be separate species. Orangutans from Sumatra and Borneo, once kept together in the same colonies, are now being separated because it has been discovered that they have distinct differences in their chromosomes.

In view of all these difficulties, it is not surprising that such a thorough investigation has actually been carried out on very few living species. And it is fairly obvious that, without a time machine, it would be very hard to make such observations on our long-dead ancestors! If it is so easy for biologists to be fooled about which species a living animal belongs to, then it must be even easier for paleontologists to be fooled, because they are forced to work only with bones. Imagine some anatomist from the far future investigating the remains of our civilization and discovering the skulls of a dachshund and a Boston terrier. These animals would certainly be put into different species (see plate 10).

The comparison is an apt one, for most of the differences between the dachshund and the Boston terrier, just as in our hominid ancestors, are concentrated in their skulls. The postcranial skeletons, aside from the great differences in the lengths of the legs, are quite similar to each other. Of course, the differences between the skulls of the dachshund and the Boston terrier have been produced by deliberate artificial selection that seized on a combination of preexisting genetic variability and new mutations and was carried out over the span of a few hundred canine generations. There has been no time for the small behavioral, developmental, and anatomical differences to accumulate that would eventually lead to a separation of species.

Was there enough time for such differences to accumulate between *Paranthropus* and *Homo habilis?* We will probably never know. But we strongly suspect, from the evidence provided by their hand bones, that the *capability* to evolve human characteristics was possessed by the robust Australopithecines, and for a very good reason—they had already managed to evolve a long way in the human direction.

There is an understandable tendency among scientists to rank various hominid lineages in importance, depending on how close they are to that elusive Direct Line of Human Descent. The most popular view is that the line runs fairly close to the oldest of the Australopithecines,

Australopithecus afarensis from the Afar plateau of Ethiopia, then through *Homo habilis, Homo erectus,* and finally *Homo sapiens.* This line is thought to stretch back perhaps four or more million years, although the oldest substantial remains that have been found, those from the Afar Triangle, are younger than that. Branching off from this line, perhaps even before four million years ago, was the lineage that gave rise to the Taung child and the other gracile Australopithecines. And branching from this in turn, rather later, arose the lineage leading to the robust Australopithecines with their strong jaws and sagittal crests.

Yet this view, which underlies so many popular and semipopular accounts of human evolution, seems strangely confining and one-dimensional. The Australopithecines, even the robust ones, were more like humans than they were like apes. They walked upright. Possibly they used tools that, although crude, were far more sophisticated than the branches or twigs employed occasionally by chimpanzees—and the pattern of wear on some of the bone tools indicates that unlike chimpanzees they also used them repeatedly. They may even have been able to kill baboons, fierce animals that are incredibly vicious when threatened, and this ability would in turn imply that they were organized into hunting bands and had developed strategies for dealing with strong, fast-moving, and dangerous prey. And perhaps most remarkably, many of the characteristics of the Australopithecines' skulls appear to have evolved in parallel with similar changes that were taking place in the direct hominid lineage.

Rather than relegate the Australopithecines to the ash heap of history, it might make more sense to ask why they and we have so much in common. What was it about the genes of these near ancestors of ours that permitted them to evolve in this remarkable way, and perhaps to evolve humanlike characters independently of our own lineage? Imagine that the hominids of the Rift Valley of one and a half million years ago had been capable of speculating about their evolution—and indeed, perhaps they were. Then would a more hirsute equivalent of Carleton Coon have looked about at the great variety of upright, tool-using hominids living at the time and proposed a multiple-origins theory of hominid evolution?

Further, if a brain-culture feedback loop was already driving the evolution of our ancestors at that distant time, it would not be too surprising if history has repeated itself closer to our own time. As we will see, perhaps it has.

6

The Latest Steps

We laid out our finds on the large table in Weidenreich's modern laboratory: on one side the Chinese, on the other the Javanese skulls. The former were bright yellow and not nearly so strongly fossilized as our Javanese material; this is no doubt partly owing to the fact that they were much better protected in their cave than our Javanese finds. . . . Every detail of the originals was compared: in every respect they showed a considerable degree of correspondence. . . . The two types of fossil man are undoubtedly closely allied, and Davidson Black's original conjecture that Sinanthropus *and* Pithecanthropus *are related forms—against which Dubois threw the whole weight of his authority—was fully confirmed by our detailed comparison.*

—G. H. R. von Koenigswald, *Meeting Prehistoric Man* (1956)

Trapped Light

In 1891 a painfully shy Dutch physician, Eugène Dubois, found a skullcap embedded in a bank of the Solo River at Trinil in central Java. It was immediately apparent that it was far less like ourselves than any other possible human ancestor, even the most primitive-looking Neanderthal, that had been discovered up to that time in Europe. A femur found in subsequent excavations along the river bank was, in contrast, very modern in appearance.

The finds were utter luck on Dubois's part. As a young student he had determined, quite on his own, to track down human origins. Where were they likely to be? He ignored Darwin's prediction that human ancestors would probably be found in Africa but chose instead to search in Indonesia because that part of the world was open to him. He had easily been able to obtain an appointment as an army physician in the Dutch East Indies.

Dubois's discovery received a mixed reaction in Europe. Some scientists were convinced that the bones really represented primitive relatives of humans, but most thought they were the remains of fossil apes. Dubois himself, thinking at first that they were ancestors of humans, named them *Pithecanthropus*—and then changed his mind (apparently in a fit of pique) and claimed that they were really apes. Incensed by how his fossils had been received, he hid them for many years under the floorboards of his house, and it took a deputation from the Royal Dutch Academy of Sciences to rescue them for science.

The real importance of Dubois's finds had to await the discovery, more than thirty years later, of Peking man, which turned out to be remarkably similar. It was now certain that Dubois's Java finds were something other than fossil apes. Then, in the 1930s, G. H. R. von Koenigswald found another Java man skull at Sangiran, 50 kilometers upstream from Trinil. Many more skulls, showing great variation in their cranial capacity but all very primitive in appearance, were soon discovered in the area. Still other skulls were found at Ngandong, another site below Trinil on the same river.

Compared with the huge number of finds in Europe, the fossil record of southern Asia is fragmentary and the whole vast area with its thousands of islands has hardly been surveyed. Tempting sites certainly abound—for example, a little-explored mountain range in the Padang Highlands of Sumatra is known as the "mountains of a thousand caves." And even much of Java itself is still virgin paleontological territory. Information from caves and other datable sites is desperately needed, for the Java remains that have been found so far have one great drawback: it is impossible to tell whether the bones had always been at the riverbank sites where they were found or whether they had been washed out of earlier deposits by floods, carried to their present sites, and redeposited. Because of this uncertainty about how old they are, endless argument has ensued.

The fiercest argument has to do with the kind of evolutionary change Java man and other Asian *Homo erectus* underwent—was the change gradual or did it take place in spurts of the kind that the paleontologists Niles Eldredge and Stephen Jay Gould have named punctuated equilibrium? There certainly are differences among the fossils. The Ngandong crania have on the average a substantially larger brain volume than the finds at Sangiran or at Trinil. Because they are found in deposits that are substantially younger than those farther upstream they also *appear* to be younger.

Milford Wolpoff has found that, when he pools measurements of the Java man skulls together with those from other Asian sites and *Homo erectus* from Africa and then ranges all the data in order from oldest to youngest, the skulls show a gradual increase in cranial capacity, from about 900 to about 1,200 cubic centimeters. Thus, it would appear that *Homo erectus* shows a continual, gradualistic change in brain size over time. But to obtain this relationship he had to assume that the Ngandong crania are very recent, perhaps only two hundred thousand years old. G. Philip Rightmire of the State University of New York at Binghamton has analyzed the same data and finds that, when he leaves out the Ngandong fossils, the trend disappears. Rightmire, a punctuationist, is convinced that there was very little change in the million and a half years during which *Homo erectus* first appeared in Africa, spread through the Old World, and was eventually replaced by *Homo sapiens*.

Wolpoff's anxiety to assume a recent age for the Ngandong remains is understandable when you recall that he is currently the chief proponent of the multiple-origins theory of humanity. If the Ngandong fossils really are relatively recent, say two hundred thousand years old, then they are likely to be younger than the mitochondrial Eve. Perhaps they will eventually be shown to form a bridge in time with the peoples of Australasia. If so, then those peoples would represent continuity, not with some recent invasion of true humans from Africa, but with the far more ancient *Homo erectus*.

All scientists agree that present-day Australasians represent the farthest reach of a great expansion of peoples through the Old World. But which expansion was it? Are they close relatives of Java man and therefore the final stages of the great migration of *Homo erectus* that began in Africa half a million to a million years ago? Or did their ancestors leapfrog recently from Africa through the length of Asia, leaving few traces behind, and drive Java man to extinction without issue?

To try to answer this question good dates for the hominid remains of Southeast Asia are desperately needed. Because the fossils often lie in disturbed strata, it is essential to date the fossils themselves. Until recently there were very few ways to do this, and they were very limited. But now new techniques are beginning to emerge.

One very useful but limited method involves carbon 14. This radioactive isotope of carbon is continually being generated by cosmic-ray bombardment in the upper atmosphere, drifting down in the form of carbon dioxide, and slipping into the food chain. When an animal or plant dies, it stops replacing old carbon with new, so that it no longer accumulates carbon 14. As the carbon 14 decays, its ever-diminishing amounts relative to nonradioactive carbon mark, with great precision, the time since the organism's demise.

Unfortunately, carbon 14 decays rapidly. After five thousand years, half of it is gone. Within ten thousand years, three-quarters of it is gone. It can really be used only to date fossils from the last forty thousand years, by which time only half of 1 percent of the original carbon 14 is left. The older the fossil, the greater the uncertainties, because so little carbon 14 is left to be measured.

Then, in the 1980s, two new techniques were introduced that for the first time allowed dates beyond the forty-thousand-year carbon 14 limit. Both these new methods depend on some odd properties of mineral crystals that are found in bone and tooth enamel, in many rocks, and even in the small particles of rock that make up sand.

Crystals in the real world are never perfect arrays of atoms or molecules. They contain discontinuities and impurities. And, even though they are nonliving parts of the mineral world, they are not completely unchanging. They contain traces of radioactive elements, which slowly alter the crystals over time as they decay. The crystals are also bombarded by cosmic radiation and by particles from radioactive elements in the soil or rock in which they are embedded. When energetic particles or quanta of radiation pass through a crystal, they excite a few of its atoms into a higher energy state. In most cases, the electrons that have been shifted to a higher orbital soon return to their ground state, but some atoms along the edges of discontinuities and impurities remain excited. Remarkably, in some minerals such as flint, or in crystalline materials such as the enamel of teeth, these atoms can persist in their excited state for millions of years. An undisturbed piece of flint or a tooth accumulates more and more excited atoms over time, at a rate governed by the amount of natural radioactivity to which they are exposed. The older the material, the more accurate this clock becomes, so it is just the reverse of the carbon 14 clock.

When a piece of flint is heated or a grain of sand is exposed to intense sunlight, this clock is reset because the heating makes all the excited atoms give up this energy. As soon as the material cools, it begins to build up a store of excited atoms again. Similarly, when an animal or human lays down the crystals making up its tooth enamel, they initially contain few excited atoms. As time goes on, and even after the animal's death, more and more excited atoms accumulate.

There are two ways in which geochronologists can determine how many excited atoms there are. One works well for minerals, the other for teeth. If a bit of flint from an ancient hearth or a grain of sand that was once exposed to sunlight is reheated, the excited atoms fall to their ground state and emit a tiny burst of light called thermoluminescence. And if a bit of enamel from an ancient tooth is exposed to intense microwave radiation, the way it absorbs the radiation gives a measure of the number of excited atoms it possesses that are in resonance with the radiation. This absorption, called electron spin resonance, can be measured with great precision.

Like other dating methods these techniques have their problems, but with proper calibration consistent results can be obtained. At many archeological sites both methods can be used, giving quite independent estimates of the age. Little has been done as yet to investigate the Java fossils using these techniques, but some tantalizing hints about the patterns of human migration have emerged from other places nearby.

The Farthest Shore

The ability of sand to accumulate excited atoms with time when it is shielded from sunlight has enabled a team of researchers from Wollongong and Australian National universities to date some very old aboriginal occupation sites on the north coast of Australia. The sites are on the western slopes of the central plateau of Arnhem Land, the blunt peninsula that separates the Gulf of Carpenteria from the Timor Sea. Buried in the first 2.5 meters of the sand that has eroded from the uplands are traces of human occupation—layers of charcoal accompanied by quartz fragments and chunks of various pigments such as ochres that were probably used as body paint.

The dates obtained from the sand grains surrounding the deepest of these deposits lie between fifty thousand and sixty thousand years ago, and make them the oldest yet determined for human occupation of Australia. It seems that people first arrived in Australia at a time when Neanderthals occupied Europe. And the traces of early occupation are at precisely the place where one would expect to find them.

The Australians are not seafaring peoples. Before the arrival of Europeans, the peoples of the northern Australian coast did make rafts of mangrove or driftwood, but these could not float for more than a few hours before becoming waterlogged. These peoples lost, or never acquired, the technology needed to hollow out tree trunks to make canoes. So how did they get to Australia in the first place? Perhaps on bamboo rafts, which are as easy to make as bark boats but float for a good deal longer. There is plentiful bamboo on the tropical island of Timor, just to the northeast of the Australian mainland. And, sixty thousand years ago, soon after the start of the last major Ice Age, the sea level was at least 200 meters lower than it is now. Today, Timor is separated from Australia by 600 kilometers of open ocean. During the Ice Age maximum, a huge Australian coastal plain extended almost all the way to the island. Migrants from Timor would have been checked only by a narrow stretch of water some 70–90 kilometers wide.

This strait could easily have been crossed, by accident or design, by peoples on bamboo rafts. But, if there were few or no bamboos in the area when they arrived, it would have been a one-way voyage. After their rafts had cracked and split in the sun on the hot Australian shore, there was no way back.

We have no way of knowing how long the first migrants to Australia stayed on the enormous coastal plain exposed by the Ice Age, because that plain now lies beneath the Timor Sea. But we do know that nearly sixty thousand years ago, totally committed to a new existence on that dry, flat pancake of a continent at the edge of the world, they were living on the slopes of the Arnhem Land plateau. Soon their descendants were scattered throughout the continent, forming hundreds of small tribes of at the most four or five hundred individuals. At the time of the arrival of the white man they spoke at least two hundred different languages, many with several dialects. It has been claimed that most of these can be traced back to a single protolanguage, although it is easier to find relationships among the south Australian languages than among those of the north. The heterogeneity of the northern languages may reflect several waves of migration from more than one area in Southeast Asia.

These preliterate tribes developed elaborate cultures in spite of their forbidding surroundings. They kept track, in great detail, of complex familial relationships and endless rules of behavior and taboos. The featurelessness of the landscape was a great challenge. In order for the tribespeople to find their way from one area to another in search of food, the tiniest distinctions had to be made about the terrain, down to the sound of footsteps on different types of soil. Complex songs guided both tribes and lone individuals around the continent from one sacred site to the next—and most sites were sacred. The late Bruce Chatwin in *The Songlines* describes the way the music of the song reflects the landscape:

> Regardless of the words, it seems the melodic contour of the song describes the nature of the land over which the song passes. So, if the

> Lizard Man were dragging his heels across the salt pans of Lake Eyre, you could expect a succession of long flats, like Chopin's "Funeral March." If he were skipping up and down the MacDonnell escarpments, you'd have a series of arpeggios and glissandos, like Liszt's Hungarian Rhapsodies.

Chatwin told the story of one aboriginal elder who traveled to London and Amsterdam for a meeting and on his return sang a long and complex song about it. The song was in part an account of his adventures, in part designed to guide any other member of the tribe who might follow in his footsteps through the complexities of London's Heathrow!

Undisturbed by the upheavals that took place elsewhere in the world, Australian tribes have enshrined in their legends a remarkable cultural continuity. Local legends commemorate the drying of inland lakes that might have taken place as long as fifteen thousand years ago. The tribes of the Atherton Tablelands in northern Queensland still tell tales about fire coming from the earth, even though it has been ten thousand years since there was last any active volcanism in the area. In Arnhem Land, the oldest settled part of Australia, the first Europeans found tribes worshipping the Rainbow Serpent, who had supposedly shaped the sinuous rivers of the region with her writhing body. Rock paintings of the Rainbow Serpent have been dated to nine thousand years ago, so that the period of her worship exceeds by at least a factor of three any other religions that have been handed down to the present time elsewhere in the world.

Flood legends abounded among the tribes of the coast, and it is not unreasonable to suppose that they reflect memories of a distant time when the sea itself rose with astonishing speed. The last great sea level rise started around twelve to fifteen thousand years ago, reaching roughly present levels between five and seven thousand years ago. The rise was enormous in magnitude, about 150 meters. During the most rapid period of change, the sea must have been rising 3 centimeters a year, so that over a single long human lifetime it could have risen 2.5 meters. This is the stuff of legend indeed. The flood legends of the Middle East have been confused and distorted by subsequent events. But an elder of an aboriginal tribe could point to the specific features of the landscape that had been drowned by the flood that was actually witnessed by his remote ancestors.

Fossil skeletons show that the peoples who arrived in Australia were no more homogeneous than the peoples of Paleolithic Europe or the peoples who inhabited Africa's Great Rift Valley before the arrival of Europeans. A number of important finds have been made throughout Australia, but the most striking have been those at Willandra Lakes in western New South Wales and Kow Swamp in northern Victoria.

The Willandra lakes are a series of dry lake beds that last had appreciable water in them fifteen thousand years ago. A remarkable series of finds, particularly at Lake Mungo, have given glimpses of the people who inhabited the lake area during the last Ice Age, when the lakes themselves were extensive and filled with fish. Before forty-five thousand years ago the lake beds were as dry as they are today. Not long after they filled with water, the first people arrived, bringing with them stone and bone tools and a culture of complex rituals. During one ritual burial red ochre was scattered over the body of a male whose well-preserved skeleton has been dated to thirty thousand years ago. And an even more elaborate burial of a female, involving ritual cremation, has been dated to twenty-six thousand years ago.

Most of these people of the Willandra lakes region were slender, with high-domed, thin-walled skulls. They appear, in fact, to have been more slender-boned than the present-day inhabitants of the region, the Bagundji tribe. There is one exception in the form of a single find that was given the designation W. L. H. 50, a skull and some postcranial fragments of what was apparently a very large and heavy-boned male. These fragments were discovered on the surface of one of the lakes where they had been exposed by a flash flood, so it is not possible to say how old they are. But they are heavily mineralized, even more so than any of the other finds in the area. The skull is enormously thick with a pronounced and continuous browridge and a low and sloping forehead. When it is viewed from above there is a severe postorbital constriction, so that it looks from that angle like a typically flask-shaped skull of *Homo erectus*. Yet the skull is much larger, with a cranial capacity of about 1,300 cubic centimeters.

At Kow Swamp, the skeletal remains of about fifty individuals with features like those of W. L. H. 50, although not as extreme, have been found. These people died and were buried, in a variety of ways, at various times from about fifteen thousand to nine thousand years ago. Larger than most present-day aborigines, they typically had thick-walled, sugar-loaf-shaped skulls, a shape also found in some present-day tribes. There has been a great deal of argument about whether their unusual skull shape could have been due to deliberate deformation, and it seems likely that some of it may be. Yet, these people were strikingly different from the majority of the peoples of the Willandra lakes and the other gracile peoples whose remains have been found elsewhere in Australia. Even if possible cranial deformations are discounted, they provide strong evidence for an extreme and long-preserved heterogeneity of physical types among the tribes of Australia during that period before the arrival of Europeans that the aborigines knew as the dreamtime. How much of this heterogeneity persists today is an open question. The anthropologist Alan Thorne has compared the robust Kow Swamp and gracile Willandra Lakes skulls to skulls of present-day aborigines. He finds the recent skulls to lie somewhere in between those extremes. Unfortunately, all the recent skulls used in his study came from one small part of Australia, so they are probably not representative of present-day variation.

(The local Echuca community of aboriginals, furious at the desecration of their ancient burial grounds, has recently forced the reburial of the Kow Swamp remains. Their move, however damaging to anthropology, has received wide support among the Australian public.)

Argument continues about whether the variety of physical types among the migrants to Australia were the result of two or more migrations or whether the first arrivals were themselves *phenotypically* heterogeneous—that is, whether they had markedly different appearances. Alternatively, the phenotypic variation seen among present-day aboriginals might have been generated by selection and tribal fragmentation after people arrived in Australia. It seems possible that some combination of all three might be true.

Milford Wolpoff and Alan Thorne have used the *Homo erectus*-like characteristics of some of the Australian fossils to argue for a continuity of local evolution in the area. They visualize the Australians as being descended from ancestors like the people of Ngandong in Java, who were certainly *Homo erectus.* The aborigines would then have made the transition to *Homo sapiens,* either before or after their arrival on the Australian mainland. Wolpoff and Thorne also postulate that a role in this transition was played by gene exchange with other, more advanced groups. You will remember that this idea of extensive gene exchange solves the problem of how similar evolutionary events could have happened in different places. Of course, it does not explain how genes that conferred a more *Homo sapiens*-like character could have migrated into the aboriginal population without at the same time bringing in other genes that would have diluted their distinctive racial characteristics—unless the aboriginals were at the same time being strongly selected for the various characteristics that mark them off from other racial groups.

The basis of the multiple-origins theory has changed very little since the time of Carleton Coon. Wolpoff and Thorne follow rather closely Coon's original views on the origin of Australians. Coon had suggested that they were still "sloughing off the genetic traits which distinguish *Homo erectus* from *Homo sapiens.*" (Coon's vagueness about the mechanics of genetics shows through rather clearly here!) Coon also suggested that there might have been a "slow trickle" of genes into Australasia from the Mongoloid populations to the northeast, but he was silent on whether this could have aided in the sloughing off of those invidious *Homo erectus* characteristics.

Wolpoff has repeatedly argued against the view so strongly espoused by Allan Wilson that a recent migration of *Homo sapiens* completely displaced the earlier peoples of Asia, a migration so sweeping that it allowed no gene exchange at all. Indeed, there are puzzling features of the paleoanthropology of Australia that suggest that Wilson's scenario is too simplistic. There are enough similarities between the fossils of Java on the one hand and some (but not all!) of the fossils of Australia on the other to suggest that something much more complicated must have gone on.

Could the transition from *Homo erectus* to *Homo sapiens* have taken place in Australia without the aid of hypothetical gene flow from elsewhere? Perhaps it could, but only if the early migrants to Australia brought enough genetic variation with them. The high degree of phenotypic variability among the Australian fossils seems to suggest that there was also a great deal of genetic variation. Selection pressures were certainly not lacking. The complex and subtle songlines of the aborigines, and the many other unique features of their culture, were able to develop even though the physical environment seems (to our eyes) a very uncomplicated one. If substantial human evolution did take place in Australia, then it is a testament to the ability of runaway brain evolution to drive our ancestors in the direction of human abilities even in the absence of advanced technology. Elaborate human culture, rather than an elaborate external environment, may turn out to be the most important determiner of our accelerating evolution.

An Evolutionary Sideshow?

In 1922 Teilhard de Chardin met the adventurer Henry de Monfried on the boat as he was coming home from China. De Monfried, who joined Teilhard's boat at Djibouti, had explored the coast of Abyssinia in search of treasure and knew the area intimately. It would, he suggested to Teilhard, be a great place to search for signs of early man.

Teilhard did not manage to do any exploring of the area until 1928, but after a brief sortie ashore he agreed with de Monfried. He found some late Stone Age tools near Obock, a little way up the coast from the town of Djibouti, on the edges of the grim lava region where the Great Rift Valley branches into the Gulf of Aden and the Red Sea. Although Teilhard did not realize it, the lunar landscape he was trudging across was new land, recently created through the process of seafloor spreading. Gubet Karah, the small bay of the Gulf of Aden that lies between the towns of Djibouti and Obock, is the tiny beginning of the great future cleft that will one day separate the Horn of Africa from the rest of the continent. Signs of the cleft are already there, only yards from the shore, in the form of the Assal Depression, which plunges to 600 feet below sea level. A tiny shift of the underlying plates is all that will be needed to let the hot waters of the Gulf of Aden pour in, drown the depression, and instantly double the size of Gubet Karah.

The area is now so hot and dry that there is virtually no vegetation. But it was not always so inhospitable. Teilhard ventured inland some 250 kilometers to the edge of the Harar Plateau and the town of Diré Dawa. In the hills nearby he was taken to the high Cave of the Porcupine, where he saw pictures and incised patterns on the walls.

Teilhard was no authority on cave paintings, but he knew someone who was. His coreligionist, the Abbé Henri Breuil, had spent his life crawling through the caves of France and Spain and meticulously recording the painted and scratched pictures left there by Paleolithic peoples. Breuil was also an authority on the various industries of stone tools that were found in the caves. He had made sense out of these tools, arranging them in order of increasing sophistication.

Teilhard, Breuil, and de Monfried returned to the area in 1933. When Breuil entered the cave in order to copy the paintings, his attention was attracted to a ledge near the entrance filled with encrusted material. When he brushed it away, the jaw of a primitive man was revealed sitting on the ledge as if it had been deliberately placed there.

The jaw showed many of the characteristics of the Neanderthals of northern Europe (though it appears now that it was probably older than the Neanderthals). It had an undeveloped chin, unusually well-worn front teeth, and a large space between the last molar and the angle at which the jaw turns upward to form the ascending ramus. All these clearly distinguished it from the jaws of modern humans (see plate 11).

Had the owner of the jaw made the paintings in the cave? If he or she had, it would be astonishing, for in none of the caves of Europe inhabited by Neanderthals or older races is there any sign of painting. Neanderthals sometimes made handprints on the walls of their caves, but real cave paintings first appeared in Europe about eighteen thousand years ago, some time after the appearance of modern humans, and probably some time after San tribespeople began making equally advanced paintings in southern Africa. The paintings in the Cave of the Porcupine are old, so old that they are now covered with a layer of transparent calcite. They are also sophisticated, showing a number of human figures, elephants, and a lion. The likelihood, however, is that they were made by peoples who occupied the cave long after the owner of the jaw.

Breuil's discovery is only one of many remarkable finds that have been made in the century and a half since the first remains of peoples that predate modern humans were discovered. In that short span of time, the scientific world has gone from utter disbelief that there could have been human ancestors at all, to an uncritical embrace of Neanderthals as fellow human beings, to the current tendency to release the embrace and hold them perhaps at arm's length. The Neanderthals form an important part of the story of our species, but we are not quite sure what their part in this evolutionary saga actually was. Were they the immediate ancestors of present-day Europeans? Were they country bumpkins who were eventually given the privilege of joining modern humans around the campfire and who perhaps left a few of their genes behind in the process? Or did our ancestors see them as nothing but savage brutal animals to be hunted down and destroyed?

The origins of the Neanderthals are unclear. When the Atlantic poured through the Pillars of Hercules some five million years ago to fill the Mediterranean basin, there were no prehumans in Europe. After the sudden imposition of this new sea between Europe and Africa, there were only two routes by which our ancestors could have invaded the European peninsula from Africa. These routes were the obvious one through the Middle East and the far more difficult one across the narrow span of water that separates Spain from Africa. Both routes seem to have been taken. The Middle East migration began perhaps a million years ago and left traces of stone tools at Ubeidiyah in the Jordan valley. At the other end of the Mediterranean, a very few simple stone tools that have been dated to almost as long ago have been found at Soleihac and Vallonet near Nice. More tools, again nearly a million years old, have been found at Isernia in central Italy. (Claims for even older tools have been made, but they appear to be rather doubtful.)

But who were the people who made the tools? The presumption is that the first invaders of Europe were *Homo erectus,* close relatives of Peking man, who were of course known to have been living in Africa at the time. But the oldest hominids whose traces have been found in Europe seem to have been rather more humanlike than the typical *Homo erectus* of eastern Africa or Zhoukoudian. The most complete skulls have been turned up at Petralona in northern Greece, Steinheim in Germany, and Arago in the Spanish Pyrenees, with a very recent exciting find made in a cave near Burgos in northern Spain. All of these finds have been dated to two hundred thousand to three hundred thousand years ago, with Steinheim and Burgos apparently the oldest. In all of these four cases the cranial capacities of these skulls are greater than those of most *Homo erectus*, and the Burgos finds in particular resemble Neanderthals. Thus there is an almost completely blank period between the million-year-old tools at Ubeidiyah and Vallonet and these four substantial skeletal remains of early European hominids. Virtually nothing is known about the inhabitants of Europe during this immense span of time, a span that begins well before Peking man began to fill up the caves at Zhoukoudian and ends when Peking man disappeared. The few hominid fossils that seem to date from this great blank period are mere fragments of jaw or skull.

If the first hominids to find their way into the remote and forbidding European peninsula were *Homo erectus,* then the skulls from Petralona, Steinheim, Arago, and Burgos show that either they did a substantial amount of evolving or later and more advanced hominids followed and perhaps displaced them. The skulls are distinctly different from those of modern humans, and even show clear differences from each other. The Steinheim skull, despite its great age, has a vertical face very like that of modern humans. The Petralona and Arago skulls have more projecting faces. All have browridges, but none are as extremely developed and shelflike as the browridges of *H. erectus,* and none show the pronounced narrowing of the skull behind the browridge that gives the *H. erectus* skull viewed from the top its characteristic flask shape. Various authors have discerned features in these skulls that suggest similarities to *H. erectus,* to Neanderthals, and to modern humans. There is great confusion in the literature that has grown up around these fossils. Nonetheless, it is generally agreed that they were all more humanlike than *Homo erectus.*

Subsequent to these finds there is another blank in the fossil record, so we do not know which if any of these hominids might have given rise to the Neanderthals. Traces of the Neanderthals, who flourished between perhaps a hundred thousand and thirty-five thousand years ago, have been found over a region extending from Spain and northern Europe through the Middle East to Iraq. They, too, showed a good deal of variability, with the most extreme types being found in western Europe during the latter part of this span of time.

This latter part of the Neanderthal period coincided with the peak of the Würm glaciation in Europe. Although the Würm ice sheet was not as extensive as that of some of the earlier glaciations, it still advanced as far as southern England, Berlin, and nearly to Moscow. All of Scandinavia was buried in ice during much of that time, and the ocean level was perhaps 100 meters lower than at present. As a result, the English Channel was almost entirely dry land. The vegetation of Middle Europe was dwarf forest and tundra, separated from the Mediterranean coast by much thicker forest that must have posed a substantial barrier to travel. The climate of northern Europe was savage—at the peak of the glaciation the mean July temperature in the Netherlands was less than 5 degrees Celsius. The demands on the peoples who lived in this grim region must have been severe, perhaps even more severe than those made on Eskimos at the present time. The Neanderthals, apparently, did not have the benefits of the highly developed technology that allows the Eskimos to live on the very northern fringes of the tundra.

Because the Neanderthals were the first prehumans to be discovered, and because virtually every likely cavern and crevice in western Europe has been explored in the search for their remains, we know a great deal about them. If a similarly intensive effort had been brought to bear in other fossil-rich regions such as China or Java, we would certainly know a great deal more about the hominids inhabiting those areas. What is striking is that in spite of all the discoveries and all the intensive study, there is still profound disagreement over exactly where Neanderthals stand in our family tree.

The first Neanderthal remains of which we have records were found at the beginning of the eighteenth century at Cannstatt in Germany, and a fairly complete skull was discovered in 1848 in Gibraltar. But these puzzling remains ended up tucked away virtually unnoticed in museums. The first find that aroused real scientific interest was made in the Neander Valley near Düsseldorf. The valley, named after the seventeenth-century poet Joachim Neander who used to repair there to commune with his muse, had fallen victim by the mid-nineteenth century to the construction industry's voracious appetite for limestone. In 1856 quarry workers found the top of a long skull with a sloping forehead and low vault, and with substantial browridges still intact. The workers dumped these fossils aside, but luckily they fell into the hands of Carl Fuhlrott, a local amateur historian.

The first explanations that were proposed for the skull reflected the cramped and narrow view that most scientists of the time had about human history. Perhaps, it was suggested, it was the skull of a tribesman killed during the Roman invasion, or of a recent mental defective. Or perhaps the quarry workers had found one of Napoleon's cossacks returning from Moscow, who had crawled up the cliff into the cave to die—managing to cover himself with a deep layer of earth in the process.

Similar debates about this and other fragmentary Neanderthal remains raged through the rest of the century, but as more and more were found, particularly two skeletons at Spy in southern Belgium, it became less and less likely that they were some recent burial or pathological aberration.

The Spy skeletons were so crushed and distorted that it was not possible to reconstruct them with accuracy. It was not until 1908 that the first really well-preserved Neanderthal skeletons were discovered, in the course of a systematic excavation of a cave near La Chapelle-aux-Saintes in central France. There is a trace of irony in the fact that the discovery, which certainly helped to call into question the doctrine of special creation, was made by three priests. They carefully packed and shipped the bones to Paris.

There the most complete of the skeletons was reconstructed by Marcellin Boule. Because the skeleton was that of an old man, showing the degenerative changes of arthritis, Boule made it into everyone's idea of an ape-man, knees bent and head thrust forward.

Boule's primitive-appearing reconstruction was published at about the same time as the "discovery" of the far more modern-seeming Piltdown fake. These two events had the effect of relegating Neanderthals to an evolutionary sideshow, surely not anywhere near that Holy Grail of physical anthropologists of the time, the Direct Line of Human Descent. It was not until the 1950s that more careful reconstructions of the skeletons, particularly by Camille Arambourg, showed that Neanderthals were fully erect and really not very different from modern humans. Some of the many reconstructions of Neanderthals are shown in plate 12.

Now opinion is swinging back again, with the discovery by Erik Trinkaus of the University of New Mexico and others that there really are distinct differences between Neanderthals and ourselves. Two of these are particularly striking.

First, Neanderthals had enormous upper body strength. From the thickness of the bones, the large size of the muscle attachments, and the short bones of the upper arms, it has been calculated that they must have been twice as strong as modern humans. This fact in itself is not too surprising—chimpanzees are also considerably stronger than humans. It appears that we, not they, are the aberrant ones, which makes one wonder why we have such a weak musculature compared with Neanderthals and chimpanzees. Perhaps there has been some evolutionary trade-off—overall weakness may have been compensated for by increased fine motor control.

Second, Neanderthals had forward-thrusting faces with very receding chins, faces that were memorably described by Carleton Coon as beaklike. Their front teeth tended to be badly worn down. Neanderthals seem to have been able to use their front teeth like a strong pair of pliers, gripping and manipulating something—hides? tendons?—as if their mouths were a third hand. This ability was surely in their genes, rather than acquired, for the youngest Neanderthals had the same unusual beaklike faces. There was also, as we saw earlier, a large gap between the wisdom teeth and the ascending ramus of the jaw. Neanderthals were presumably not troubled by the impacted wisdom teeth that are so common among modern humans and that seem to be a consequence of the movement of our whole face into a vertical position that has crowded the teeth in the back of our jaws.

Of course, Neanderthals had sloping foreheads and beetling brows (so do some of us). Their brains, however, were actually substantially larger than our own on the average. There is no indication that their brain functions differed noticeably from our own, but there are some substantial pieces of evidence that they were fully capable of speech.

Human beings can produce an enormous range of sounds, although some of us, such as opera singers, do a much better job than others. This variety of sounds comes from many sources. The tone and timbre of our speech is largely shaped by the voice box or larynx, an expanded region of the upper part of the trachea, or windpipe. The top and bottom parts of the larynx can move relative to each other, stretching or relaxing the folds of tissue within the voice box that make up the vocal cords. You can feel the top part of the larynx move up and down by placing your fingers lightly on your throat and singing a high and a low note.

In the apes, the larynx is relatively narrow in cross section and is pushed up toward the back of the throat, which greatly restricts its movement and the number of sounds that can be made. In humans, the larynx is wider and positioned further down the throat, which gives it much more freedom to move. This freedom of movement has been, along with the less advantageous crowding of our wisdom teeth, one of the many results of our acquisition of upright posture, which has shifted the head back relative to the spinal column. The front of the throat has lengthened, freeing the larynx to grow larger and more mobile. As a result, a human has far more control over the movement of air through the voice box and throat than does an ape. We also have more freedom to change the positions of the soft palate and the tongue, a freedom that adds even greater variety to the sounds.

Unfortunately, all these important structures are made of flesh or cartilage and have left no trace in the fossil record. The size, shape, and position of the larynx in our hominid ancestors have to be guessed at on the basis of the morphology of the bottom of the skull and of the jaws, bones that are some distance away.

However, there is some fossilizable material in this mass of tissue. Just above the larynx is the hyoid bone, which provides support for the back of the tongue. It is a freely suspended U of bone, with the sides of the U pointing backward. In humans, the sides of the U splay outward, reflecting the great width of the larynx directly below it. In the apes, the sides of the U are parallel. So it was a very exciting discovery when in 1989 Yoel Rak and his colleagues at Tel-Aviv University found a fossil hyoid bone among Neanderthal remains in Kebara Cave on Mount Carmel in northern Israel. The hyoid bone was fully human in morphology, even though the cave deposits had been shown to be sixty thousand years old. As a result of this find, there is at the moment no compelling reason to suppose that Neanderthals were incapable of making the same range of sounds as modern humans.

The hyoid bone is not the only remarkable find to emerge from the many caves on or near the slopes of Mount Carmel, near the Mediterranean coast. The caves have been used for generations by goatherders, and before that they were occupied by Bronze Age and Neolithic peoples. Their history of human occupancy, however, goes back much further. First excavated extensively in the 1930s, they have over the succeeding decades yielded a treasure trove of human and animal skeletons and thousands of stone implements.

They have also been at the center of enormous controversy. The Paleolithic human skeletons that were found vary remarkably in their appearance, from extreme Neanderthal to quite modern. This fact suggested, to Dorothy Garrod and the others who were involved in the first excavations, that the peoples of Mount Carmel were a mixture of Neanderthals and modern humans. Possibly they were able to exchange genes.

However, although the skeletons are indeed variable within a cave, there do seem to be differences from one cave to another. Skeletons from the caves of Tabun and Kebara, south of Haifa, are predominantly Neanderthal, with one skeleton from Kebara looking remarkably like the classical Neanderthals of western Europe. Skeletons from the cave of Skhul, only a few hundred yards from Tabun, and from Qafzeh some 20 kilometers to the east, seem on the whole to be less Neanderthal and more human. Two skulls in particular, Skhul 5 (plate 13) and Qafzeh 6, have often been used to argue that these people were essentially modern, even though their slightly projecting browridges and slightly forward-thrusting faces argue that they still retained some primitive characteristics. Other skulls from the same layers in the same caves are less modern looking and suggest a complex pattern of gene flow between Neanderthals and modern humans. Initially, many fragmentary bits of dating evidence were put together to show that all these peoples, modern and Neanderthal, lived together in cozy proximity some forty thousand years ago. This fit the assumption that Neanderthals predated modern humans and perhaps evolved into them.

There were problems with the dates, however. They were heavily dependent on the carbon 14 method, which you will remember is unable to provide reliable dates for times greater than forty thousand years ago. The material in the caves of Mount Carmel fell into this uncertain time. Was the forty-thousand-year date real, or was it merely a product of the imaginations of scientists pushing a technique to its limit?

Recently, material from hearthstones and teeth found in the caves has been reanalyzed using the thermoluminescence and electron spin resonance methods. When bits of flint fell into the hearths of the inhabitants of the caves, their excited-atom clocks were reset. And similar clocks had also begun to tick in the teeth of the animal remains that were brought into the caves—the enamel of the teeth, grown during the animals' lifetimes, had no excited atoms when it was first laid down but gradually began to accumulate them.

The father-daughter team of Georges and Hélène Valladas at France's Centre national de la recherche scientifique found that their thermoluminescence dates were very different from those obtained with carbon 14. Hearths in the Neanderthal caves of Tabun and Kebara gave consistent dates of about sixty thousand years ago, so they fell squarely in the range of dates for Neanderthal remains from western Europe. But hearths from the caves of Qafzeh and Skhul, with their more modern-seeming inhabitants, gave much older dates—ninety thousand years or more. These new dates set the world of archeology on its ear.

Did they mean that modern humans actually predated Neanderthals and gave rise to them as a kind of evolutionary offshoot? If so, then the apparently primitive features of Neanderthal skeletons were not primitive at all—Neanderthals were actually a highly adaptive branch of the evolutionary tree that merely happened to go off in a slightly different direction from modern humans.

Most recently, however, this remarkable story from the caves of Mount Carmel is beginning to dissolve in the welter of confusion that seems to characterize all our attempts to understand our fossil record whenever we examine it closely. The more humanlike remains from Qafzeh and Skhul turn out on close examination not to be exactly like us. They show a great range of characteristics, some of them like modern humans and some more like Neanderthals. The most humanlike skulls from each of the caves are not really representative of the majority of the hominids that lived there. And the Neanderthals of the area, from the caves of Tabun and Kebara, really seem to be rather different from those who lived far to their west in Europe. In the Middle Eastern Neanderthals the extreme beaklike appearance of the face is softened, with a more pronounced chin and less aggressively jutting browridges. In this they are not alone, for many Neanderthal remains have been found nearby in eastern Europe that also share these characteristics.

To compound the confusion further, even the thermoluminescence dates have been cast into some doubt. Rainer Grün of Cambridge University has dated teeth from the fossil-bearing layers of the Neanderthal cave of Kebara using electron spin resonance and obtained dates twice as old as the thermoluminescence values (though his dates for Tabun agree with the thermoluminescence numbers). Electron spin resonance has also given very old dates for some Neanderthal finds in Germany. It may be that the Neanderthals of Mount Carmel really will turn out to be older than the modern humans from the nearby caves. Indeed, to judge from the recent find at Burgos, the origins of Neanderthals may lie 300,000 years or more in the past.

Finally, the idea that Neanderthals were incapable of constructing an elaborate culture, which would explain why *Homo sapiens* presumably displaced them, has also recently received a setback. In 1980 quite advanced stone tools were discovered along with undoubtedly Neanderthal remains in a cave at Saint-Césaire in western France. These tools are much more modern than any others that have yet been found associated with Neanderthals. This remarkable find throws the whole Neanderthal picture into confusion. Were Neanderthals simply a highly specialized offshoot of premodern humans, evolving on their own separate path in the tundras of glacier-bound Europe while those premodern humans evolved toward us elsewhere? Or did they, as C. Loring Brace of the University of Michigan has passionately insisted for years, contribute their genes to the emergence of modern peoples? Brace claimed that Neanderthals could actually have evolved into modern humans directly, as they invented advanced tools and better ways of cooking their food. These cultural changes would cause them to lose their extreme facial characteristics as they ceased to rely on their front teeth. Not many people side with Brace, but the finds from Saint Césaire strongly suggest that there was at least some cultural exchange between Neanderthals and modern humans. If there was cultural exchange, then genetic exchange is a real possibility.

No matter how much leapfrogging the various dates may do in the future, it seems certain that Neanderthals arose long after the time of the mitochondrial Eve. And there is no gainsaying the facts that Neanderthals lived in close proximity to modern-appearing humans in the Middle East and that there is a gradient from less to more markedly Neanderthal-like features as one moves from east to west through the European peninsula and from the more remote times of the Mount Carmel populations to the more recent populations of Spy and La Chapelle-aux-Saintes. These facts make arguments that Neanderthals always remained genetically distinct from other human groups lose much of their force.

This is why I predicted earlier that, if a Neanderthal thaws out of an Alpine glacier in the next few years, he or she will probably be found to have mitochondrial genes very like those of the rest of us. If Neanderthal genes do turn out to be similar to our own, then scientists will have been led astray all these years by skeletal differences that are quite superficial, differences that are perhaps caused by a relatively small number of genes. And the Neanderthals will turn out to be nothing more than a small variation on a theme.

What is that theme? Niles Eldredge and Stephen Jay Gould have made a powerful argument that evolution, as it can be followed in the fossil record, proceeds by fits and starts rather than through the smooth gradual changes that had been emphasized by Charles Darwin. They call this process punctuated equilibrium, and indeed it seems to have figured large in the evolutionary history of many, but not all, organisms. But in the history of humans in Europe, fragmentary as the fossil record is, there is nothing that suggests any such sudden jumps. The somewhat small-brained but physically very diverse peoples of Steinheim, Arago, and Petralona, who lived two to three hundred thousand years ago, might indeed have been replaced by larger-brained premodern humans and then by Neanderthals arriving from somewhere else about a hundred thousand years ago. On the other hand, they might simply have evolved into them. The difference in brain size between these peoples and ourselves is after all only some 20 percent. And there is a good chance that the Neanderthals, with their beetling brows and robust skeletons, will be found to be part of our gene pool, so the possibility that they were completely replaced by modern humans thirty-five thousand years ago is not as likely as it once was. No matter what the details of this complex story eventually turn out to be, runaway brain evolution was acting on all those primitive Europeans, driving them in the direction of big-brained, weak-muscled, highly acculturated *Homo sapiens*.

We have seen that there are other stories similar to that of the Neanderthals in our prehistory. These are marked by a recurring pattern, in which several different types of hominid, distinguished by clear-cut differences but also sharing many features, coexisted in one place. The aboriginals who arrived in Australia were made up of at least two, and probably more than two, distinct types. And, long before the mitochondrial Eve, the robust and gracile Australopithecines were showing parallel evolution in many of their features, although not in their jaw muscles or dentition. One cannot help but wonder whether the robust Australopithecines as well as the Neanderthals will turn out to have been victims of a bad press, needlessly banished from the Direct Line of Human Descent.

It is lucky that our knowledge of genetics has not remained static from the time of Carleton Coon to now. If we are ever going to resolve these puzzles about our origins, we must understand how our genes work, and what the *genetic* differences and similarities between ourselves and our ancestors might be. If the mitochondrial Eve is old rather than young, this would of course allow more time for the flow of new genes for human attributes suggested by Wolpoff to take place. But how many of the genetic differences that separate us from our ancestors really are new, the result of recent mutations of the type that Coon imagined, and how many were already present in our ancestors, perhaps at low frequency? How much of our evolution is due to selection for new genes, and how much to selection for genetic variants that were already present as our runaway brain evolution began to gather speed? The exploding world of molecular genetics is about to give answers to these central questions.

III

THE GENES

7

Two Kinds of Geneticist—Two Kinds of Gene

Man is not just an overgrown Drosophila *[fly]. Some laws of biology apply, however, to men as well as to flies.*

—Theodosius Dobzhansky, *Mankind Evolving* (1962)

The genes that control the development of our runaway brains have obviously undergone a great deal of evolution from the time of our earliest hominid ancestors to the present—but just how much and what kind of evolution was it? What properties of our gene pools could have allowed such an astonishing process to take place? Now is the time to confront this problem squarely, by asking precisely what evolutionists mean by genes and by genetic change. There are, as we will see, a great many different opinions about these matters, but there is also a slowly emerging consensus.

You will recall that Charles Lumsden and E. O. Wilson gave the name *culturgens* to new inventions and other novel aspects of our culture. A typical culturgen might be a spear thrower, a device for increasing a spear's force and accuracy. After the spear thrower was invented, its presence in the cultural environment immediately began to select for new genes that allowed the invention to be used effectively. These new genes then spread through the population, so that eventually most people had the genetic capacity to be expert with a spear thrower. This is what might be called the Lucky Dip approach to the gene pool—somehow, the right gene is always around when a new culturgen comes along.

But what are these new genes? Where did they come from?

In one sense, of course, there are no new genes. Each of our genes has an evolutionary history covering at least three and a half billion years, a history stretching back to the time when the first DNA-based living organisms appeared. Although some extremely important genes have remained remarkably well preserved after all this time, most genes of today have little if any resemblance to their distant ancestors of billions of years ago. Much mutation and selection and many chance events have supervened. In spite of all these evolutionary changes, this long history means that most of the proteins made by our genes are extremely versatile and very good indeed at what they do. Hemoglobin, the protein that carries oxygen from our lungs to other parts of the body, is a good example. Even though many scientific lifetimes have gone into its study, new and extraordinary properties of hemoglobin are still being discovered.

In various members of every species, slight variants of such superbly well-adapted genes are appearing all the time as the result of mutations. These mutations, however, are not infinite in their variety, simply because there are not an infinite number of new kinds of genes that can be produced from an old one by a single mutational change. It is simply not possible to change a hemoglobin gene into an antibody gene in one step. Similarly, it is difficult to imagine a single mutation in a particular gene that produces the capacity for plantigrade walking or for speech—or indeed for being expert at using a spear thrower. If our evolution had to await such mutations, our ancestors might never have descended from the trees.

To understand how evolution *really* works, we have to abandon the notion that such mutations can happen. Instead, we must think of new mutations as small changes affecting the functions of preexisting genes that already have long and complex histories. Usually, new mutations tend to damage the genes in which they occur because they upset their precise functioning and their finely honed interactions with other genes. But sometimes they change them in ways that increase the fitness of their carriers or in ways that *might* increase the fitness of their carriers farther down the line if the environment should alter in a particular way.

It turns out, as we will see, that the gene pool of the human population is filled with various mutant allelic forms of preexisting genes, mutants that have arisen at various times in the past. Even harmful mutant alleles can lurk in the population for many generations, if they happen to be recessive or nearly so—and so can more benign mutant alleles that are not advantageous now but may be at some time in the future.

Because some of this already-existing variation has the potential to play an evolutionary role, our species would be able to go on evolving even if mutations themselves suddenly stopped. Of course, in the absence of new mutations, evolution would eventually begin to sputter and finally come to a halt. But it could go on running for a while because the great quantity of variation in every gene pool has provided each species with the evolutionary equivalent of a full tank of gas.

This stored potential for evolutionary change helped to power events like the explosive adaptive radiation of the mammals after the sudden demise of the dinosaurs. The tiny insectivore mammals that survived the end of the age of dinosaurs were not forced to start at ground zero, waiting around until new advantageous mutations appeared, because they could draw on preexisting mutations that had already accumulated in their gene pools. In short, *the rate of evolution is not directly dependent on mutation rate.*

Surely, you say, if this is true then all it does is make the Lucky Dip model of evolution more likely. If there are many mutant alleles already lurking in the gene pool, this simply increases the likelihood that when a culturgen like a spear thrower is invented there will already be a mutant allele present in the population that enables its carriers to take advantage of this new technique. Pack enough variation into the population, and there is likely to be, somewhere, an allele for the ability to use a spear thrower, ready to be selected for when somebody invents one.

Not quite, because the *existence* of all this variation in the gene pool is only part of the story. The variability in the human gene pool is more complicated than a collection of small recent mutational changes. There is no genetic variation for the ability to use a spear thrower per se, but rather variation for balance, for strength, for speed of reflexes, for the distribution of proprioceptors in the muscles of the arm that allow the brain to determine exactly where the arm *is* at any given instant. Just as importantly, there is variation in the ability of the brain to integrate environmental information, an integration process that allows some of us to predict the exact moment in the future when the frightened leap of the prey meets the arc of the spear.

Before spear throwers were introduced, hunters with the strongest throwing arms, a tiny fraction of the tribe, were at an advantage in the hunt. When the new technology appeared, it provided far more hunters, by proxy, with a strong arm, so that success in the hunt now depended on subtler and less brutal distinctions. As a result, the evolutionary process could draw on a wider sampling of the gene pool. Alleles of genes that had made little difference earlier could now become very important in determining hunting success.

So, what is the nature of all these remarkable genes, and what kinds of mutations affect them? We now know a surprising amount about these matters, and our information is growing at a dizzying pace. To see where we now stand, it is necessary to go back in time and look at how our knowledge of the gene pool has grown over the decades. The story is a strange mixture of genetics, politics, and colliding value systems.

The Classical View of Population Structure

In March 1910 a nineteen-year-old undergraduate at Columbia University named Hermann J. Muller gave a talk to the little biology club he had founded on campus the year before. He called his talk "Revelations of Biology and Their Significance," and it was more than a mere example of juvenile precocity. It forecast Muller's future career and the central role he was to play in both genetics and evolutionary theory.

Unafraid to tackle the big issues even then, Muller dealt in his talk with the role of genetics in the perfection of humankind. He did not fall into the trap of assuming that evolution itself had a higher goal, but he did state very clearly that we as human beings ought to have such a goal. Our growing understanding of biology would enable us to perfect ourselves:

> Science, in the form especially of psychology and sociology, will discover what qualities are desirable for the most efficient cooperation and for the best enjoyment of life; and science, in the form especially of physiology and genetics . . . will discover what the elementary bases for these qualities are, and how to procure them for man.

In order to achieve this goal, the youthful Muller was very clear about what had to be done with physically and mentally defective people. They had to be sterilized, to prevent their harmful genes from being passed on to the next generation. In this he was expanding on the ideas of Francis Galton, Charles Darwin's cousin, who in 1883 had introduced the term *eugenics* (from the Greek *eugenēs,* "wellborn") out of his concern for preserving what he perceived to be humankind's higher qualities. Galton founded the Eugenics Society to further this idea. Muller, however, went a step beyond Galton when he espoused sterilization of genetic defectives.

Muller's formative years were spent during the heyday of the eugenics movement in the United States. At the same time as eugenics laws providing for the compulsory sterilization of mental incompetents were being placed on the books in a number of states, the eugenics movement itself was divided about the origin of mental defects. Many eugenicists, knowing and caring nothing about genetics, assumed that the environment was at fault, that a degenerate environment would somehow result in a degenerate race of humans. All sorts of confused ideas were mixed up in their thinking, among them the idea that acquired characteristics were inherited. Their confusion even extended to the cure they proposed, for surely if the environment was at fault the sensible thing to do was to attack the defective environment rather than to sterilize people who had been adversely affected by it.

Muller, along with other more genetically oriented eugenicists, thought more clearly about the problem. He had been greatly influenced by remarkable discoveries concerning the inheritance of chromosomes and the inner workings of the cell, which he had recently learned about from his distinguished professor of biology Edmund B. Wilson. To Muller it was obvious that the ills of humankind could be traced to defective genes, not to defective environments. A defective environment could be altered, but defective genes were permanent and irreversible changes. The young Muller, severe and uncompromising in his approach to eugenics, divided the genetic universe up into good and bad genes.

The excesses of the eugenics movement soon produced a reaction. Even though Francis Galton was English, the idea of eugenics never caught on in England to the same extent that it did in Calvinist and perfection-obsessed America. A number of eugenics bills were introduced into the English parliament, but none was passed, partly because of a well-founded fear that they would be only the first step toward further curtailment of civil liberties. In a 1922 book titled *Eugenics and Other Evils,* G. K. Chesterton expressed concern that, if the government could order people not to have babies, it could interfere even further in one's private life—perhaps forcing one to give up smoking! In fact, Chesterton was right to worry. Some American eugenicist literature was as adamant about the evils of smoking as about the evils of mental defectives having children. The claim was made that excessive use of tobacco could cause the sexual organs to shrivel—which if true would make a very effective addition to the surgeon general's list of warnings.

At the same time, however, there was a widely followed discussion among intellectuals about the problems that eugenics posed. Even a casual examination of the idea of eugenics revealed some of its flaws. In the same book, Chesterton expanded on some folk wisdom about inheritance. He remarked, "Seeing, therefore, that there are apparently healthy people of all types, it is obvious that if you mate two of them, you may even then produce a discord out of two inconsistent harmonies. It is obvious that you can no more be certain of a good offspring than you can be certain of a good tune if you play two fine airs at once on the same piano."

Chesterton certainly went too far, for there are usually some resemblances between parents and offspring! Any differences, and there are sometimes many, can be traced to several genetic phenomena. Sexual reproduction mixes genes together in new combinations. As a result, these new mixtures of genes may interact with each other in quite different ways in the offspring. You will recall that at each genetic locus on our chromosomes, each of us carries two versions, or alleles, of a gene, one inherited from our mother and one from our father. At such a locus, an allele that was masked by the other allele when it was in the parent (and was therefore recessive) may sometimes turn out to mask its matching allele in the offspring (thus, in its new genetic milieu, it is now dominant). And often genes that behave in one way when they interact with certain other genes in the parent may behave in quite a different way when they interact with different alleles of these genes in the offspring. This phenomenon is called *epistasis* (from the Greek for "stepping on"). In epistasis, one gene steps on the effects of another, sometimes suppressing and sometimes enhancing it. Epistasis can be even more complicated than dominance, for the various genes involved in epistasis may be on quite different chromosomes and the number of possible interactions are legion. Finally, there is the effect of the environment. Often, when organisms with the same genotype are raised in different environments, the results are very different.

In short, there are a multitude of possible ways that the thousands of genes in any given animal or plant can influence each other and interact with the environment. Luckily for geneticists, not all genes are so complicated, for some show clear-cut modes of inheritance and nearly always produce the same effect regardless of what the other genes in the plant or animal are doing. Gregor Mendel worked with such genes when he carried out his famous crosses between different strains of pea plant. It was a good thing for the nascent science of genetics that he had picked such clear-cut genes to work with, for if he had been forced to work with less clear-cut characteristics he would never have discovered his famous laws.

When plant and animal breeders carry out crosses, however, they are usually trying to enhance characteristics with very complicated patterns of inheritance. Careful breeding experiments have shown that unexpected results in such crosses must be due to all three factors—dominance, epistasis, and the environment—but in most cases unraveling the complexity of the numerous genetic and biochemical interactions involved has proved to be quite beyond our capabilities.

In the early part of the century, of course, little was known about these various effects, and they played little part in the eugenics debate except to provide some basis for a vague feeling that the genetics of human beings was probably very complicated. Nonetheless, Chesterton's concern about the impossibility of predicting the outcome of marriages between people with apparently desirable characteristics foreshadowed the direction the debate would eventually take.

This debate had certainly not reached any level of genetic sophistication in 1910, when the young Hermann Muller felt self-confident enough to be able to prescribe genetic cures for the human race. His confidence in himself was not misplaced, since his later work would provide much of the genetic basis for far more sophisticated debates about the true nature of genes. This in turn would give rise to a long and continuing debate about the nature of evolution, particularly human evolution.

A short, intense, and severe man, Muller did not make friends easily—although the few friendships he made, such as that with the geneticist Edgar Altenberg, were lifelong. His first real scientific contributions took place in one of the twentieth century's most exciting scientific environments, the tiny "fly room" on the sixth floor of Columbia University's Schermerhorn Hall.

During the teens of this century, under the often bemused leadership of Thomas Hunt Morgan, a crowd of excited graduate students uncovered the secrets of genetics in 613 Schermerhorn. There they filled thousands of milk bottles with the progeny of the little fruit fly *Drosophila melanogaster*, an organism that Morgan had originally picked to study development but that turned out to be perfect for studying genetics. In the space of a few short years brilliant young men like Muller, Calvin Bridges, and Alfred Henry Sturtevant turned this modest little fly into the geneticist's organism of choice.

In the course of examining hundreds of thousands of flies, they discovered dozens of different mutations affecting eye color, wing shape, and many other characteristics. These occasional discoveries showed that such mutations could arise spontaneously and that, once they had arisen, they could be passed unchanged to the next generation like any other genes. The ones they chose to concentrate on, some dominant and some recessive, had the same kind of clear-cut and obvious inheritance that Mendel had found among his pea plants. Next, they were able to demonstrate that these genes were located on, or at least attached to, the sausage-shaped structures called chromosomes that were found in the cell nucleus. Finally, and perhaps most dramatically, the chromosomes themselves were not immune to giant mutational events. Whole chromosomes could sometimes go to the wrong daughter cell during cell division, and entire chunks of chromosomes carrying many genes could sometimes be lost or rearranged, with large genetic consequences.

Over the course of that decade, the fly-room workers showed that the mutant genes of *Drosophila* could be classified into four groups based on their inheritance. Generally, mutants in the same group tended to be inherited together. They already knew that, when the cells of *Drosophila* were examined under the microscope, they could be seen to contain four pairs of chromosomes, one member of each pair derived from each parent (the members of each pair were named the maternal and paternal chromosomes, to indicate their origin). Surely the fact that there were four groups of genes and four pairs of chromosomes must be more than a coincidence. Indeed, this correspondence was one of the strongest pieces of evidence that genes really did lie on chromosomes. However, they also realized that, if genes always remained tied to each other on the chromosomes, then all those genes should be inherited as a unit from one generation to the next. It would be as if the flies had only four genes, since each chromosome would act like a giant gene. But they already knew that the flies' inheritance was not that simple.

Studies on plants had shown that the nuclei of cells that were about to give rise to gametes contained chromosomes that underwent strange contortions. First, maternal and paternal chromosomes paired up, and then they appeared to exchange matching pieces with each other. Of course it was impossible to be sure that this exchange really took place simply by looking at the chromosomes under the microscope, but if it really did happen then it would have the effect of exchanging genes between the maternal and paternal chromosomes of the pair. One generation of such exchanges would not completely scramble the maternal and paternal chromosomes, because rather few of these crossover events seemed to take place each generation. However, if the crossing-over process were repeated over many generations each time sex cells were formed, and if the crossovers happened in different parts of the chromosomes each generation, the genes on the chromosomes would become more and more thoroughly scrambled with time.

Morgan and his students could not see the chromosomes of

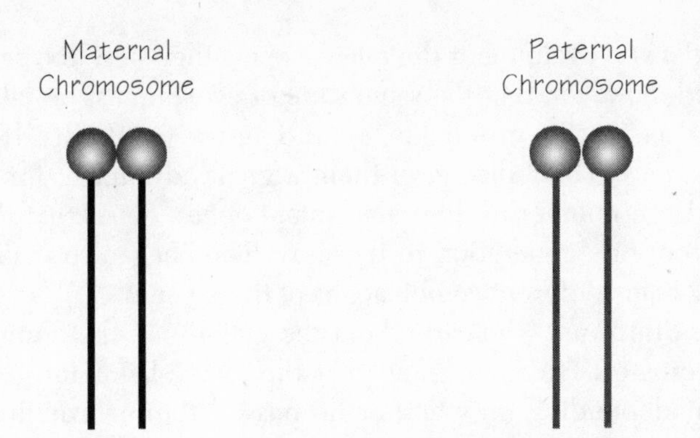

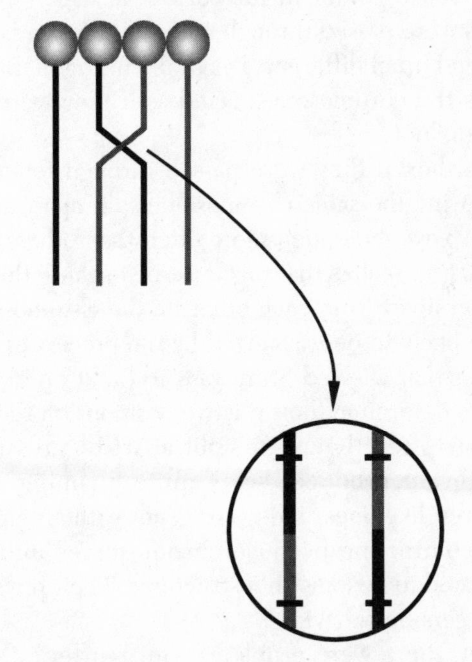

FIGURE 7.1. A simple diagram of crossing-over. Maternal and paternal chromosomes, which have already duplicated before cell division, pair and recombine. If the maternal and paternal chromosomes carry different alleles of a gene, and the recombination point happens to fall within the gene, then a new allele that is made up of parts of the old ones can be generated.

Drosophila very well under the microscope—they were too small. They suspected, however, that the same sorts of events must be taking place in them as in the much larger and more easily studied plant chromosomes. Their flies gave them a great advantage, for they had found a large number of different mutant genes, genes that they could follow from one generation to the next. They began to make crosses involving many different combinations of these genes.

It was Sturtevant who carried out the majority of these innumerable genetic crosses. He soon found that the fates of mutant genes were different depending on whether he passed them through males or through females to the next generation. If two different mutant genes on the same chromosome were passed through males, they always appeared together in flies of the next generation. When these same genes were passed through a female, however, they would often split up and end up in different flies in the next generation. For some reason, in males the chromosomes acted as if they were giant genes. In females they did not.*

Further, if they were passed through females, some pairs of genes occupying the same chromosome split apart and went to different flies in the next generation more often than others did. Morgan himself was the first to realize that this must be because these genes happened to lie further apart from each other on the chromosomes and therefore were more likely to be exchanged by the process of crossing-over. It was this insight that allowed Sturtevant to build up detailed maps of the genes and to determine their relative positions on the chromosomes according to how often they were split apart by crossing-over. The maps that Sturtevant produced were quite unambiguous—the chromosomes behaved like linear strings of genes without any loops or branches. This was comforting because chromosomes under the microscope also appeared to be long, thin structures. Their physical appearance matched their genetic behavior.

Yet, there were problems with Sturtevant's maps. When the genes were very far apart from each other on the chromosomes, his numbers did not add up properly. For his Ph.D. thesis, Muller took Sturtevant's data and showed that the reason was a simple but subtle one. Once a crossover took place at a particular point on a chromosome, something prevented other crossovers from taking place nearby. The discovery of

*This situation is rather unusual, and it is one that fruit fly geneticists have turned to their advantage. In many other organisms, including man, more crossing-over takes place in the females than in the males, but only in the fruit flies and their relatives is the difference so extreme.

this phenomenon of *interference* made Muller begin to wonder whether there were genes that controlled it. If there were, perhaps he could lock together groups of genes in females as well as males and prevent them from recombining at all.

In 1914 Muller's thesis work was incorporated into a book that he wrote jointly with Morgan and Sturtevant, a book that set out the marvelous genetic discoveries of the previous few hectic years. By this time, however, Muller was no longer getting along very well with the other members of Morgan's group. He felt, with some justification, that the many ideas he had contributed to the fly group had not received the acknowledgment they deserved. His biographer Elof Axel Carlson points out that Muller was blessed—or cursed—with an excellent memory and was able to recall conversations in great detail after a lapse of years. Unable to understand that not everyone else had the same ability, he could not comprehend why Bridges and Sturtevant were always failing to recognize his contributions.

In that same year, Muller moved to the Rice Institute in Houston. There he flung himself into the study of the great unanswered question that had been posed by the frenetic years at Columbia. What was the nature of the gene? What were these mysterious characters that passed, almost always unchanged, from one generation to the next?

He tried several ways to approach that central problem of genetics, but the most successful by far was his attempt to produce mutant genes in the laboratory.

X Rays—Bringers of Life and Death

If genes changed spontaneously at a low rate, as the occasional discovery of new mutants by the fly-room geneticists had shown, then surely there should be some way to change them at will, ideally at a much higher rate. If this could be done, then the means by which this could be accomplished might give a clue to the nature of the genes themselves.

Of course, this was such an obvious idea that it had occurred to every other geneticist on the planet. As a result, experimental organisms were being subjected to every imaginable environmental insult, ranging from radioactivity through heat, cold, and chemicals to loud sounds and other vibrations. All these approaches seemed to fail, or at least to give ambiguous results, which is especially surprising in view of the fact that these legions of scientists were often using things like X rays that we now know *can* change genes.

Muller had returned to Columbia in 1917, but his position was terminated in 1920. He was sure that it was Morgan's machinations that had prevented his appointment from being converted into a permanent one, though by that time Morgan had left Columbia for the California Institute of Technology (and Sturtevant later denied that Morgan had anything to do with it). Angry and disappointed, Muller took a job at the University of Texas at Austin. Before going there, he spent the summer of 1920 with Edgar Altenberg at Woods Hole, doing research. It was there that he found the strain of flies that would let him detect, unambiguously, whether a given treatment could produce mutations.

The sex chromosomes of flies, like the sex chromosomes of humans, have an odd inheritance. As in humans, females have two X chromosomes (the X has nothing to do with X rays, by the way), while males have an X and a Y. The Y chromosome is almost empty of genes, and the result is that male flies, like male humans, express any recessive mutant genes that they happen to carry on their single copy of the X chromosome. If these recessive genes are passed to a female instead, they are likely to be masked by the dominant wild-type allele so that the female is unaffected. This is why, in humans, the recessive condition hemophilia is found predominantly in males.

Muller began by constructing strains of *Drosophila* that carried many different mutant genes on the X chromosome. Using these strains, he had already isolated what appeared to be a large number of spontaneously arising lethal genes, but the results were ambiguous and could be interpreted in several ways. So he began the tedious job of genetically mapping his new lethal mutations in order to prove that they really existed. Then something odd happened in one of his fly bottles.

The flies in one of the crosses he was working on carried several genes on their X chromosomes, genes that had been put together in the course of many previous crosses. One of these was the recessive lethal gene he was trying to map, and another was a dominant mutant gene called Bar eyes, which shrank the large oval compound eyes of the flies to a slitlike shape. Accompanying these was an assortment of recessive mutant genes scattered along the length of the X chromosome. This dense population of mutant genes could be used as markers for mapping. Normally, when he mated females carrying one of these chromosomes to unmarked males, the males of the next generation would show a great variety of different combinations of the mutant genes because of crossing-over in the female parent, which broke up the multiply marked X chromosome in many different ways. These different X chromosomes would be passed on to the next generation, where their

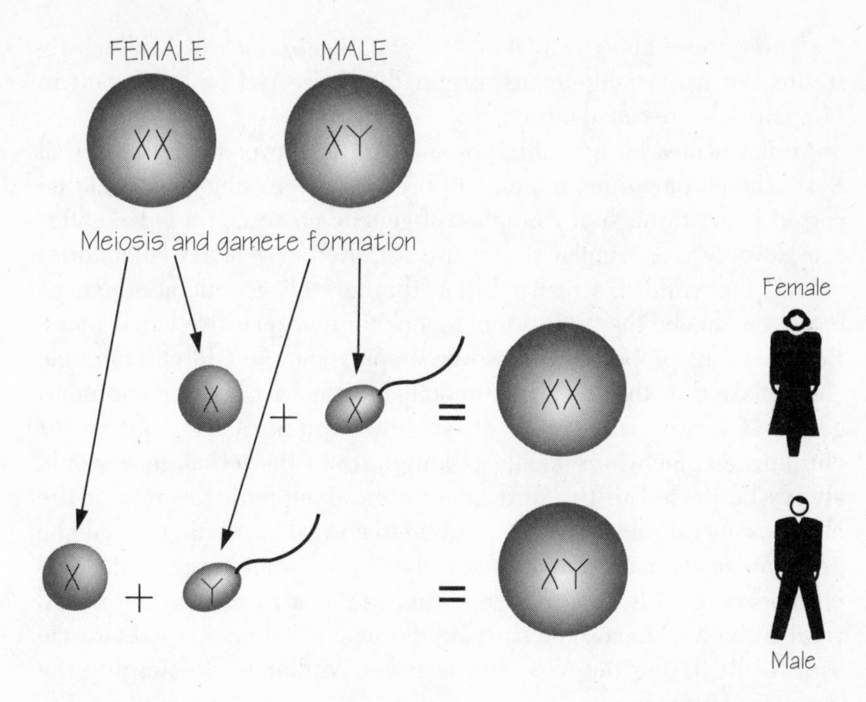

FIGURE 7.2. The inheritance of sex is very similar in humans and fruit flies.

various genetic cargoes would immediately be made visible in the males. If he really had made a lethal gene on this chromosome, then he could locate it by looking for combinations of marker genes that should have been there but were now missing or rare.

In one of these crosses, however, he did not get his usual collection of different classes of flies. The only flies in the next generation were Bar-eyed females and males that did not carry any of the genes of the original chromosome. Muller instantly realized what had happened. Something had suppressed crossing-over along the entire length of the female's X chromosome, so that males received either the multiply marked chromosome with its lethal gene and died or the other unmarked X chromosome and lived. Whatever this influence was that prevented crossing-over, Muller realized that it must be like his phenomenon of interference, but far stronger and extending along the whole length of the chromosome. (It was not until some years later that Muller and Sturtevant simultaneously discovered the reason Muller's chromosome behaved the way it did—a massive rearrangement of the order of the genes had taken place in one of his flies, a rearrangement that effectively prevented crossing-over between this new and aberrant

X chromosome and the rest of the X chromosomes in Muller's fly strains. Such rearrangements, originally discovered by Sturtevant in Morgan's lab, are called *inversions.*)

Muller named his new chromosome *ClB* (crossover-suppressor lethal Bar). The chromosome, immune to breakup by crossing-over, could be passed intact through any number of genetic crosses, and indeed after over seventy years similar strains are still used in genetics laboratories around the world. He realized that, through this genetic accident, he had been handed the perfect tool to look for new recessive lethal genes. Because of its property of crossover suppression, the *ClB* chromosome could make the other X chromosome in a female act like one enormous gene. If a new recessive lethal gene appeared on that other chromosome anywhere along its length, then the lethal gene would always be passed to the next generation along with the rest of the chromosome. If such a fly were crossed to normal males, there would be *no males in the next generation!* Either males would receive the *ClB* chromosome with its lethal gene and die as a result or they would receive the X chromosome carrying the new lethal gene—and also die as a result. Using the *ClB* chromosome, Muller could simplify the genetics of his flies.

It was not until 1926 that Muller carried out what was to be the critical experiment, irradiating normal males with X rays and mating them to *ClB* flies. Half the sperm from these males carried a Y chromosome and would give rise to males. The other half carried an X chromosome and would give rise to females. Among these X-chromosome-carrying sperm, he suspected, some would carry newly induced recessive lethal genes. (The sperm carrying them would not die because these genes were not expressed in sperm. They would exert their effects only as the flies began to develop.)

He then separated the *ClB* females of the next generation and mated each one individually to males. Those that were carrying lethal-bearing X chromosomes from their fathers would produce only female progeny. There was no need to count thousands of flies—a quick look at a vial of flies was enough to see whether there were males among them. The results were stunning—150 times as many new lethals appeared in his irradiated flies as in the unirradiated controls.

When Muller presented his detailed results to a meeting in Ithaca, New York, in 1928, the scientific world was instantly convinced that he had proved it was possible to alter genes in the laboratory. The results, widely reported, made the world at large aware that X rays and other forms of radiation were not simply benign tools of medicine—they

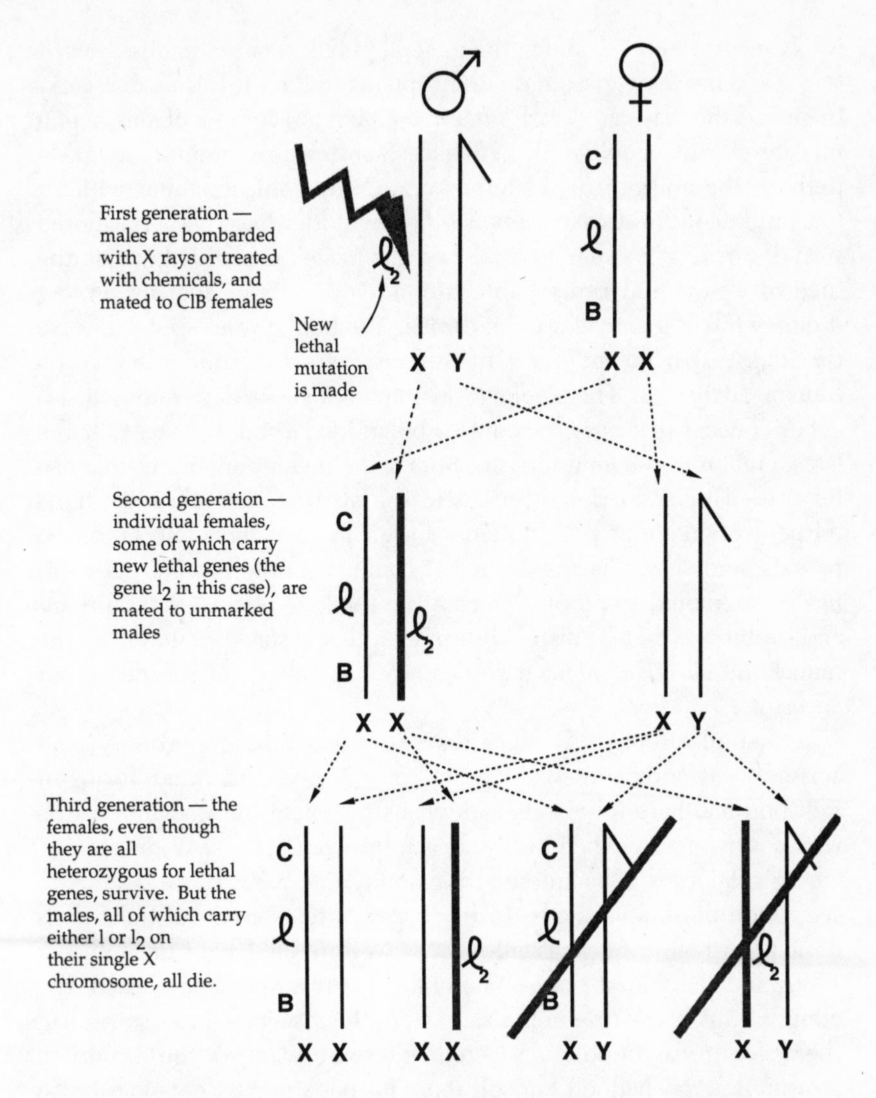

FIGURE 7.3. How Muller found his X ray–induced lethal genes. The figure shows the set of three crosses he carried out using his *ClB* crossover-suppressing strain of genetically marked flies. If a lethal gene was induced in one of his X ray-treated flies, there would be no males in the third generation.

could have dangerous genetic effects as well. Scary stories appeared in the press about the effects of irradiation, for it proved impossible to communicate to the public the other important result of Muller's work, that he was not making monsters with his X rays. He was making lethal genes that usually killed their carriers very early in development.

Muller's experiment set a number of things in motion. One was a wildly inaccurate idea among the general public about mutations—what they are, how they arise, and what happens to the organisms that carry them. To this day, the word *mutant* conjures up images of misshapen monsters or of people with new and sometimes frightening talents—perhaps the ability to read minds or to move objects about without touching them. Movies and science-fiction stories have blithely ignored scientific reality in order to reinforce this image, primarily because the idea of a new and remarkable human suddenly appearing among ordinary folk is a marvelous plot device. The public swallowed with ease the idea that even members of other species can suddenly take on human attributes. The Teenage Mutant Ninja Turtles, a tongue-in-cheek concept that has grown into a billion-dollar industry, are only the latest creatures in a long series of humanlike mutant organisms that can be traced back to the science-fiction stories of the 1930s. This misperception about what mutations are and what their effects can be persists, and, as we have seen, is not confined to the general public. It has even colored some of the scientific thinking about human origins and evolution—recall Allan Wilson's idea that a single mutation in the mitochondrial DNA might have conferred the ability of speech on our ancestors.

A second effect of Muller's discovery was a long-continued and acrimonious controversy about whether X rays and other forms of radiation had harmful effects and what they might be. Radiation had a very positive image when Muller began his work because of the cancer-curing properties of radium and the ability that X rays gave doctors to see inside the human body. In the hypercautious world of today, it is difficult to realize how casually X rays were treated right up until the 1950s and the fallout scare. As a child, I was fascinated to watch the bones of my feet wiggle in the shoe-store fluoroscope—most stores had them, for it was impossible to tell otherwise whether the armorlike shoes of the day had left enough room for one's toes. A radiologist who practiced during the 1930s and 1940s told me how he and his fellow radiologists had entertained each other during their lunch breaks by eating their lunches behind the fluoroscope screen. The accumulated dose was enormous and had left him unable to have children. Indeed, massive doses of radiation were a common way to induce temporary male sterility. Until the late 1950s, it was standard practice to X ray every visitor to San Quentin Prison from head to foot. The relatives of long-term inmates received frighteningly high cumulative doses. Although Muller's work had begun to raise concerns about the health effects of these cavalier uses of radiation, it was the specter of atomic fallout that finally turned public opinion away from the image of radiation as a savior and replaced it with that of a threat.

Even at the scientific level, Muller's work was immediately embroiled in controversy. It was difficult to detect the genetic effect of X rays without Muller's sophisticated approach because most of the mutations caused by X rays seemed to be lethals. And it was this fact that began to worry other geneticists. Only rarely, it seemed, did X rays produce mutants that had a visible effect on the flies, and usually these effects were very different from the kinds of mutations that had been discovered in Columbia's fly room. Why did Muller find so few of these other kinds of mutations? Was he actually changing genes with his X rays or was he only destroying them? If the latter, then was it some other influence that had caused the genes of the flies in the fly room to mutate? The process of mutation seemed to many geneticists to be far more complicated, and mutations themselves appeared to have many more origins, than Muller's original experiments had suggested.

Muller's brilliant experiments had given him a tremendous reputation and provided a platform for his views. Luckily, his earlier severe version of eugenics was now tempered by genetic commonsense. At the third (and last) International Congress of Eugenics that took place in New York in August 1932, his address—which the organizers of the congress had tried to prevent—pointed out that the simple-minded eugenic approach of dividing humans into fit and unfit was nonsense. "Genetic worth is a practically continuous variant, and there is no hard and fast line between the fit and the unfit, nor does relative fitness in the great majority of individuals depend on one or a few pre-specified genes," he pointed out.

Indeed, a giant social experiment that demonstrated the truth of Muller's words had just been carried out. Millions of people who had been model, hardworking citizens a few short years earlier were now standing in breadlines or selling apples on street corners. Their genes had not changed—society had. Therefore, said Muller, society had to change yet again. Capitalism provided no incentives for the genetically fit to reproduce. "The social direction of human evolution . . . can occur only under a socially directed economic system."

It was not long before Muller found out for himself what the consequences to genetics of a socially directed economic system could be. He had helped to edit and distribute a left-wing newspaper on the University of Texas campus. This, along with his unabashedly socialist speeches, had made him persona non grata with the university administration. In the fall of 1932 he left for what was to be a sabbatical year in Germany, a year that turned into a lengthy period of exile abroad. The rise of the Nazis was swift and merciless and caught both Muller and his German socialist friends completely by surprise. He watched storm troopers break into the Kaiser Wilhelm Institute for Brain Research, where Muller had been working with the Russian geneticist Nikolai Timoféeff-Ressovsky.

Muller was anxious to get his family out of Germany, and he seized on what appeared to be a marvelous opportunity. The distinguished Russian plant breeder Nikolai Vavilov wanted Muller to be the director of the genetics laboratory in Vavilov's Institute for Applied Botany in Leningrad. He offered Muller superb facilities and many co-workers and assistants at a time when American universities were striving desperately to stay afloat.

Muller's years in the Soviet Union began well, and he appeared to be welcomed by the Soviet government. In 1935 he even marched in the Red Square parade honoring the anniversary of the Bolshevik Revolution, perhaps the only American geneticist ever to have done so. But soon he was again catapulted into political controversy and physical danger because of the rise of a pseudoscientist who came to dominate and eventually to destroy Soviet biology.

Lenin, and later Stalin, inherited a medieval agricultural system. In order to force the industrialization of the country, they uprooted peasants from the farms and sent them to work in the factories and steel mills. These former peasants had to be fed, so it was essential that food production be increased in turn. The dictators tried to turn the primitive agricultural system into a collectivized one.

After Stalin consolidated his power, the apparatchiks in charge of agriculture had every reason to panic, for Stalin brought a new artistry to the application of terror. He exercised it with a large element of randomness, keeping every level of society off balance and preventing the formation of any opposition. He rarely let the apparatchiks know whether he was pleased or displeased by their efforts. So, like primitive tribesmen frantically trying to propitiate an unpredictable volcano, they turned to shamans for help.

One of these was Trofim Denisovich Lysenko, who perfectly fit George Orwell's description of a Soviet apparatchik as "half gangster, half gramophone." He made extravagant and totally unfounded claims for a number of doubtful agricultural techniques, some old and some invented by himself, and shamelessly faked data to show that these methods increased yields enormously. He expertly wove a network of fear and intimidation around the scientific establishment, through rabid accusations that mainstream scientists were trying to wreck the Soviet state.

As a result, Lysenko effectively destroyed Soviet biology in the 1930s and 1940s, so thoroughly that his malign influence is still being felt. Ironically, his power grew primarily through a misapprehension on the part of scientists who assumed that, because Lysenko's supposed triumphs were trumpeted by the press, he must have the ear of Stalin. In fact, Stalin and the politburo only finally came out clearly for Lysenko and against genetics in 1948, when the damage to biological science had largely been done. Perhaps the greatest irony, and the clearest indication of the ingeniously random nature of Stalin's terror, is that many of those who stood up against Lysenko during those terrible years were not arrested. David Joravsky has documented this in his remarkable book *The Lysenko Affair,* and has even listed a number of Lysenkoists who *were* arrested! Still, the heaviest hand of terror seemed to fall both on those who tried to appease Lysenko and on those who tried to ignore the whole thing and get on with their work.

In the middle of all this, Muller was as confused as everybody else. At one moment, Lysenko was shouting that "Mendelian-Morganian" genetics was vicious nonsense and must be stamped out. At the next, Muller was given to understand that the party was solidly behind Vavilov's plan to hold an International Congress of Genetics in Moscow in 1938. He was heartened by rumors that Stalin was thinking of establishing Muller's dream—a eugenics program! In what must rank as a true milestone of naïveté, Muller actually sent a letter to Stalin about the program and about the little book Muller had written advocating it. Icily, word came back from the Kremlin that Stalin was displeased and that all such work must stop.

As Stalin's terror proceeded and Lysenko's attacks on genetics increased in stridency, as some of Muller's Russian friends began to disappear and even Vavilov was threatened (he was later imprisoned and died), the scales began to fall from Muller's eyes. In retrospect his innocence about matters Soviet seems astonishing, but he was not alone. It was extremely difficult for him and his socialist contemporaries to admit that their cherished dream had turned into a nightmare. At last, reluctantly, Muller realized that he must get out of the Soviet Union, and do so in a way that would do the least harm to his friends. His ploy was ingenious—he left for Spain to serve a brief stint in a transfusion unit of the International Brigade. Then, before the transient glory of this socialist exploit wore off, he returned hastily to Moscow, packed, and left for Edinburgh.

There, in the brief period before the war broke out, Muller could finally get back to his science. Perhaps his most important accomplishment was to encourage and channel the research of a young hardworking geneticist, Charlotte Auerbach, as she began the difficult and dangerous task of finding out whether chemicals, as well as X rays, could cause mutations. She eventually showed that mutations could indeed be induced in *Drosophila* by frightful chemicals such as mustard gas. It is now known that mustard gas damages DNA, and indeed the lesions caused in humans and experimental animals by this deadly chemical have many of the same properties as radiation burns. Auerbach's first mutations were very similar to the lethal genes Muller had induced with X rays, but since that time many other chemicals have been shown to cause a much wider variety of mutational changes. Muller, initially supportive of Auerbach's work, later became unhappy with the way mutation research was going because he believed that it distracted from the most important problem, the harmful effects of radiation.

On his return to the United States in 1940, Muller found himself an academic pariah. During the war he eked out a year-to-year existence in a temporary appointment at Amherst College, where his work was hampered by the lack of graduate students. Finally, through the good offices of friends, particularly at the Rockefeller Foundation, he got a permanent job at Indiana University in 1945. His friends' faith in him was fully justified when he was awarded the Nobel Prize in 1946, the first year the prizes were resumed after the war.

After the war, Muller's interests in politics and eugenics continued unabated. Disillusioned with the Soviet perversion of socialism, he resigned from the Soviet Academy of Sciences in 1948 and blasted the Lysenkoist movement at every opportunity. So far as eugenics was concerned he had moved to a more free-enterprise approach, espousing the idea of *germinal choice.* A woman whose husband was sterile could, he envisioned, choose from an assortment of outstanding sperm donors. Pressed to provide a list, he included Lenin's name on it, to the embarrassment of his colleagues.*

*A Center for Germinal Choice was actually established in Escondido, California, in 1981 by a wealthy retired optometrist and industrialist named Robert Graham, and a similar one was set up in Pasadena in 1984 by a former employee of Graham's. The Escondido freezers originally included sperm from three Nobel laureates, including the controversial physicist William Shockley, although those particularly distinguished sperm have now grown too old to use. Over 150 babies (an unknown number from Nobelists) have resulted, although Graham has been unable to find any evidence to show that the children are anything out of the ordinary.

Perhaps the most important of Muller's contributions after the war was an address given in December 1949 to a meeting of the American Society of Human Genetics in New York. Titled "Our Load of Mutations," it was in part a response to the worldwide acceleration in the rate of atomic testing with its attendant fallout, in part an apocalyptic vision of what might happen to the world if the human race did not begin to practice eugenics.

The paper was written at a time when the nature of genes was still essentially unknown. Muller, like most geneticists of the time, was sure that genes were made of proteins, for only protein molecules seemed to be complicated enough to perform such a multitude of different tasks. In his view such elaborate structures were far more likely to be damaged than improved, and any influence such as fallout that increased the rate of damage was a serious problem. He pointed out that, although dominant harmful mutations would tend to be removed rapidly from the human population by selection, recessive ones could lurk concealed there for generations. Further, the descendants of a single harmful mutation could, like the descendants of a single human being, become very numerous in the future. An uninterrupted flow of new harmful mutations would be sure to damage the genetic structure of the human species.

Muller's argument that radiation was bad for our species is generally accepted today, but it did not jibe with the official government position in the 1950s. In 1955 he was prevented by the Atomic Energy Commission from giving a paper on the effects of radiation to a United Nations conference, a case of censorship that caused a great outcry in the press. Unfortunately, there is no doubt that Muller's political history made him an easy target for government censorship and interfered with his effectiveness as a spokesperson for the perils of radiation.

I met Muller briefly in 1965, two years before his death. Still intellectually very active, he gave an informal evening seminar about the probable number of genes in humans and other higher organisms. By that time, of course, Muller along with all other geneticists had been converted to the view that DNA was the genetic material. With impeccable logic, Muller showed that most of the DNA in higher organisms could not be genes. His argument was simple: Suppose that genes are arranged along the DNA with no spaces between them, so that all the DNA is made up of genes. Then there would be so many genes that they would pose a great danger to the organism carrying them. Given the known mutation rate to harmful alleles, dozens of new harmful mutations would arise in each individual in the course of his or her lifetime, instead of the one or two that actually appear. This would place an insupportable genetic load on the population. To his audience of largely disbelieving students, Muller concluded that most of the DNA must not be genes but must have other functions.

A year after Muller's death, Roy Britten and David Kohne of CalTech proved him right by showing that a large fraction of the DNA in higher organisms has a simple repeated structure, so simple that it cannot possibly code for genetic information. It is now known, as Muller had predicted, that most of the DNA in humans and other complex organisms is indeed not made up of genes but has some other function or functions—perhaps to do with gene regulation or with the structure of the chromosomes themselves.

On the way to the airport the day after his talk, I told Muller about some experiments of my own in which I had used his favorite organism *Drosophila*. The straightforward interpretation of my experimental results was that Muller was right about the unconditionally harmful effects of most mutations. Using various crosses, I had examined the effects of alleles that I knew were harmful when the fly carried two copies—that is, when the fly was *homozygous* for the harmful allele. I then found that some of these harmful genes could also affect a fly adversely even if the other allele carried by the fly had no obvious effect—that is, if the fly was *heterozygous* for the harmful allele.

However, when I carried out a more detailed analysis, I found that the heterozygous effect of these harmful alleles was so slight, and so dependent on the environment, that the alleles could not have been unrelievedly bad. I found that under some environmental conditions they could actually have positive effects on the flies' survival, and that in other environments their effects might be negative.

Unlike Muller, I had not produced these harmful genes by X rays or chemicals but rather had found them in natural populations of flies. Possibly, as I suggested to him, they might be different in their effects from the newly induced harmful mutations that Muller had spent his lifetime investigating. My genes, after all, had a much longer history of natural selection behind them than his did.

Muller was undisturbed by this idea. He remarked that any harmful genes that did manage to persist for many generations in natural populations would very likely be the ones that had the most involved epistatic interactions with other genes and that were also the most affected by the environment. As a result, any harm that they did would be felt under some genetic and environmental circumstances and not others. So it was not surprising that small changes in the environment could make their effects disappear. I was most impressed—his view of genes and how they operated was far more complex and subtle than a superficial glance at his more polemical writings would imply.

Muller left a powerful legacy. His early investigations into the nature of the gene eventually led to answers to the questions of what genes were made of and how they might be manipulated in the laboratory. He alerted the world to the dangers of radiation and other mutagenic agents and worked tirelessly to publicize these dangers. There is no doubt, however, that his published views of the structure of the genetic complements, or genomes, of organisms were too simplistic. He had clearly shown that all genomes are littered with bad genes, the result of X rays, chemicals, and other mutagenic agents. But he had a strong and lifelong conviction that there were, at the other end of the spectrum, good genotypes in the human population that could be selected for through programs of eugenics. This polarization of genes into good and bad made his name synonymous with what came to be called the classical view of population genetics and evolution.

As I have tried to suggest, Muller's own views were more complex than that, but unfortunately his career lent itself to simplification. He embraced eugenics throughout its whole stormy twentieth-century history, in the face of the excesses of the American eugenics movement, the monstrous laws imposed by the Nazis, and the rise of a liberal viewpoint in the United States that downplayed any genetic role in human differences. His passionate embrace of eugenics seems puzzling now. It put him at odds with the rest of the scientific establishment and indeed with much of the public in general. There is no doubt that from the dualistic point of view his logic was impeccable. If there were bad genes, then why should there not be genes of an unalloyed goodness? But was this simplifying dichotomy a reasonable one? A whole school of population geneticists did not think so.

A Balance of Forces

I have met only two really charismatic people in my life. One was the Maharishi Mahesh Yoga, who radiated goodwill and kindliness at the meeting I attended and whose eyes sparkled with joie de vivre as he surveyed his adoring followers from behind the heaps of flowers they piled around him. The other was the geneticist Theodosius Dobzhansky, whose eyes held exactly the same kindly sparkle and who inspired a similar adoration in the many students and collaborators he gathered around him during his long scientific career. Dobzhansky was enormously gregarious, infinitely patient with students, and always interested in their work. His impish sense of humor, combined with his emphatic mode of speaking and his unique accent, gave his every utterance the force of a bon mot. All his students soon collected their favorite Dobzhansky stories, which they still, almost twenty years after his death, exchange fondly at meetings.

Dobzhansky's path crossed Muller's many times, but their views on genetics were poles apart. During my own brief collaboration with Dobzhansky in the early 1960s, I was struck by how his views of both evolution and the human condition were shaped by his own complex and contradictory character.

He was born in 1900 in the little town of Nemirov in southern Ukraine. As a child he became intensely interested in collecting butterflies and other insects and decided very early that he would be a biologist. When he entered university in St. Petersburg, he took courses in a great variety of biological subjects ranging from cytology to anatomy. But soon the war, and later the revolution, disrupted his life as it did that of so many others. His faculty mentor was forced to go into hiding for political reasons. In 1920, during the dreadful starvation period before Lenin's new economic policy began, his mother choked on a piece of ersatz bread that he had brought home, and died in his arms.

Forced to learn his science where and when he could, Dobzhansky finished his undergraduate studies at Leningrad without obtaining a formal degree. He landed a job as an assistant instructor at the University of Kiev. During the early 1920s he began an intense study of ladybug beetles, collecting them over a wide area and discovering a number of new species. From the beginning he was fascinated not only by differences among the many ladybug species but also by the many differences that he saw within what taxonomists would agree was a single species. The wing cases, particularly variable, showed wide variation in color and in the number, shape, size, and position of the black spots. The ladybug populations were phenotypically *polymorphic*—that is, there were several different phenotypes, or distinct sets of physical characteristics, coexisting in the same population.

These studies, which he swiftly summarized in a series of papers, introduced him to a central question in evolution, the species problem. Dobzhansky was soon attracted to the ideas of Darwinian evolution and in particular to Darwin's view that most evolution occurs by gradual, almost imperceptible steps. At what point in this gradual process does one species become two? He had an intense desire to learn more about the steps by which various species had become different and whether all the variation that he had found within his species of ladybugs might play a role. When he made crosses between the different varieties of ladybug that he had found within each of his species, however, he obtained confusing results. There were too many genes contributing to each of the characters he looked at, and too many epistatic interactions among them, to produce the kind of simple Mendelian ratios that made genetic interpretation easy.

Because only fragments of information were penetrating Russia from the scientific world outside, he knew virtually nothing of the remarkable advances that had taken place in Morgan's fly room. When Nikolai Vavilov returned to Leningrad from a trip to the West with a load of books and reprints of papers, Dobzhansky learned about their contents and was dazzled. He determined to become a geneticist and to use the powerful new techniques of genetics to solve the species problem.

Then, in 1927, came an opportunity to spend a year in Morgan's lab at Columbia. This altered his scientific career and probably also saved his life—if he had been in Russia during the rise of Lysenko, he would certainly have been too vocal and devastating a critic of Lysenko's brand of pseudoscience to have lasted for long. Soon after Dobzhansky's arrival in New York, the fly group moved permanently to CalTech in Pasadena, and Dobzhansky accompanied them.

Here, during a highly productive but often stormy collaboration with Alfred Sturtevant, Dobzhansky found the perfect organisms with which to attack the species problem. Donald Lancefield, who had been a student of Morgan's at Columbia, had been working with a remarkable species of *Drosophila,* very different from the highly inbred and thoroughly domesticated *Drosophila melanogaster* that inhabited the Columbia fly room. These flies were larger and darker in color, and held the extra attraction for Dobzhansky that they inhabited some of the most picturesque regions of the American West—he loved our great national parks, and was never happier than when he was trapping flies, often illegally, within their precincts.

Lancefield had discovered that these flies could be divided into two groups on the basis of mating experiments, even though they were essentially indistinguishable from each other in their physical appearance. Within each group the flies bred normally among themselves, but if flies of one group were crossed with flies of the other something went wrong. The females were fertile or partly so, but the males were sterile because their testes were small and unable to function.

Dobzhansky and Sturtevant realized that these two groups of flies were exactly what they needed to study the species problem. They suspected that Lancefield's two races, partially genetically isolated from each other, must lie at some fairly early point along the process of species formation. At the beginning of that process all the flies had belonged to one species, freely able to exchange genes. By the end, their descendants would form two distinct groups, unable to exchange genes because of all the differences that had accumulated between them in the course of their evolution. Lancefield's two races of flies must form a kind of missing link in the process by which species arise.

From the moment they began to study these flies, Dobzhansky and Sturtevant had different views about them. Because hybrids between the two races had such difficulty in producing offspring, Dobzhansky soon felt compelled to raise the races to the status of species. The flies had been named *Drosophila pseudoobscura* because for a long time they had been confused with a similar European fly named *Drosophila obscura.* Dobzhansky gave one race that name and awarded the other the name *Drosophila persimilis.*

Sturtevant strongly disagreed with the way Dobhzansky was taking liberties with taxonomy. He felt that the differences between the two races were simply not great enough to justify splitting them into species. Yet Dobhzansky felt that he was warranted in doing so—after all, he already had infinitely more information about the genetics of his flies than paleoanthropologists have about the genetics of those various human ancestors that they have blithely split into different species or even different genera.

Then Dobzhansky and Sturtevant were provided with a new genetic tool that they thought might resolve their disagreement. The giant chromosomes of *Drosophila* had recently been found by Theophilus Painter at the University of Texas. These chromosomes, located in the salivary glands of the larvae of the flies, were a thousand times as large as normal chromosomes.

In most of the cells in a fly's body (or in ours, for that matter), the chromosomes in the nucleus are usually invisible even under a microscope. They are stretched out to form long, thin threads, filling the nucleus with a featureless tangle of DNA and various proteins. Only when the cells are dividing do these threads coil and condense to form the clearly visible sausagelike structures that we usually associate with the term *chromosome.*

The salivary glands consist of a few hundred huge and metabolically very active cells, so active that they require many copies of certain genes to carry out their functions. Rather than make copies only of those genes, the cells have taken the simpler route of duplicating all their chromosomes a thousand times or so. As the cells of the gland mature and their chromosomes double again and again, the chromosomes become visible, first as ribbons and then as giant bundles of threads. In the mature cells of the salivary glands these huge bundles are easily visible even under a low-power microscope.

These immense chromosomes show complex and distinctive patterns of bands along their length. Painter, and later Calvin Bridges, used these bands to draw detailed maps of the chromosomes of *Drosophila melanogaster.* Each region was distinct and unique. Bridges was the better mapmaker. He assigned numbers and letters to individual regions, and his maps have been used ever since to allow geneticists to find their way easily around the chromosomes (see plate 14).

Dobzhansky and Sturtevant found to their delight that *Drosophila pseudoobscura* also had excellent giant chromosomes, with equally complicated patterns of bands that allowed a detailed map to be made. The pattern of bands on the chromosomes of *pseudoobscura,* however, showed almost no resemblance to the pattern found on those of *melanogaster.* This was surprising because obviously the flies must have many similar *genes*—they are both species of *Drosophila,* after all. Why shouldn't their chromosomes resemble each other? Yet they did not. Too many differences, large and small, had accumulated in the DNA of these two species in the course of their long evolutionary separation. The many differences in the giant chromosomes of the two species were a visible demonstration of the millions of years of evolution that separate them.

This was not the case when Dobzhansky compared the two races that he wanted to call new species. *D. pseudoobscura* and *D. persimilis* turned out to have chromosomes that were remarkably similar. The order, shape, and size of the bands on the chromosomes matched perfectly—or almost perfectly. There were only a few places where there were detectable differences, and these were not in the appearance of the bands but rather in their order. Sometimes there was an inversion, in which part of a chromosome had been snipped out, turned 180 degrees, and reinserted into its old place in the chromosome. These inversions were like the crossover-suppressing inversion of Muller's *ClB* flies, but they had occurred at some point in the flies' distant past rather than in a laboratory stock.

The close resemblance between the chromosomes of Dobzhansky's two species, Sturtevant felt, cast into even more doubt Dobzhansky's insistence on turning the single species into two. Were these two types of fly so closely related genetically that the only differences between them were a few variations in the order of the bands on their giant chromosomes? This hardly seemed enough to justify giving them different names.

Dobzhansky knew that there had to be other differences, invisible even at the relatively detailed level of the giant chromosomes. He made crosses between the two species, crosses that he was able to carry out for more than one generation by mating the fertile female hybrids back to males of one or the other species. These crosses enabled him to show that the sterility of the hybrid males was due not to a single gene but to many genes scattered over all the chromosomes, although he was not able to show exactly where the genes were located or how many of them there were. As with his ladybug beetles, the genetics turned out to be too complicated for this kind of precise analysis, with too many interactions among the genes and too many environmental influences.

He was also able to measure subtle behavioral differences between the species, the result of genes that were even more difficult to track down. These differences were not enough to prevent males and females of the two species from mating when they were confined to a bottle on the laboratory shelf and presumably had nothing else with which to occupy their time. In nature, however, where it was quite possible for a female fly to escape from the unwanted attentions of a male of the other species simply by flying away, such slight behavioral differences made an effective barrier to gene exchange. He was able to provide a direct demonstration of the effectiveness of these behavioral differences, for after years spent trapping tens of thousands of flies in places where the two species lived together, he had found only two hybrid flies. *D. pseudoobscura* and *D. persimilis* really were, in spite of Sturtevant's hesitation, different species. Their common ancestor had existed very recently, perhaps only a few hundred thousand years ago, but now they formed two separate gene pools, almost incapable of exchanging genes except through the rough and insensitive matchmaking of geneticists in the laboratory.

Dobzhansky had caught, as if in a snapshot, two species in the early stages of diverging. He had shown that they had some genetic differences that were actually visible at the level of the giant chromosomes, along with many more differences that were detectable only through changes in the flies' fertility and behavior. The genes that caused these latter differences could not be tracked down by any technique available to scientists at the time. Nonetheless, he knew that they had to be there because he saw their effects.

In 1940 Dobzhansky left CalTech for Columbia. Over the next thirty years, he and his co-workers extended their studies to many other pairs of species of *Drosophila* in different parts of the world. Some of these pairs had proceeded further along the speciation process than *D. pseudoobscura* and *D. persimilis,* and other pairs had just barely begun to separate. It really looked as if speciation were a gradual and continuous process, just as Charles Darwin had predicted it would be. Dobzhansky was able to reconstruct it by taking a series of "snapshots" of the various stages of speciation that had been achieved among flies that lived in places such as Yosemite Valley, the rain forest of Brazil, and the cordillera of the Andes. These studies cast a new and brilliant light on evolution, for Dobzhansky was working at a genetic level that had not yet been explored.

Of course, while this level is quite inaccessible to paleontologists, who can only guess how many species there might be among their fossils, at least their fossils can be dated. There were no fossils of Dobzhansky's fruit flies, so he had to make educated guesses about how long ago the various speciation events that he had caught in his snapshots had begun to take place.

As he continued these studies, Dobhzansky gained more hints about how the speciation process occurs. Does a species simply split into two large groups, perhaps through some geographical accident, and then do these two groups, now able to diverge slowly from each other, give rise to two new species? Or does a new species begin its career through a kind of birth trauma, resulting in the sort of size bottleneck that we saw earlier might have taken place at the time of our mitochondrial Eve? He thought he could see traces of such a birth trauma in the giant chromosomes.

There are two complete sets of giant chromosomes in each of the cells of the salivary glands, one set from the fly's mother and one from its father. They do not remain aloof from each other in the nuclei of these cells, however, as the maternal and paternal chromosomes do in the nuclei of other cells. Early in the salivary gland cell's career, the corresponding maternal and paternal chromosomes find each other and pair up gene for gene along their entire lengths. Each giant chromosome seen under the microscope is really two joined together along their length. Normally, this pairing results in a single thick bundle of threads rather than two thinner ones. But if there has been any change in the order of the genes, such as an inversion, and if one of the chromosomes carries the inversion while the other does not, then this immediately becomes obvious under the microscope. As the matching genes try to find each other, the chromosomes form a loop. You can see why this happens from the simple diagram on the next page, in which the genes along the chromosomes have been labeled with the letters A, B, C, and so on.

Geneticists like Muller found such inversions very useful because they lock genes together into groups and prevent them from crossing over. Dobzhansky's inversions, however, were different from the ones that had appeared suddenly in the Columbia fly room and Muller's laboratory. He found that some of them seemed to have arisen millions of years ago and to have persisted in the fly population right down to the present time. During all that time they had neither been lost nor managed to spread to every fly in the population. Dobzhansky's ladybug populations had been *phenotypically* polymorphic. In contrast, in his *Drosophila* populations all the flies looked the same—until their giant chromosomes were examined. Then they were discovered to be *genotypically* polymorphic. There was a great deal of evolutionary activity going on in his flies at a level that was quite invisible to the naked eye.

When he looked at the giant chromosomes of flies taken from natural populations, Dobzhansky found that one of them in particular, designated the third chromosome, was very genotypically polymorphic. Examining the nuclei of salivary glands from flies heterozygous for two different kinds of these third chromosomes, one inherited from each parent, he found that these chromosomes exhibited a remarkable variety of inversion loops.

In most cases the chromosomes twisted themselves into more than a single loop because at various times in the past inversions had happened on top of other inversions. If the maternal and paternal chromosomes of a fly happened to differ by a number of these overlapping inversions, they would form an astonishing tangle when they tried to pair up. Some of these complex chromosomal configurations are shown in plate 15.

Dobzhansky and his students quickly became skilled at reading the tangles they saw under the microscope and at telling which chromosomes were involved. As they explored the huge region over which the two species lived, extending from British Columbia south to Central America and from California east to Nebraska, they found that some of these chromosomal types were very common and very widespread. There must, they realized, be hundreds of millions or even billions of copies of these various types of chromosome in the fly populations.

The ranges of the two species *D. pseudoobscura* and *D. persimilis* overlap in a wide area of northern California and Oregon. In this overlap region, a particularly common chromosome is shared by both

Normal chromosome

A B C D E F G H I J K L M N O P

Chromosome with an inversion

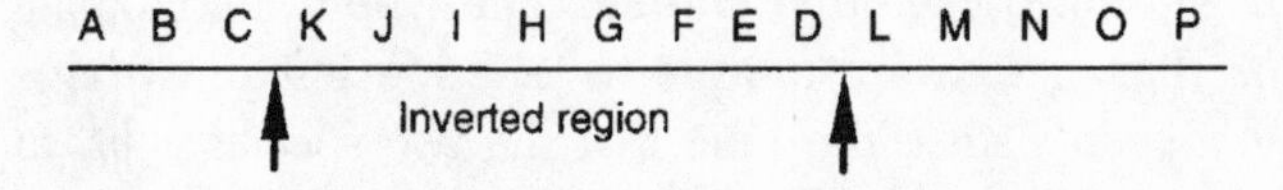

What happens when these chromosomes pair up in the nuclei of the salivary gland cells:

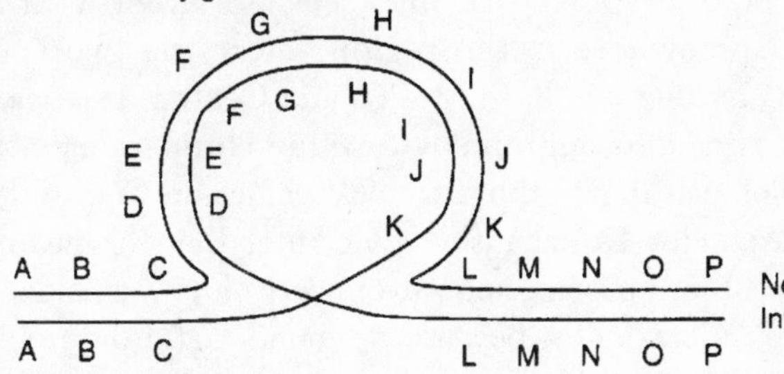

species. So far as could be told from the pattern of bands, the genes were arranged in exactly the same order in these chromosomes regardless of which species they came from. Dobzhansky used this chromosome as the standard against which to compare other chromosomes. Although many other kinds of chromosome were found in each of the two species, the only kind that both species had in common was the one that he named the Standard sequence of genes.

This finding gave Dobhzansky a good idea of how the process of speciation had actually begun. Earlier we saw a similar situation among the human mitochondrial chromosomes. An ancestral chromosome carried by the mitochondrial Eve has, over hundreds of thousands of years, accumulated changes so that almost everybody today carries a chromosome slightly different from his or her neighbors. Among Dobzhansky's flies, an ancestral fly carrying the Standard chromosome had given rise to many other chromosomal types. Undoubtedly, the Standard chromosomes in the population today are genetically somewhat different from that ancestral Standard chromosome, but the accumulating genetic differences have not become great enough to be visible under the microscope.

The story was becoming clear. *D. persimilis* is a species that, from many lines of evidence, had a more recent origin than *D.*

pseudoobscura. It has a smaller and more ecologically restricted range, even though it has acquired new abilities that distinguish it from *D. pseudoobscura*—it is able to live farther north and at higher altitudes. It shares the Standard chromosome and no other with *D. pseudoobscura.* Therefore, the common ancestor of the two species must have carried the Standard chromosome. The other chromosome types found among the *D. persimilis* flies, which are unique to that species, must have arisen by inversion events since the time that the two species split, in the same way that changes have happened in our mitochondrial chromosomes since the time of the mitochondrial Eve.

Dobzhansky could also infer that the speciation event itself must have happened somewhere in the region where the two species currently overlap. Farther south, in Mexico and Central America, it is too warm for *D. persimilis,* and in any case the chromosomes of the *pseudoobscura* flies that live in those tropical climes are very different from Standard. Actually, Dobzhansky could infer that the inversions carried by these southern chromosomes were very old, much older than the entire species *D. persimilis,* because he found that a different and much older species had arisen from an ancestor of these inversions. Therefore, *D. persimilis* could not have arisen from flies carrying those ancient southern chromosomes.

The chromosomes of Dobzhansky's flies have preserved vivid traces of the speciation event itself. It seemed likely that there were very few flies involved in the actual splitting process, for otherwise there would be a very good chance that the *D. pseudoobscura* and *D. persimilis* populations today would share more kinds of chromosomes than the Standard one. This suggested that the speciation event may have involved a size bottleneck, in which the population of flies that was destined to become *D. persimilis* was temporarily restricted in numbers and all its chromosomal types except Standard were accidentally lost.

So the Standard chromosome might have been very like the mitochondrial chromosome carried by our mitochondrial Eve. Unfortunately, we do not have giant chromosomes like those of *Drosophila,* and the ancestors of our closest living relatives, the chimpanzees, parted genetic company with our own long before Eve. As a result, we cannot now say whether that bottleneck (if it happened) was part of a speciation process or merely an accidental reduction in our ancestors' numbers. Dobzhansky could make a stronger connection between a size bottleneck in his flies' distant past and the first stages of speciation. What he could not tell, because there was no fossil record, was just how quickly all the many differences between *D.*

pseudoobscura and *D. persimilis* had accumulated after their gene pools had parted company.

Even after the lapse of decades, Dobzhansky's work on the two sibling species *D. pseudoobscura* and *D. persimilis* remains one of the best and most detailed studies of speciation ever carried out. These flies, with their giant chromosomes and their easily seen inversions, were the most ideal material imaginable. He wrote a seminal book on the subject, *Genetics and the Origin of Species,* a book that influenced evolutionary studies for decades afterward. But he was not content with having revealed how speciation worked. He wanted to know more.

By the end of the 1930s the collaboration between Sturtevant and Dobzhansky had come to an end. Sturtevant had become more and more unhappy with Dobzhansky's hurried—even sloppy—experimental approach. It is certainly true that Dobzhansky's experiments, hastily conceived in a first flush of enthusiasm, were often poorly designed. Dobzhansky began with a strong conviction about how the real world should operate, but sometimes the phenomena he was trying to measure were so difficult to detect and the results open to so many different interpretations that his experiments—and there were many of them—stirred up confusion and controversy.

In part this was simply haste—Dobzhansky was always a man in a hurry. But in part it was because he had begun to ask more fundamental questions about the nature of genes themselves and the ways in which they work together. These questions turned out to be far more difficult to answer than the problem of how his various *Drosophila* species had arisen, so difficult that he never obtained results that satisfied the majority of his fellow geneticists. This did not stop him from trying, and indeed he went on performing experiments with the same boundless enthusiasm up until a few days before he died in 1975.

Dobzhansky was convinced by his early work on ladybugs that most genes did not act alone. They acted in groups, and indeed he even went so far as to suppose that every gene affected every character to some degree. All these complex interrelationships must themselves, he strongly believed, be the product of evolution. It was the genotype that evolved, not the individual genes.

His conviction was reinforced by his work on *Drosophila.* The inversions that told him so much about the history of his flies were remarkable genetic events. At the time that they first happened, they had locked together whole blocks of genes on the chromosomes and effectively prevented them from recombining. These blocks, he knew, consisted of hundreds or perhaps thousands of genes, and they persisted for very long times in the fly populations. Some of them, such as the Standard chromosome, predated the genetic split between *D. pseudoobscura* and *D. persimilis,* and some, like the chromosomes of the Mexican and Central American *D. pseudoobscura* populations, seemed to be even older than that.

Such long persistence, he thought, must mean that the genes that happened to be locked together on these chromosomes must work particularly well together. Further, they must interact in interesting ways with the genes carried by other inversions in the population because otherwise an inversion either would be lost or would spread through the whole population and displace all the other different types of chromosomes.

To get some idea of why these fly populations had persisted in their genetically polymorphic state for so long, he made extensive studies of little populations that he set up in the laboratory. When such a population had flies carrying only two kinds of chromosome, he found that both types would persist for many generations, neither one being lost. But when he changed the environmental conditions, or added flies carrying another inversion, the various inversions would rise or fall in numbers in the population over time in unpredictable ways. These complex, environment- and genotype-related effects, he felt, must somehow explain the long persistence of these inversions in natural populations of the flies.

Dobzhansky spent his life studying such natural populations, either observing them in the field or working with flies that had been brought recently into the laboratory. The genes that he brought in from the wild seemed to behave in far less predictable ways than those newly arisen mutant genes, lacking any evolutionary history, that were dealt with by more laboratory-oriented geneticists like Muller. He was quite convinced of this, even though he was never able to prove it directly.

Just as his flies seemed ideally designed to illustrate the process of speciation, Dobzhansky himself seemed ideally designed for the rigors of life as an evolutionary biologist. He loved working in the Sierras of California and was quite immune to the effects of the poison oak that grew everywhere in the sites where he trapped his flies. His forays into the Brazilian rain forest left his companions covered from head to foot with mosquito bites—Dobzhansky himself, however, was always unscathed. He also had a sublime confidence in his own immune system. Once, during a summer I spent with him in the Sierras in 1963, I came back from a trip to Berkeley bearing a treasure trove of overripe bananas that I had found discarded on the docks of the Port of Oakland.

Dobzhansky mashed them up that evening to replenish the bananas in his fly traps, and I was horrified to see him licking his fingers as he worked. "Stop!" I cried. "Don't you remember that I found those bananas in a garbage can in Oakland?" "Yes, very resourceful of you," he replied, continuing to lick his fingers.

Dobzhansky was always interested in the unpredictable, the mysterious, and the extraordinary. He had strong and lifelong religious beliefs and was fascinated by, and very sympathetic toward, Teilhard de Chardin's efforts to link Christianity and evolution. A very active member of the American Philosophical Society, he published a number of papers on philosophy and evolution. His students had no doubt that Dobzhansky was a man of mythic proportions. Unfortunately, he espoused a view of evolutionary biology that also seemed to have a large mythic element—that most of the genetic variation in natural populations has pronounced and complex effects on the organisms carrying it. Myths do not last very long in science.

8

Pulling the Plug

With the ascension of Charles I to the throne we come at last to the Central Period of English History . . . consisting in the utterly memorable struggle between the Cavaliers (Wrong but Wromantic) and the Roundheads (Right but Repulsive).

—W. C. Sellar and R. J. Yeatman, *1066 and All That* (1930)

Imagine that you are watching a film. It is a high-tech production, in glorious color and 3-D, projected onto a giant Cinemascope screen and bathing the audience in stereophonic sound. The film is filled with fascinating people and complicated plot lines, crackling with brilliant dialogue as the characters fall in and out of love, battle, make up, have children, grow old, and die.

Suddenly, as you watch, the film changes. The screen shrinks and becomes two-dimensional. The color disappears, to be replaced by dingy black and white. The sound, too, vanishes. There are just as many people crowding the screen as before, but their tiny images are now almost indistinguishable. Their mouths open and close meaninglessly. If there is any plot motivating the characters milling about on the little screen, it is now impossible for the audience to follow it.

Something as jarring as this transition has happened to Darwinian evolutionists, among them the many followers of Theodosius Dobhzansky, who until a few decades ago had supposed that the great diversity of genotypes carried by the millions of species of organisms on the planet could be explained primarily by the forces of natural selection. Now, a new and very different explanation of the process of evolution has appeared. It demonstrates by cold and incontrovertible mathematics that most evolutionary change is not driven by natural selection at all. Instead, it can be explained more simply, cleanly, and convincingly by random events—mere throws of genetic dice. The innocence of the premathematical Darwinian era has been replaced by a very different world, a world of mathematical models in which selection plays a very much reduced role.

You can imagine the frustration of the Darwinians, who are sure to the depths of their beings that there is something wrong with this new worldview but who are quite unable to prove it. The mathematicians, triumphant, toss the Darwinians an occasional crumb. Yes, they say, of course Darwinian selection plays a role in evolution. They cheerfully admit that a few genes are without a doubt subject to selection, and they usually mention the gene for sickle-cell anemia. This gene, widespread in African populations, protects its heterozygous carriers against the worst effects of malaria. Homozygotes for the gene, however, become dangerously ill. In Africa the heterozygote advantage just balances the homozygote disadvantage, so the gene is retained at high frequencies in the population. So strong are these two kinds of selection acting on the sickle-cell gene, simultaneously increasing the survival of heterozygotes and decreasing the survival of homozygotes, that these selective effects can easily be detected in a variety of ways. Such genes are rare, however, as the mathematicians point out. Most genetic change, particularly when it is viewed at the level of DNA, must simply be the result of alleles being buffeted by the winds of chance.

The Darwinians look at the incredible variety of living organisms in the world around them, organisms that are linked and interrelated by marvelous webs of adaptation, and they wring their hands in despair. Unfortunately, the connections between all this phenotypic variation and the underlying genes that produce it are for the most part quite unknown. Ever since the founding of genetics as a science, Darwinians have speculated freely about the nature of these connections. They have speculated about how many genes might be involved, what they do, and what their evolutionary histories might be. Until recently, there have been very few data to help answer these questions. As a result, all the Darwinians have been able to do, when confronted by the mathematicians' bloodless view of the natural world, is croak, "Where's the beef?"

If things are as simple as the mathematicians say, then evolution must somehow have produced the infinite variety of the living world through changes in a handful of genes that are as obvious in their effects as the gene for sickle-cell anemia. It is, the Darwinians feel, like trying to produce the Sistine Chapel with a bucket of flat white paint and a roller. If evolution is hamstrung in this way, then how can it work at all? But in the hardball world of science, sentimental appeals to presumed complexity are not good enough. The mathematicians have, without mercy, called the Darwinians' bluff.

How has evolutionary theory gotten itself into such an impasse? In fact, this conflict has been present in one form or another since the earliest days of genetics, but only in the last few decades has it been possible to formulate the disagreements in a way that is testable.

Genetic Neutrinos

One summer day in 1953 a slender young student of genetics named Motoo Kimura stepped ashore in Seattle after a two-week voyage from Yokohama. His ship, the Hikawa-Maru, was the only Japanese passenger vessel that had been left afloat at the end of the war. For the first time in his life, young Kimura, eating decent food and playing deck games, had experienced luxury. So pleasant was the voyage that a persistent stomach problem, a legacy of the grim life in postwar Japan, disappeared.

During the previous few years Kimura had been an assistant at the new National Institute of Genetics at Mishima, south of Tokyo. Quite isolated there, with no one else to share his interests, he had begun to study the papers of the population geneticist Sewall Wright, working step by step through Wright's complex mathematics. He was so consumed by his studies that one night Wright (whom he had of course never met) appeared to him in a dream and asked him kindly how many of his papers he had read!

Wright was one of the founders of mathematical population genetics. This apparently arcane subject deals with the way that various alleles of genes change in frequency in populations over time. It can describe with some precision, for example, how a strongly selected allele such as the mutant allele for sickle-cell anemia can increase in frequency in places where malaria is present and fall in frequency in places where it is not. It can also, as Wright and others discovered, make predictions about the behaviors of many other kinds of genes over time. This branch of mathematical biology is very important to the theory of evolution because it can be argued that evolution itself is nothing more than changes in frequency of alleles over time—if there were no such change, then the genotypes of organisms would be immutable and evolution would come to a stop.

When dealing with biology, however, mathematicians are forced to simplify. To begin with, it is obvious that a narrow view of evolution as nothing but the change of allele frequencies over time must be a simplification. Most organisms are complex, made up of many different kinds of cell. They grow and develop and eventually die, and along the way they interact with other organisms and with their environment in many unpredictable ways. The genes that dictate how they develop are numbered in the thousands or in the tens of thousands, and they, too, interact with each other in many different ways. Each generation a few of these genes, among the many millions in the population, alter as a result of mutation, and they can do so in many different ways because many different kinds of mutation can happen. How can a mathematician make sense out of all this complexity?

Since there is as yet no mathematical theory that can handle all these variables, one way is to deal with one gene at a time and not worry too much about the details of how that gene works, which other genes it might interact with, and how it might be affected by various kinds of mutational change. The mathematician might begin the process of simplification by labeling genes with letters, calling a wild-type allele *A*, for example, and a mutant allele at the same genetic locus *a*. Then equations can be constructed that can be used to predict the frequencies of these alleles in the population as they change from one generation to the next, on the assumption that the mutant allele has some selective advantage or disadvantage.

Such equations were first proposed during the early days of population genetics at the beginning of the century. They showed that, if a mutant allele has an advantage, then even if it is initially rare in the population and the advantage is very small, it eventually rises in frequency in the population and displaces the original wild-type allele. Similar equations were derived to describe the properties of heterozygous organisms, those with the genotype *Aa*, that carried one allele from their mother and a different one from their father. If such heterozygous organisms were the most genetically fit in the population, then both the *A* and *a* alleles could coexist in the population indefinitely. Exactly such a mechanism had been found by Dobzhansky to be operating in his *Drosophila* populations—heterozygotes for two different chromosome inversions tended to leave more offspring than homozygotes for either inversion, and as a result both inversions could persist in the population without one displacing the other.

Beginning in the 1920s, Sewall Wright, R. A. Fisher and J. B. S. Haldane began to go beyond these simple equations. Wright, in particular, began to ask what would happen if he abandoned one great simplifying assumption of these equations, the assumption that populations are infinite in size.

Now, such an assumption is obviously not correct. Wright himself worked with guinea pigs, which for financial reasons if nothing else had to be kept in the laboratory in small numbers. He knew that it was not possible even after a few generations to find animals in such tiny populations that were not closely related to one another. The unavoidable inbreeding that resulted was harmful because injurious recessive genes, the very genes that Hermann Muller was so concerned about, lurked in his guinea pig populations as they did in every other population. They were usually masked, or nearly masked, by the wild-type alleles, but they were revealed in their true destructiveness when related parents each carried the same harmful allele and passed it on to their offspring.

Wright constructed clever equations that allowed the amount of inbreeding to be calculated, equations that quickly became essential for plant and animal breeders. Then he began to construct more general equations dealing with inbreeding in small populations. Essentially, his equations followed the fate of imaginary populations in which there were two alleles initially present at many different genetic loci, or places on the chromosome. As one generation succeeded another in these small populations, some members of a given generation would by chance leave many offspring. Others by chance would leave none. As a result, allele frequencies would fluctuate wildly from one generation to the next in small populations. Large populations, by their sheer size, would damp out these fluctuations.

Because the frequencies in one generation depended only on the frequencies in the previous generation and on the vagaries of chance, Wright could treat his alleles in the same way that physicists treat the positions of molecules colliding with each other in a chamber full of gas or liquid. Molecules have no memory of where they used to be two or three collisions ago, and their latest collision is just as likely to batter them in one direction as in another. Similarly, alleles have no memory of their frequencies two or three generations ago, and, unless selection plays a role, they are equally likely to rise or fall in frequency in the next generation.

Even though both the alleles at each of the genetic loci might start at a frequency of 50 percent, as time goes on the frequencies at the various loci form a kind of spreading cloud as some alleles become more common and others become rarer. Eventually, after many generations have gone by, one allele becomes *fixed* in the population at every locus and replaces the other allele.

It is as if the molecules of the physicist's gas, diffusing from the center of the chamber, were to adhere to the chamber's walls once they

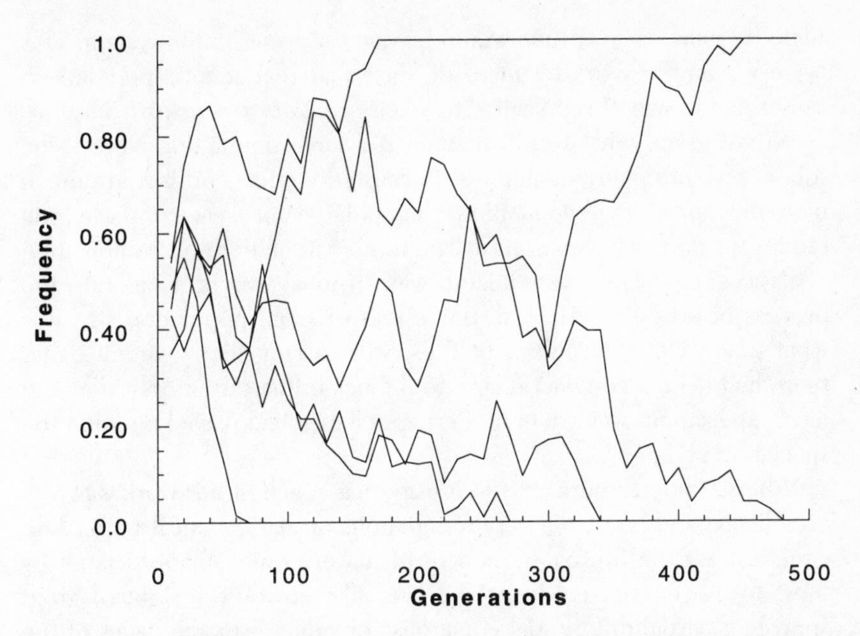

FIGURE 8.1. To show the effect of random genetic drift, I directed the computer to create six populations, each made up of one hundred organisms, and to follow them for five hundred generations of sexual reproduction. At the outset, each population was provided with two alleles, each at equal frequency. Over time, and quite by chance, four of the populations became "fixed" for the first allele and two became fixed for the second. Some of the populations reached fixation very quickly; others took much longer. Eventually, all became fixed for one allele or the other. If the populations had been made larger, it would have taken many more generations for all of them to reach fixation.

bumped into it. Eventually there would be no molecules left floating about in the chamber, although the process of "fixing" the last few molecules might take a very long time.

Wright knew, of course, that selection and other factors that were not due to these chance events could dramatically affect the outcome of this diffusion-like process. Selection acting in various ways on alleles of different loci might increase or decrease the rate at which the cloud of allele frequencies would spread. If new alleles arose by mutation, or were carried in from other populations by migrants, this would also have a great effect, particularly if such alleles appeared in substantial numbers. This would be the equivalent of introducing a new supply of molecules into the physicist's chamber. Even if all these non-random processes were operating, however, at least some of the diffusion of allele frequencies over time would be due to "noise" in the system. This noise was entirely the result of the fact that real genetic populations were finite in size. Wright called this noise *random genetic drift.*

Wright asked what would happen if a large population were to be subdivided into many small ones. Because the effects of drift are much more obvious in small populations than in large ones, each of these little groups would by chance eventually end up with a different combination of fixed genes. This, he realized, was an ideal way to accelerate the process of evolution. Indeed, this effect of small population size was what made Dobzhansky's fruit flies, still carying their evidence that there had been a population size bottleneck in the distant past that had led to speciation, so intriguing. Perhaps chance had played a role in the speciation process.

When Motoo Kimura arrived in America, much of this work was well established and lay at the very foundations of mathematical population genetics. Yet the full consequences of it were quite unappreciated by most of us geneticists, daunted as we all were by the sight of great sprawling partial differential equations spreading over the pages of the journals. Were these equations relevant to the real world? Surely not, we all thought. So many nonrandom things were happening in real populations, especially the highly nonrandom process of selection, that drift must play a subordinate role. It seemed ridiculous to try to describe all the genetic goings on in those messy, pullulating, incredibly complex populations of real organisms by cold and abstract equations.

Kimura, however, did not think it was ridiculous at all. Soon after arriving in the United States, he moved to the University of Wisconsin at Madison. There he became the Ph.D. student of a courtly, viola-playing population geneticist with the unlikely name of Jim Crow.

Crow has a deep understanding of the ways in which genes rise and fall in frequency as populations evolve over time. He posed a deceptively simple problem to Kimura: Suppose we forget about selection, he said, and simply imagine a population in which new alleles are arising all the time by mutation. None of these alleles affect the organisms carrying them in any detectable way, and each of the new mutant alleles can in turn give rise to yet more unique but still selectively neutral alleles. Thus, there is a potentially infinite supply of these neutral alleles, each in turn capable of giving rise to others. Suppose further that this process has been going on for many generations, so that the rate at which these neutral alleles arise now balances the rate at which they are lost or fixed—that is, suppose that the population has reached a steady state. Now, imagine that we could somehow detect the presence of all these neutral alleles and measure their frequencies. What would such a population look like?

Kimura used elegant mathematical methods to explore this situation and found a most surprising result. If a population was large, then even if the mutation rate was low, when it eventually reached Crow's steady-state situation it should be full of neutral alleles! Suppose that one in a hundred thousand genes mutates to a new neutral allele each generation. Then, if a population of one million organisms is allowed to reach Crow's steady state, there should be about forty different alleles, at a variety of frequencies, at *each* genetic locus. So, if this process were going on in the real world, there ought to be a huge number of neutral alleles in large populations, resulting in a great deal of genetic variation that is essentially invisible to the process of natural selection. This variation obeys the law of diffusion, not the law of the jungle. These neutral alleles are the genetic equivalent of the chargeless and massless particles known to physicists as neutrinos, perhaps occasionally detectable by the geneticist but nonetheless invisible to natural selection.

Back in chapter 1, we met just such a steady-state population full of apparently neutral alleles in the form of the population of human mitochondrial DNA molecules. John Avise's model for the history of human mitochondrial DNA, you will remember, assumed that the human population has stayed constant in size for a very long period, since well before the time of the mitochondrial Eve. This has enabled it, he suggested, to reach the equivalent of Crow's steady state for neutral alleles. Over any given period of time, about as many alleles should be lost or fixed as are introduced by mutation.

However, most large populations do not remain large indefinitely—even the immense populations of fruit flies, as we have seen, can go through size bottlenecks. These bottlenecks can cause many neutral alleles to be lost or fixed. When this happens, even if the population were to expand back to its old size immediately, it would have to begin the process of accumulating its standing crop of neutral variants all over again. As Kimura also showed, this can take a very long time. A population that loses all its variation and then expands to a size of one million and stays there takes on average about two million generations to get back to steady state—not an inconsiderable span of time if each generation takes several years!

Wesley Brown's alternative version of the history of human mitochondrial DNA provides us a glimpse of just such a series of events. Suppose that Brown's view of human mitochondrial history is closer to the truth than Avise's and that there was a short but severe bottleneck in numbers back at about the time of the mitochondrial Eve. Then it is quite possible that the mitochondrial chromosomes of the human species as a whole might not yet have reached Crow's steady state. We might still be building up our crop of neutral mitochondrial alleles. After all, if Eve lived two hundred thousand years ago and the average length of a human generation is twenty years, then a mere ten thousand human generations have gone by since then. If our species had a population size larger than twenty thousand* during most of that time, which does not seem unlikely, then we would not have had time to reach steady state. Indeed, as it turns out, more and more evidence is emerging that we are not, at least for our mitochondrial chromosomes, at Avise's (or Kimura's) steady state.

Kimura pointed out another simple relationship that emerged from the dynamics of his neutrino-like neutral alleles. Most neutral alleles are lost rather than fixed because for many generations after they have been introduced by mutation they remain very rare. Only very occasionally does a new neutral allele replace the old allele in the population and become fixed.

Remarkably, if the population remains stable in size, the likelihood of fixation turns out to be the same as the likelihood of mutation. It is easy to understand this simple relationship once it is pointed out, and indeed it can be done through intuition as well as mathematics. Imagine a population of half a million organisms, each (because they are diploid) with two alleles at a particular genetic locus (we are talking here about nuclear genes, not mitochondrial genes). Thus, there are a million alleles at each locus in the population at the present time. Suppose further that the population stays at this size of half a million for many generations into the future. Now, if you could see far enough into the future, you would find that, just by chance, all the alleles in that future population would be descendants of *one* of the million alleles present today. This one allele would be like the chromosome of the mitochondrial Eve. Of course it is impossible to predict by looking at today's population which of those million alleles will be the lucky one that will become fixed in the far distant future, just as a Paleolithic

*Why twenty thousand? Remember that only females count in mitochondrial inheritance and each female passes on only one kind of mitochondrial DNA to the next generation. So there are forty thousand nuclear genes at each locus in such a population, but effectively only ten thousand copies of mitochondrial DNA. Such a population is therefore the equivalent of five thousand dipoloid individuals, and will take about ten thousand generations to get back to Crow's steady state.

geneticist would not have known which of the many hominid females alive at that time would be destined eventually to become the mitochondrial Eve.

Now, suppose that new mutant alleles arise at this genetic locus at the rate of one in a million, so that on average one allele becomes mutant each generation. There is only one chance in a million that this new mutation will occur in the same allele that is destined eventually to become fixed, which is of course the same as the chance that particular allele is the one that will be fixed! With odds like that, you can easily see why most new mutant alleles eventually become lost. On average, a million generations must go by and a million new mutations must arise before one will eventually become fixed.

If each generation is several years long, as it is for us and for many other mammals, then the intervals between such rare fixation events could easily be several million years. When Kimura first pointed out this consequence of his theory, it seemed to be nothing more than an amusing mathematical construct, involving undetectable mutations replacing each other over unimaginable spans of time. This abstract mathematical model was soon, however, to be made very concrete.

At about the same time as Kimura and Crow were working on their model, the Nobel Prize–winning chemist Linus Pauling and his student Emile Zuckerkandl at CalTech were looking at evolution in a very new way. They were gathering together slender but growing pieces of evolutionary evidence, not at the familiar level of changes in the physical appearance of organisms over time, but at the much less familiar level of changes in the molecules that make up these evolving organisms.

The many properties of living cells are largely determined by the proteins that are their most important component. Most cells contain small amounts of tens of thousands of different kinds of proteins, each shaped by evolution to perform a particular task. Proteins in their turn are very large molecules consisting of various combinations of smaller molecules called amino acids. Some twenty different kinds of these amino acids are ubiquitous in the living world, and they exhibit a wide variety of chemical properties. One thing they have in common, however, is their ability to hook together in long chains. The number of possible chains that can be built from these twenty different kinds of amino acids is immense, and evolution has drawn from this essentially infinite set of possibilities to produce the billions of different kinds of proteins made by the ten million or so species of animals and plants on earth.

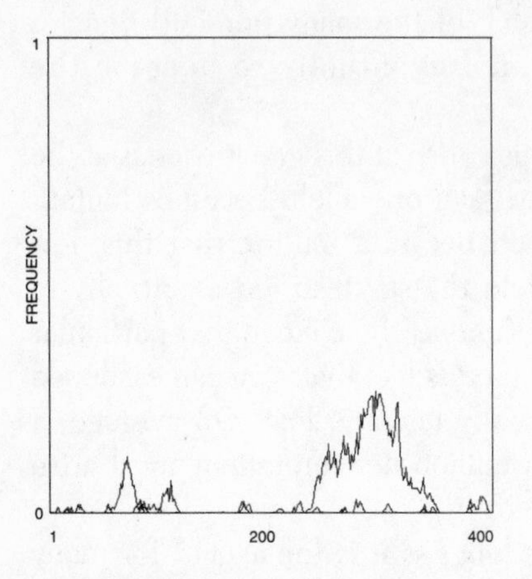

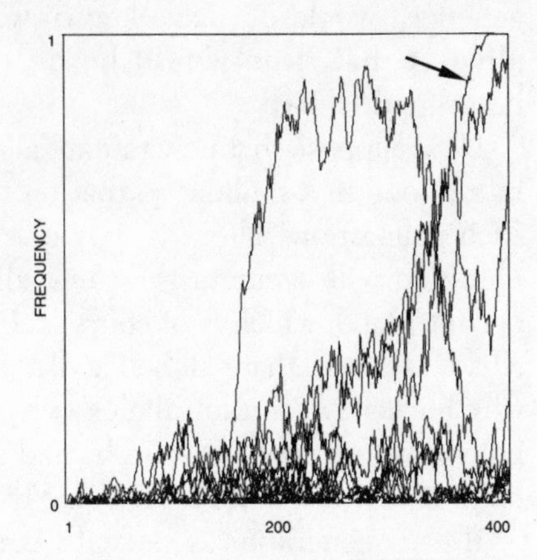

FIGURE 8.2. To show the fate of most selectively neutral mutations, I instructed the computer to follow populations made up of one hundred organisms for four hundred generations and to introduce new mutant alleles periodically. The first part of the figure shows what happens when a new mutation appears on average once every ten generations (an unusually high mutation rate for such a small population). None of the mutant alleles lasts for more than a few generations before falling back, entirely by chance, to a frequency of zero. In the second part of the figure the mutation rate is increased, to one per generation. Now, one of these four hundred mutations (marked with an arrow) has succeeded in persisting long enough that it eventually rises to a frequency of one and replaces all the other alleles at that locus. It has, in the jargon of population genetics, become fixed. Another allele almost makes it, but, buffeted entirely by chance, it falls back. Again, the overwhelming majority of the new mutant alleles are lost. Fixation of selectively neutral alleles is a rare event, but given enough time it can happen even in a large population.

Biochemists had already painfully worked out the amino acid sequences of a few proteins isolated from various species of animals. Zuckerkandl and Pauling began to wonder whether these sequences might be related to each other, just as the animals from which they came were related.

One of the easiest proteins to obtain in quantity is hemoglobin—this protein makes up much of a mature red blood cell, and it was one of the first to be sequenced. Hemoglobin molecules consist of four subunits, each about 150 amino acids long. Zuckerkandl and Pauling compared the amino acid sequences from different animals by lining them up next to each other, as in figure 8.3. It was immediately apparent that they were all very similar to each other. Although the same amino acid is usually found at the same place in all the chains they looked at, there

HUMAN

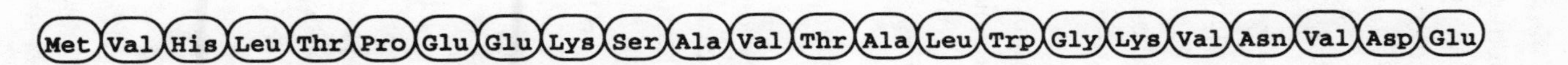

SHEEP

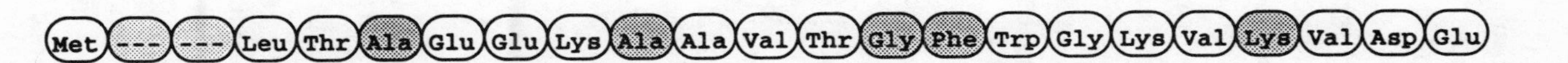

OPOSSUM

FIGURE 8.3. The amino acid sequences of the first part of the beta-hemoglobin genes of three animals—humans, sheep, and opossums. Amino acids that differ from the human sequence are shaded, and you can see that there are more differences between human and opossum than between human and sheep—this is what would be predicted from the fact that opossums are marsupials, in a completely different group of mammals from humans and sheep. You can also see that, at some time along the lineage leading to the sheep, two amino acids have been lost.

were many places where different amino acids were found in different animals—one amino acid must have been substituted for another in the course of evolution. There were also places where the chains did not match properly, but Zuckerkandl and Pauling found that they could be matched if little gaps were introduced in one or the other chain. Not only could one amino acid be substituted for another in the course of evolution, but sometimes amino acids could be lost or gained.

There was a remarkable correlation between the number of amino acid differences and the *taxonomic distance* of the animals being compared. Taxonomists are biologists who worry about the naming and the evolutionary relationships among animals and plants. Although they disagree on many things, they are in universal agreement that chimpanzees are more closely related to humans than sheep, and sheep in turn are more closely related to humans than opossums. Zuckerkandl and Pauling found that these taxonomic differences were paralleled with remarkable exactitude by differences in the proteins. Almost always, the greater the morphological differences between two species, the greater the differences between their hemoglobins.

Zuckerkandl and Pauling realized that if enough such information about proteins could be collected, it would provide a far more detailed and unequivocal view of evolution than could be gleaned by looking at the shapes, sizes, and behaviors of animals or plants. Such gross taxonomic characters are subject to the vagaries of selection and adaptation, and often give no clear idea of how much evolutionary time has separated different species. Proteins, however, can be used to build an internally consistent family tree, such as the one shown in figure 8.4. Some of the branch points of the tree can be dated, at least approximately, from the known fossil record. The ancestors of humans and horses, for example, probably parted company from each other fewer than sixty-five million years ago, during the mammalian evolutionary explosion following the demise of the dinosaurs. You will recall that it was this well-defined mammalian radiation that was later used by Sarich and Wilson as a yardstick to date the much more recent split between the ancestors of humans and chimpanzees.

Given this and other dates from the fossil record, Zuckerkandl and Pauling calculated that the rate of molecular evolution was very slow indeed. In the hemoglobins, on average, one amino acid was substituted for another every seven million years. They also observed that this rate of substitution seemed to have been remarkably constant over hundreds of millions of years.

Organisms evolve by fits and starts. Occasionally they gain complex

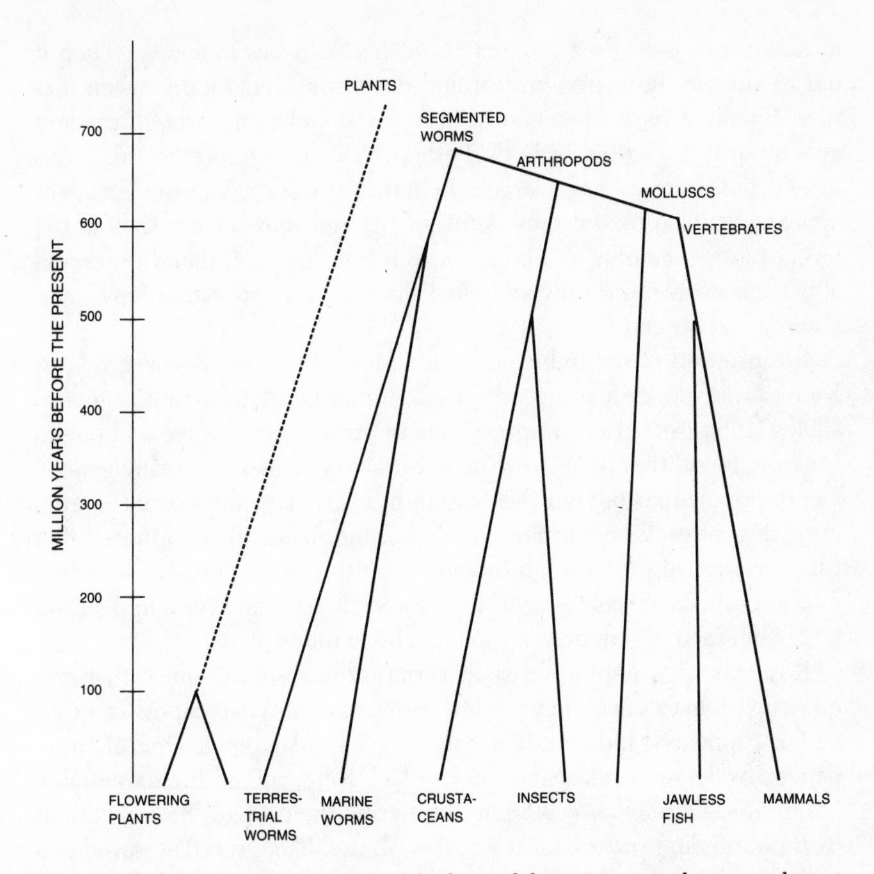

FIGURE 8.4. A simplified family tree of the hemoglobins. Surprisingly, many plants, too, have hemoglobins, although these molecules function very differently in plants and animals. They resemble animal hemoglobins so little that it is impossible to estimate when their common ancestor lived. Notice that the common ancestors of the segmented worms, arthropods, mollusks, and vertebrates are all shown as living between about six hundred and five hundred million years ago, but many educated guesses went into making this estimate. The paleontological data suggest, however, that the ancestors of these diverse groups of animals did branch off from each other over a relatively short time period.

new capabilities, such as the ability to live on land, to bear live young, to control their internal temperatures, or to fly. Although these obvious phenotypic changes take place at unpredictable times and at unpredictable speeds, it seems that proteins, sequestered inside cells, evolve in a far more sedate fashion. They alter according to the dictates of a slow, calm, and regular *molecular clock*.

This was exactly what Kimura's population model predicted. The process of substituting one amino acid for another in a protein is not an instantaneous one. First a mutation in the DNA has to appear. Then it has to survive those first critical and dangerous generations when it is rare. Finally it has to become fixed in the population, through random genetic drift, selection, or both. Perhaps, if various unpredictable kinds of selection were to play a large role in this process, then proteins might appear to evolve by the same kinds of fits and starts as are seen at the level of gross anatomy and behavior. But if it does not, then they would appear to evolve in a slow and clocklike way—just as Zuckerkandl and Pauling had observed.

Kimura and Crow published their paper in 1964. Two years later Kimura was suddenly provided with even more evidence that his neutral alleles might exist. Two groups of geneticists, one in Chicago and one in London, found that there were large numbers of alleles at many genetic loci in populations of fruit flies and of humans. This discovery caused a great deal of excitement, particularly because it was first supposed that this variation must be maintained in these populations by selective forces. Perhaps, it was thought, the source of Dobzhansky's complex and highly interactive genetic variation had been found at last.

Kimura's arguments that such variation was simply noise that was quite invisible to selection were little noticed—until two papers by other authors appeared independently making a similar point. One of these papers, by Thomas Jukes and the late Jack King, coined the memorable phrase *non-Darwinian evolution* to describe the changes brought about by these random processes. It was this phrase, like a red flag waved at a bull, that finally caught the attention of the community of evolutionists and population geneticists.

It was not long before two warring camps formed, the neutralists and the selectionists. Everybody joined one side or the other, like the Roundheads and the Cavaliers in the English Civil War. And, just as in that war, innumerable inconclusive skirmishes were fought.

Kimura soon became the leader of the neutralists. He gathered a group of students and collaborators who shared his growing conviction that neutral alleles made up the bulk of genetic variation. He also led the devastating counterattacks whenever the selectionists tried to demonstrate that the pattern of variation revealed by one technique or another could be explained by selection.

Some of the counterattacks were easy. Many papers were written suggesting that the distribution of allele frequencies in natural populations showed evidence of being shaped by selection—that there were, for example, too many common alleles and too few rare ones or that there were suspiciously regular changes in allele frequency from one part of the geographical distribution of a species to another. Kimura and his collaborators demonstrated that given the proper choice of mutation rates, population sizes, and population history, it was possible to show how virtually any distribution of allele frequencies could have arisen entirely as a result of chance. In those cases that were impossible to explain except by selection, they pointed out that the genes the experimenters were observing were probably not the genes that were actually being selected. All that was needed to give such effects was a few strongly selected genes here and there on the chromosomes. As these strongly selected genes rose or fell in frequency, they would tow in their wake many neutral alleles of other genes that simply happened to be nearby on the chromosomes.

Other counterattacks were more devious. Much effort was expended by the selectionists to show that the molecular clock was not as constant as Zuckerkandl and Pauling had originally suggested. All this was for naught, for neutral theory could easily encompass such irregularities—all that was required was fluctuations in population size that would produce waves of fixation or, if this was not sufficient, changes in the generation time or the rate of mutation to neutral alleles. These changes could slow down or speed up the rate of fixation as much as was needed to explain the data.

At times the debate veered toward paranoia. Tom Blundell, an X-ray crystallographer at the University of London, published some studies in 1975 of the structure of the small protein molecule insulin. He was able to show in detail what had happened during the evolution of this molecule. In primitive fish, each insulin molecule is a separate protein chain. The insulin of the more advanced bony fish, reptiles, and mammals is a little cluster of six chains. Blundell showed that accompanying this change there had also been a burst of evolution among precisely those amino acids that were involved in joining these molecules together. This unexceptionable work, demonstrating that selection had acted on the insulin molecule, was intensely criticized by Kimura at meetings. Blundell told me that Kimura's path had crossed his several times during a lecture tour following publication of his paper and that he had begun to feel that Kimura was pursuing him around the world like some Fury from a Greek legend.

There were many reasons why the debates were so acrimonious. Experimenters who had spent years trying to demonstrate unequivocally that selection was operating on some gene or other were furious at neutralists who simply sat at their desks and demolished their experiments. Another problem, I think, was the tone adopted by the neutralists, which triggered a similar response from the selectionists. Most people in the neutralist camp are Japanese, and a certain note of severity is an accepted part of the way a Japanese senior scientist addresses a junior colleague. This severity did not sit well when it was translated into English at a meeting or in a paper. Hackles were unnecessarily raised.

Like some recently crusted-over volcano, the debate has largely died down, in part through exhaustion on both sides, in part because unequivocal evidence has now emerged that Kimura is right. There are indeed many neutral alleles in natural populations. Now that his point has been proved, his combativeness—and that of his associates—has lessened. The clincher came from the study of DNA sequences. When the DNA that codes for proteins began to be examined, it turned out that neutral or nearly neutral mutations were everywhere.

Each succeeding set of three bases along a gene, called a *codon,* specifies a particular amino acid. There is, however, a disparity between the number of codons in the code and the number of amino acids that the code can specify. The four bases found in DNA, symbolized by the letters A, T, G, and C, can each occupy any of the three positions in a codon. As a result there are four times four times four or sixty-four different ways of arranging these bases in groups of three. But there are only twenty amino acids that need to be specified by these sixty-four codons. This means that the code is a redundant one. Various amino acids may be specified by one, two, three, four, or even six different codons. The amino acid valine, for example, can be specified by the codons CAA, CAG, CAT, or CAC. And the amino acid glycine can be specified by CCA, CCG, CCT, or CCC.

You will note that in these examples the first two bases are always the same, and only the third varies. Most of the time this is true, with the redundancy being found in the third base. So, for these and most other amino acids, it does not very much matter which base is in the third position of the codon that specifies it, for the right amino acid is still inserted into the protein chain as it grows inside the cell.

This situation provides an ideal test for the neutral theory. If, in the course of evolution, one base is substituted for another in one of the first two positions of a codon, this usually results in the substitution of one amino acid for another in the protein. Because such a substitution changes the makeup of the protein, it is likely to have some effect, usually a harmful one, on the organism that carries the mutant gene. If the substitution occurs in the third base of the codon, however, it usually does not change the protein, so it is likely to have little if any effect on the organism's survival.

By the 1970s, methods for *cloning* genes, by growing many copies of a single piece of DNA taken from any organism, had become widely available. At the end of the decade, rapid methods were found to sequence this DNA, so it became possible to sequence many different cloned genes. These methods turned out to be very easy, giving clean and unambiguous results. As a result, it is now actually easier and more accurate to sequence the gene itself than it is to sequence the protein that the gene codes for.

When the genes for the hemoglobins in figure 8.3 were sequenced, striking patterns were revealed. The DNA sequences for the pieces of hemoglobin gene from figure 8.3 are shown in figure 8.5, with little arrows marking the places where differences have accumulated between human and sheep or between human and opossum. Many, though not all, of these differences are found in the third base positions of the codons. This is exactly what Kimura predicted—the rate of evolution of these third-position bases, subject to little if any selection, turns out to be faster than the rate of evolution of the other more meaningful bases in the genes.

By far the fastest rate of evolution has been found in genes that are completely released from the constraints of natural selection. In 1979 and 1980 an astonishing discovery was made by several different laboratories almost simultaneously. A careful analysis of the DNA sequences of several hemoglobin genes showed that some of these genes had been disabled and were no longer functional. The initial disabling event had been a fairly minor change in the DNA, which prevented the genes, soon named *pseudogenes,* from being read properly. Many different kinds of pseudogenes have since been found, like the hulks of wrecked ships, scattered around the chromosomes of humans and other organisms.

Pseudogenes are completely released from the pressures of natural selection. When the sequences of pseudogenes from different species are compared, it is possible to compare what has happened to them since the time at which they became disabled with the rate of evolutionary change in similar but functioning genes that are carried by the same species. As Kimura's theory predicts, pseudogenes evolve at very high speeds. In fact, they evolve two or three times as quickly as even the third-position bases of similar but functioning genes. In the pseudogenes, unlike functioning genes, the changes are not concentrated in the third positions but are scattered throughout their length. Because these genes no longer carry information, their rate of evolution must actually be governed by the rate of mutation itself. Even in pseudogenes, most of the mutations that appear in a population are

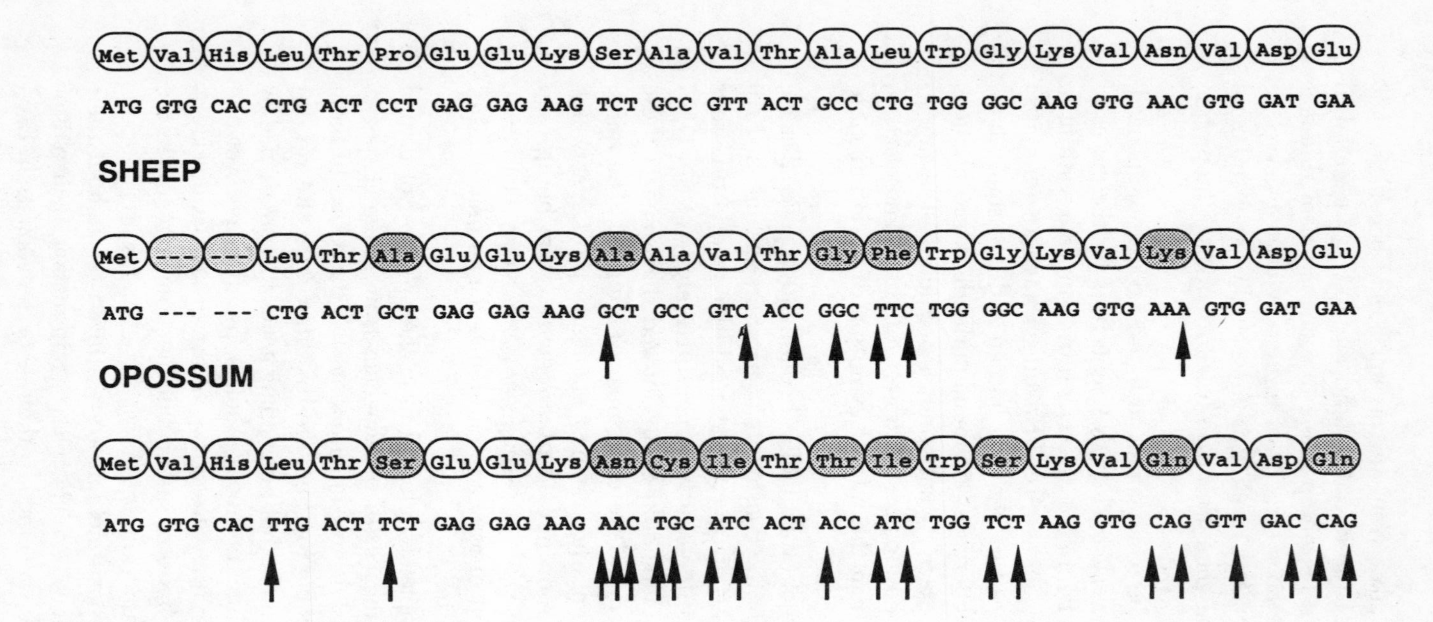

FIGURE 8.5. The DNA sequences that code for the pieces of the beta-hemoglobin protein that were shown in figure 8.3. Arrows show where the bases are different from the human DNA sequence, and again there are more differences between humans and opossums than between humans and sheep. Sometimes there has been a base change, particularly in the third position, that does not result in an amino acid change. Changes like these are, in Kimura's view, most likely to be invisible to natural selection and therefore selectively neutral. The conflict with the selectionists arose when Kimura claimed that many of the other DNA changes are also selectively neutral, even though they result in amino acid changes.

soon lost by chance, but all the mutations that arise, regardless of which bases are affected, have an equal chance of eventually becoming fixed.

Kimura's view of evolution has thus been amply confirmed. Random events are responsible for much of evolutionary change. Although it turns out that selection does play a role in these changes, it is largely a passive one, like that of a gardener who weeds but does not plant. The activity of this selective gardener can be detected from the fact that even third-position bases in functioning genes evolve more slowly than any of the bases do in pseudogenes. This suggests that new mutations in these third-position bases are lost more readily than if they were selectively neutral, which in turn implies that they are being selected against—weeded out, in effect. So it seems that even third-position bases of functioning genes must have some slight importance to the organism, because changing them appears to be disadvantageous. The selection acting on such bases seems always to be conservative, always preventing change.

Such neutral or nearly neutral mutations are suspected to be responsible for most of the changes in human mitochondrial DNA, which is why the mitochondrial clock seems to be so regular. Yet, as we saw earlier with the mitochondrial DNA, even the most clocklike evolutionary change is sometimes difficult to interpret if different parts of the chromosome evolve at different speeds.

At this point a faint murmur arises from the cowed Darwinians. Kimura's victory, they say, is a pyrrhic one. In the process of conquering the selectionists, he has swept the field clear of all the interesting genes. There must be more to evolution than mutation and random drift.

There is hope for the selectionists, for even after the neutralists have done their worst there are still plenty of interesting genes. There are about three billion DNA bases in both the human and the chimpanzee genomes. Humans and chimpanzees are now known to differ in about 1 percent of these bases. Although this does not seem like much, it represents a great many genetic changes that have taken place in the seven million years or so since our common ancestor. One percent of three billion is thirty million, so there are some thirty million base differences between ourselves and our closest living relatives. Although neither humans nor chimpanzees have gained or lost much total DNA in the course of their separate evolutionary careers, a lot of evolutionary changes have accumulated. A few of these are modestly large differences, but most of them consist of single-base substitutions and little insertions and deletions that are scattered throughout the genes—sometimes, indeed, quite a few are found in a single gene. These differences are also scattered throughout the 90 percent or so of our DNA that does not consist of genes.

How many of these tiny changes are meaningful—how many of them really contribute to the physical and mental differences between ourselves and chimpanzees? Dobzhansky would have suggested that all of them do. If nothing else, Kimura has at least simplified the picture—we do not have to track down, account for, and understand the selective reasons behind thirty million differences! Of course, we do not yet know the fraction of these differences that are neutral, but let us suppose that it is 99 percent. Then the changes that have really mattered in the course of the separate evolution of humans and chimpanzees, the changes over the last seven million years that have led among other things to our runaway brain, might number a "mere" three hundred thousand or so!

Even this reduced number is quite enough to keep selectionists busy for a very long period of time. So, in spite of the acrimonious debates of the past, it turns out that there is plenty of room for both selectionists and neutralists under the big tent of population genetics. Now we are starting to get some hints about what some of these interesting gene differences are actually like. Some of them are as richly complicated as anything Dobzhansky might have imagined. It has not yet been possible to investigate such genes in humans as thoroughly as they have been investigated in other organisms, but to give you the flavor of the enterprise I will start with a truly remarkable gene found in the fruit fly.

9

The Love Song of the Fruit Fly and Other Amazing Gene Stories

Love took up the glass of time, and turn'd it in his glowing hands;
Every moment, lightly shaken, ran itself in golden sands.

—Alfred, Lord Tennyson, "Locksley Hall" (1842)

In 1979 Charalambos Kyriacou, a postdoc in Jeffrey Hall's laboratory at Brandeis University, was trying with mounting frustration to measure the love song of the fruit fly.

In many species of *Drosophila,* a male gains the attention of the female by repeatedly approaching her from the side or behind and furiously vibrating one wing. If she does not fly away, he moves on to more ardent stages of courtship, which we will not go into because this is a PG-rated book. In any case, it is the wing vibrations that are most easily measured by scientists interested in this courtship ritual, and these vibrations turn out to have several components. The most obvious is the frequency of the beat, which a sensitive microphone picks up as a series of clicklike pulses. Then there is the length of time that the fly vibrates its wings, which can be long or short depending on the species. Finally, there is the space between the clusters of clicks.

Kyriacou was trying to affect the flies' song by making flies that had partly normal and partly genetically defective brains. In order to understand what had happened to his mutant flies, he had to measure the songs of flies that had not been manipulated. Despite repeated attempts, he could never get a consistent value for the space between the clusters of clicks. Sometimes the spacing was short and sometimes long, and there was no obvious pattern.

At last, in desperation, he sat down with dozens of yards of chart recordings and began to trace through them from one end to the other. Finally a pattern emerged. During the courtship the spaces between the clusters of clicks would gradually lengthen and shorten in a cyclic fashion. One complete cycle, from short intervals to long and back, took about a minute in *D. melanogaster*, and about forty seconds in its sibling species *D. simulans*. So it was the cycle of *intervals* between the clusters of clicks, from frantic to leisurely and back again, that characterized the song, not just the average space between clusters. He could show that the flies could tell the difference. When these songs were reproduced electronically, the long-cycle song stimulated *D.melanogaster* females, and the short-cycle song stimulated *D.simulans* females.

This little nugget of information might simply have been added to the growing lore about *Drosophila* songs, surely one of the more obscure byways of science, were it not for a very odd fact. Kyriacou measured the length of the cycle in flies kept at different temperatures, and found to his surprise that it was always the same whether the temperature was high or low.

Insensitivity to temperature happens to be an important characteristic of animal and plant circadian ("about a day") rhythms. These rhythms govern daytime and nighttime activities, and their temperature insensitivity ensures that the peak of activity of cold-blooded animals happens at the same time of the day regardless of the season. Yet the song cycle of the lovesick fruit fly was hardly a circadian rhythm. Its length was a minute or less, instead of twenty-four hours.

Both Hall and Kyriacou began to follow up on these observations. Crosses between *D. melanogaster* and *D. simulans* produce flies that survive, but both sexes are infertile (these flies are further along in the process of speciation than *D. pseudoobscura* and *D. persimilis*). By examining these hybrids, they were able to narrow down the location of the gene that controlled the differences in the cycle length between the two species. The gene turned out to be somewhere on the X chromosome, but, because their hybrids could produce no offspring, they could go no further. Here again the story might have been abandoned, were it not for a lucky guess on Hall's part.

There is a gene on the X chromosome called the *period gene*, which affects the fly's circadian rhythm. It has been found to regulate both the time of day at which the fly emerges from its pupal case and its daily activity cycle once it becomes an adult. Earlier, Hall's co-worker Ron Konopka had found mutations of this gene that lengthen or shorten these circadian rhythms to more or less than twenty-four hours, and other mutations have been found that abolish them completely. When Hall and Kyriacou measured the songs of flies carrying these period mutations, they found that they, too, were affected. The effect on the song cycle was a miniature reflection of the way the period gene affected the fly's much longer daily cycle—short daily activity cycle mutants had short song cycles, long daily activity cycle mutants had long song cycles, and aperiodic mutants had no obvious song cycle at all.

The connection between the period gene and the fly's song cycle was clinched in 1991, when Hall's group reported on a massive series of experiments that had taken them nine years to do. Using specially designed viruslike pieces of DNA, they inserted the period gene from *D. simulans* into aperiodic *D. melanogaster* mutants and found that the gene produced a *simulans*-like cycle. Then, using the delicate tools of molecular biology, they went even further and snipped the *melanogaster* and *simulans* period genes apart to make hybrid genes that were part *melanogaster* and part *simulans.*

The hybrid genes told a fascinating story. The molecular dissection showed that only a small part of the middle of the gene was responsible for the difference between the two species—if that piece from the *simulans* gene was inserted into the middle of the *melanogaster* gene, this was enough to produce a *simulans*-like cycle.

In this short stretch of gene there are only four differences between the two species. At these four places, the *melanogaster* gene specifies one amino acid and the *simulans* gene another. They were able to say with some confidence that at least one of these differences must be responsible for the different lengths of the song cycle of the two species. Kyriacou is now painstakingly dissecting the gene even further to see whether only one of the four differences is enough to produce the shift from one species' song to the other or whether some combination of two or more is required.

This remarkable story is as close as biologists have yet come to really understanding the function of an actual genetic difference between two closely related species. The difference happens to be an important one, for it dictates a behavioral alteration that helps prevent flies of the two species from mating in the wild. Note, however, how tiny this genetic distinction is. It is due to one, or at the most a handful of amino acid differences between genes that are otherwise very similar in the two species. No new gene had to arise in order to make *D. simulans* behave differently from *D. melanogaster,* or vice versa. A few amino acid substitutions, a small tweaking of the function of the gene, were enough.

How does the gene carry out this function? You will recall that the period gene influences not only the male fly's song cycle but also the time of day the adult fly emerges from its pupal case and the time of day of the fly's peak activity. It does so by influencing a clock that is buried deep in the processes of the cell. This clock is somehow immune to the usual laws of chemistry because, unlike an ordinary chemical reaction, it does not speed up or slow down as the temperature changes.

It is helpful to draw an analogy between this remarkable gene and a watch, although we must be careful not to take the analogy too far. In an old-fashioned mechanical watch, the escapement mechanism is driven by a torsion pendulum, in the form of a balance wheel. This wheel, if left to its own devices, would spin freely, but it is forced to oscillate back and forth because it is constrained by a delicate hairspring. With each oscillation, a lever attached to the balance wheel releases one tooth of a nearby escapement wheel and allows that wheel to move slightly before its next tooth catches. This in turn allows the gears that drive the watch's hands to inch forward. The jerk of the escapement wheel, which is under great tension from the mainspring, then gives the balance wheel a tiny push that keeps it oscillating.

When you adjust such a watch, using the little lever that can be seen when you take the back off, you are actually adjusting the tension on the hairspring. This makes the balance wheel oscillate more or less quickly. The changes that Hall and his co-workers detected in the period gene are like adjustments to the hairspring of a watch, adjustments that leave all the rest of the watch's elaborate mechanism untouched.

The genetic equivalent of the hairspring itself may actually have been found. Not far away from these four important changes is a part of the gene that codes for a very monotonous structure—a sequence of two amino acids, threonine and glycine, which is repeated over and over in a kind of molecular stutter. In both species there are a number of different alleles, each of which have different numbers of repeats. In *D. melanogaster,* the commonest of these alleles have seventeen, twenty, or twenty-three of these repeats.

Why should changes in the length of the gene produce this remarkable effect? Clever experiments showed that if the period gene is turned on during the fly's development but switched off in the adult, it has no effect on circadian rhythms. In contrast, if it is turned off during development and then turned on in the adult, it can govern them properly. Cell division has essentially stopped in the adult fly, so the period gene is not acting to influence some developmental process. Instead, the protein made by the period gene is apparently acting as some kind of molecular oscillator, an escapement mechanism. Hall's group, working with that of Michael Rosbash at Brandeis, constructed

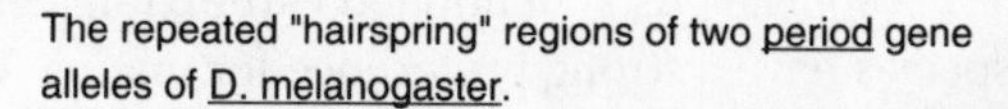

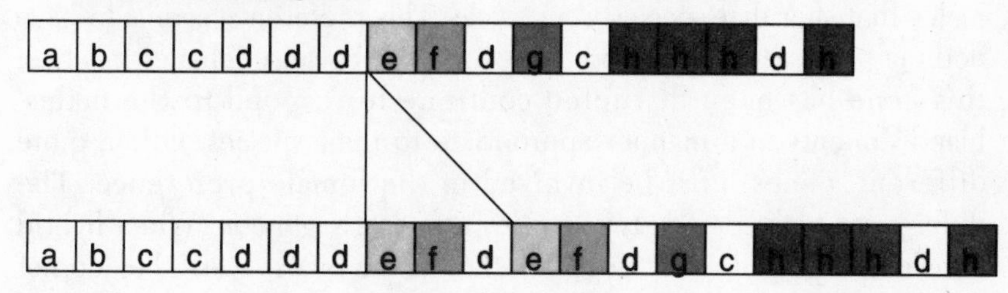

FIGURE 9.1. Adjusting the period gene hairspring. The figure shows two alleles of the period gene in *Drosophila melanogaster*, a shorter one that predominates in southern populations and a longer one that predominates in northern populations. Each of the eight kinds of lettered square represents a slightly different sequence of six bases of DNA—for example, *a* represents ACGGGC and *d* represents ACTGGA. In spite of these differences, all these eight sequences code for a threonine followed by a glycine. You can see that the difference in length between the two alleles is due to the duplication of a region *efd* that is found only once in the shorter allele and twice in the longer one. Many similar slight adjustments of the hairspring were found by Kyriacou and his co-workers when they looked at alleles belonging to related species.

mutant alleles in which the entire repeat region had been removed. Although the flies still had a circadian rhythm, the rhythm now showed some temperature dependence! The hairspring regulating the entire escapement mechanism was apparently disabled by this genetic alteration.

If the threonine-glycine repeat region is really acting as a kind of molecular hairspring, then it should be possible to detect adjustments in the hairspring itself, adjustments that would be directly analogous to increases and decreases in the tension on the hairspring of a watch. Exactly such adjustments have been found. In populations of *D. melanogaster* living in North Africa, the commonest alleles are those with a relatively short stretch of threonine-glycine repeats. In the more frigid clime of northern Europe, alleles with a larger number of repeats predominate. The colder the climate, the longer the repeated segment of the gene. These little repeats seem to form a necessary part of the escapement mechanism of this biological clock—they can regulate it even over temperature extremes. Hall and his co-workers are finding that deletions of parts of this region make the protein more likely to be destroyed by high temperatures, so perhaps it is the amount of protein that is important to the regulation of the fly's rhythms.

Most of these genetic differences affect the male's song. What about the genetics of the female's response? As I mentioned earlier, females of a particular *Drosophila* species have a strong preference for mating with males that sing that species' song cycle. This preference seems to have nothing to do with the period gene, however, because females in which this gene has been disrupted continue to respond to the males' blandishments in a manner appropriate to their species. Other, quite different, genes must be involved in the female preference. The differences between the two species are clearly genetic. When hybrid females are played male songs of either of the two species, they do not respond—but they do respond when an electronically generated hybrid song, an average of the two, is played.

A number of genes are likely to be involved in this female response, and it is certain to be many years before these genes are tracked down. They will be much more difficult to understand than the period gene, for they control a preference rather than an overt activity. They may prove as hard to find and to understand as those human genes, whatever they are, that contribute to a preference for Mozart over the Who.

Potential-Altering and Potential-Realizing Mutations

The period gene is a vivid example of a kind of gene that I wrote about a few years ago in a book called *The Wisdom of the Genes.* Much of the story of the period gene has emerged since I wrote that book, which is too bad because it is certainly the most vivid example of this genetic wisdom that I have yet come across. Genes are wise, I suggested, because most of them have had a long evolutionary history. As a result they have, over time, actually become good at evolving.

Such a statement is enough to make any evolutionist stiffen in horror, because it implies that genes can somehow anticipate future environmental changes and be ready to evolve accordingly. This is not what I meant, and I spent a good deal of time in the book trying to explain as clearly as possible what I *did* mean. It is a shame that I did not have the period gene to use as an example, for then I would have had no trouble convincing my readers.

We do not yet know the full evolutionary history of the period gene, but it is safe to assume that at some point in the past it controlled circadian rhythms in a less temperature-insensitive fashion than it does now. This might not have mattered in a world with few temperature extremes, and indeed most of the history of our planet has been marked by far more uniform temperatures from pole to equator than at present.

During less equable periods, however, insensitivity to temperature extremes might have had an advantage. It seems that the temperature insensitivity was produced by the insertion of a short stretch of DNA into the period gene, a stretch that specified a repeated run of alternating amino acids in the protein. (The original mutation may or may not have carried a run of threonine-glycine repeats—the species *Drosophila pseudoobscura* has a series of very different repeats inserted into the same part of its period gene.)

This insertion was what I have called a *potential-altering* mutation. It changed the properties of the organisms carrying it, as do many other mutations, but with a difference. This insertion, although it did largely divorce the rhythm from temperature, did not confer complete temperature independence. It needed to be "fine-tuned" to take account of local temperature extremes. This fine-tuning has happened since the potential-altering mutation first appeared, and has taken the form of subsequent mutations that have lengthened or shortened the number of repeats. This has enabled various members of the same species to keep their circadian rhythms constant even though they may happen to live in warmer or colder regions.

Now, it happens that stretches of DNA that have a monotonous repeated structure are very prone to exactly this kind of mutational change. Lengthenings or shortenings do not take place each generation, but they happen often enough that there is always, in the gene pool of the species, a supply of mutant alleles that are a little longer or shorter than the average. Such mutations are examples of what I have called *potential-realizing* mutations, because they realize the full evolutionary potential of the original mutation.

It is very likely that, during the course of its long evolutionary history, many other kinds of mutations happened to the period gene that also conferred temperature independence on its carrier's circadian rhythms. If those mutations could not easily be fine-tuned the way the repeated sequence can, however, they would not have been selected for as strongly. Over the long term, a mutation that has the capacity to give rise to subsequent potential-realizing mutations has an advantage over one that does not.

The wisdom of the genes is a very limited wisdom and can appear in the course of evolution only if organisms are faced repeatedly with oscillating environmental changes to which they must continually adapt. The ability to keep circadian rhythms steady as the ambient temperature changes from warm to cold and back again is one such situation, and the result has been the remarkably flexible period gene.

Do we know of any such situations in our own species? Certainly not in the detail that *Drosophila* geneticists have been able to provide, but genetic stories are now emerging that will, I suspect, turn out to be very similar.

Differences in skin color carry a freight of political and social problems in human societies. The darker a person's skin, independent of that person's other properties, the more negative the attitude of the rest of society tends to be. The evolutionary reason for these differences in skin color among humans is not yet clear. One popular theory suggests that before the days of vitamin D supplements, we absolutely depended on sunlight for the production of this vitamin, essential to proper bone formation. The cells of our skins, like the cells of the rest of our bodies, contain a cholesterol-like compound that can be converted to vitamin D. In order for this to happen, one of the six-carbon rings in this compound must be broken in our exposed skin cells through the action of ultraviolet light. Once this ring is broken, and only then, the compound enters our circulation and enzymes in our liver and kidneys can complete the conversion to active vitamin D. Even though they make large quantities of the precursor compound, children deprived of ultraviolet light develop abnormally, with the bowed legs and deformed chests characteristic of rickets.

On little evidence, it was assumed until recently that large amounts of pigment in the skin would slow this conversion. This would not matter in tropical regions where the ultraviolet flux is strong, but it would have had an effect in the cloudy north. The invention of clothing should also have increased the intensity of selection on northern peoples, giving a strong advantage to individuals with less skin pigment.

Unfortunately for this theory, recent studies have shown that people with heavily pigmented skins make about as much vitamin D precursor as people with little pigment. So, did northern Europeans lose their skin pigment accidentally, or are there subtle selective pressures that we have not yet discovered that dictated its loss?

Whether by selection or chance, how could these genetic changes have come about? There is some anthropological evidence to indicate that dark skin color has been lost and gained more than once in the course of the evolution of the various human races, evidence suggesting that it is fairly easy for organisms like ourselves to switch from dark to light skin and back again. Perhaps there are a small number of potential-altering genes, with properties like those of the *Drosophila* period gene, that can easily give rise to potential-realizing mutations adjusting the amount of pigment in the skin. Alternatively, there could be several separate genes, each of which has an allele that produces a small amount of pigment and an alternative allele that does not. Then the amount of pigment an individual has would depend on the number of "black" and "white" alleles he or she possesses. Either way, more or less pigment could easily be selected for, but if our skin color is controlled simply by a collection of black and white genes there is not much evolutionary subtlety in such a system. Which of these models is more nearly correct?

The German-born American geneticist Curt Stern, working in the 1950s, was among the first to try to determine the number of genes that might be involved in skin color. He began with the second and less complex of the two models, which assumes that there is a set of genes, each with black and white alleles, regulating the amount of pigment. He used skin color information from a group of African-Americans, many of whom had some white ancestors. He found that most of the group had quite dark skins but that, when the data were graphed, there was a pronounced tail on the distribution in the form of people with distinctly lighter skins. This tail, he thought, could have appeared only if there were rather few genes involved. If there were many genes, then the African-American population would have a much narrower range of pigment intensity because everybody would tend to have about the same mixture of black and white genes. But if there were few genes, then merely by chance some people would have more than the expected number of black genes and others fewer than the expected number. Those with fewer would occupy the tail at the light end of the pigment distribution and would be easy to detect.

Stern assumed that all the black genes had small equal effects and would add up, so that heterozygotes for black and white alleles at any genetic locus would be intermediate in color between the two homozygotes. He also made the rather large assumption that the people in the group had mated at random with respect to skin color. Using these and an assortment of other assumptions, he was able to estimate that skin color is determined by about three or four genes.

Since then, many other people have quarreled with his simplifying assumptions. It is possible, indeed likely, that the genes do not simply add up in their effects but may have large or small effects and may interact with each other. It is also highly improbable that the African-American population mates randomly with respect to skin color, although deviations from nonrandom mating would need to be quite large to have much of an impact on the analysis. Still, nobody has completely disproved his original conclusion that the number of genes involved is likely to be relatively small. Regardless of whether or not his original model is right, so far as skin color is concerned it can be concluded that there is not much of a genetic difference between people who are black and people who are white.

Do these genetic differences simply somehow add up in their effects, as Stern suggested, so that the color of one's skin depends only on the number of black genes one carries? The full story of skin color genes has not yet been worked out, but already it is apparent that Stern's simple model is not correct.

The color of your skin is chiefly determined by the amount of a pigment called melanin, which is formed from the amino acid tyrosine by the action of the enzyme tyrosinase and by some other processes. Melanocytes, the cells chiefly responsible for making melanin, first appear on the dorsal surface of the developing embryo. They then migrate everywhere on the embryo's surface, taking up residence in the dermal layers of the skin. Once they arrive in the skin, they rarely divide.

Melanocytes are being intensively studied at the moment because these cells can give rise to the devastating and usually fatal skin cancers called melanomas. As they mature, they produce branches called dendrites, and in this they are rather reminiscent of the neurons that make up much of the brain—indeed, they are very closely related to neurons in other ways and actually respond to some of the same growth factors that neurons do. Melanocytes, however, use their dendrites not to communicate with the surrounding skin cells but to produce packets of pigment called melanosomes. These packets are then engulfed by nearby skin cells. A very active melanocyte that produces many pigment-rich melanosomes can darken the skin in a large area.

All kinds of things influence the way melanocytes grow and change their shape. Low levels of long-wave UV radiation stimulate them to put out more extensive dendrites—the result, of course, is tanning of the skin. This appears to be due to the UV-stimulated release of a so-called second messenger that activates many different systems in the cell. More subtle influences are at work as well. Melanocytes are also stimulated by factors secreted by the surrounding skin cells, although the nature of these factors has yet to be discovered. This can clearly be demonstrated by culturing melanocytes in the laboratory and adding to them some filtered medium from a different culture in which other kinds of skin cells have been growing. This causes the melanocytes to multiply much more quickly.

The melanin pigment itself can be black, brown, or even red. The

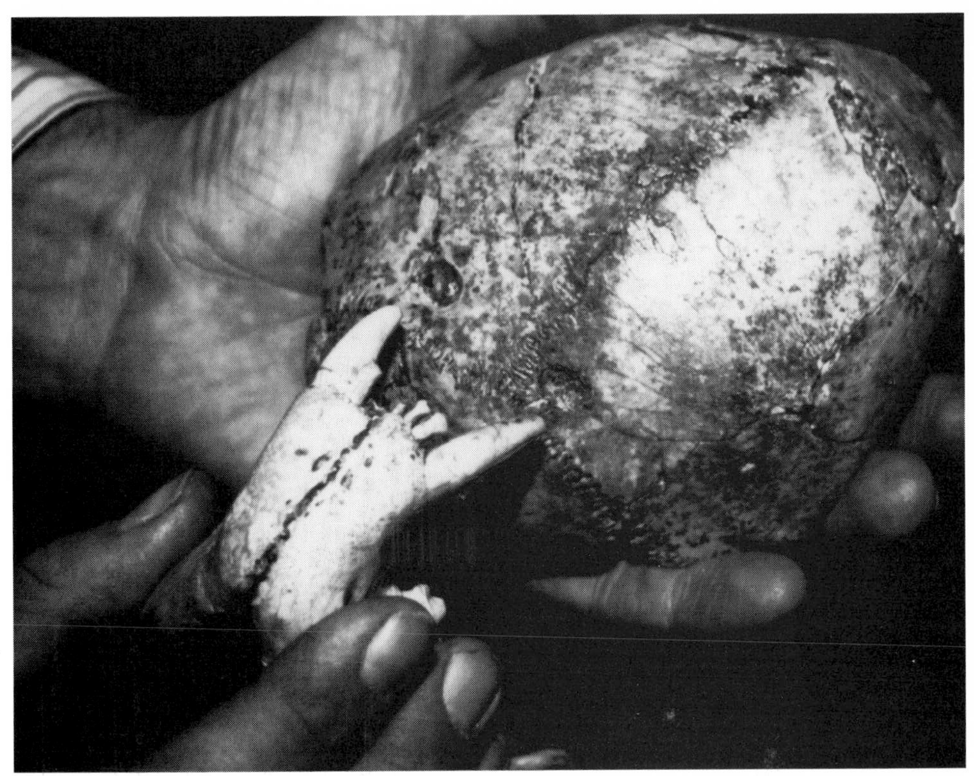

PLATE 1. Gracile Australopithecine skull showing a pair of indentations that matches perfectly to the canines of a cheetah. The gracile Australopithecines in the cave deposits of South Africa seem more to have been the victims than the hunters.

PLATE 2. Casts of some of the Peking man crania that were found in the cave at Zhoukoudian. Only parts of the top region of the face were actually found—Weidenreich's reconstruction of an entire Peking man skull is based on a good deal of guesswork.

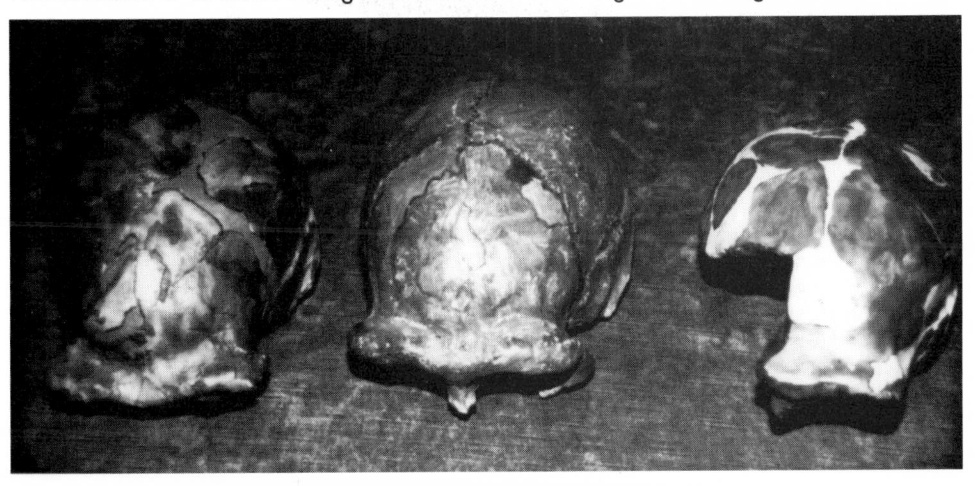

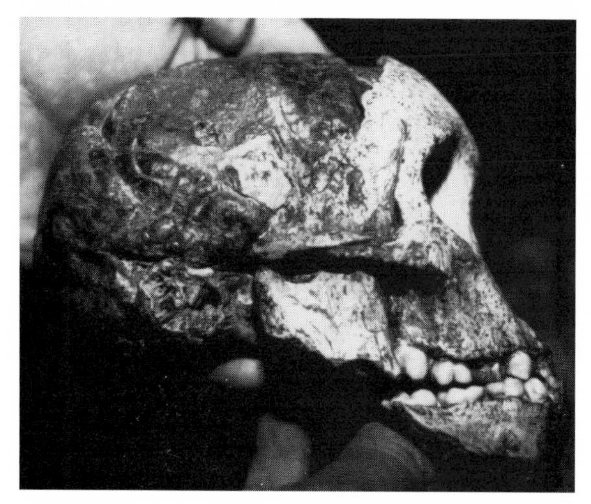

PLATE 3. The right side of the Taung baby's cranial endocast, showing how the cavity within the skull did not fill completely with limestone, so that calcium carbonate crystals could form within it like crystals in a geode. Note, too, the remarkably humanlike dentition, with its large flat molars and small canines.

PLATE 4. A chimpanzee and a human skull. The human skull is that of a three-year-old, and the chimpanzee skull is of approximately the same physiological age. The Taung baby more closely resembles the chimpanzee—except for its remarkably humanlike dentition.

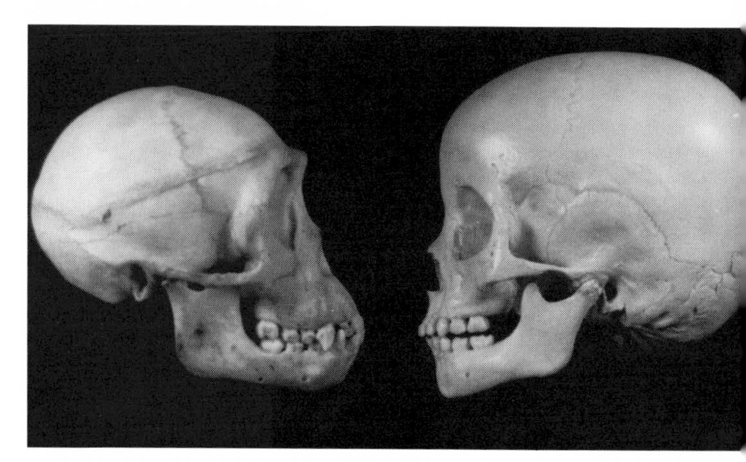

PLATE 5. The talus of (A) a chimpanzee, (B) a gorilla, (C) an *Australopithecus afarensis* similar to Lucy, and (D) a modern human. The *afarensis* talus was not quite as cuboidal as that of a modern human—perhaps her stance was not yet completely upright. Still, even though it is more than three million years old, it resembles that of a modern human much more closely than that of a chimpanzee.

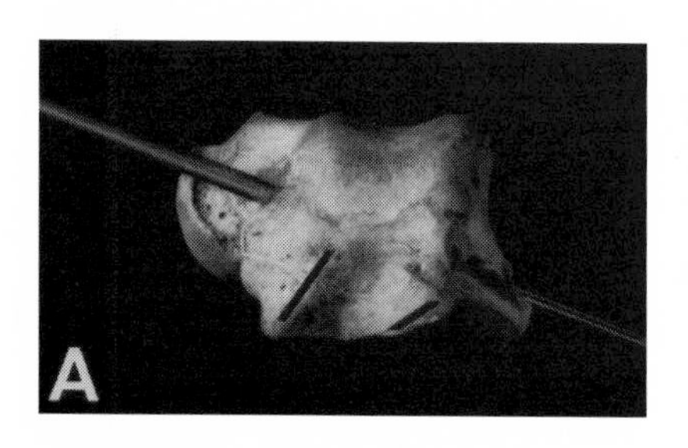

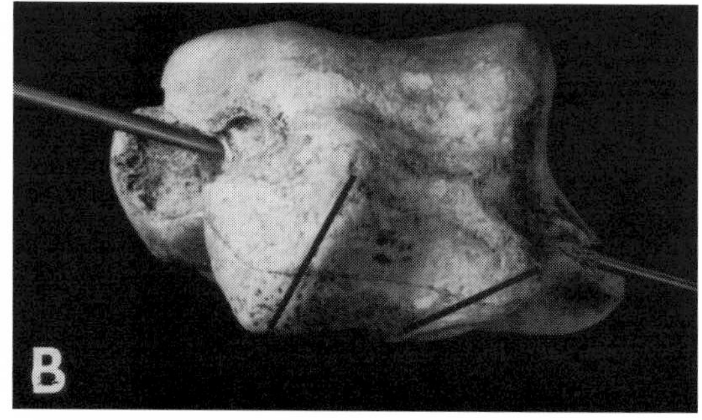

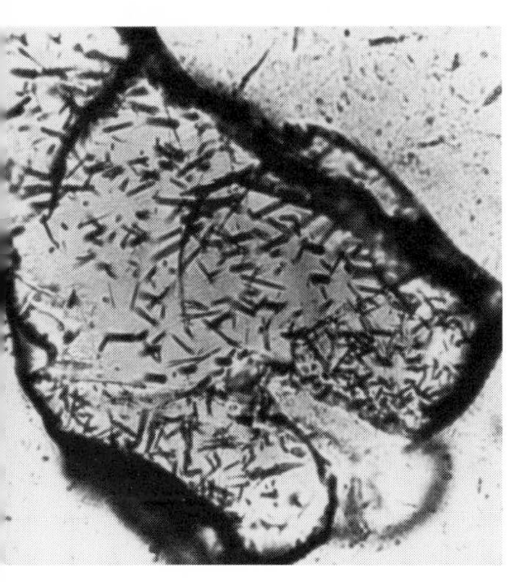

PLATE 6. Etched fission tracks.

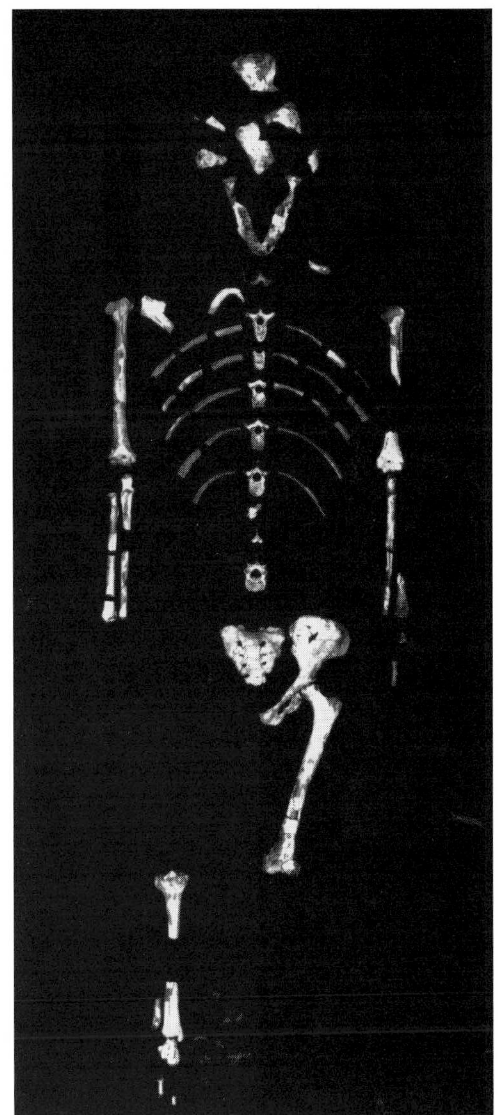

PLATE 7. The skeleton of Lucy.

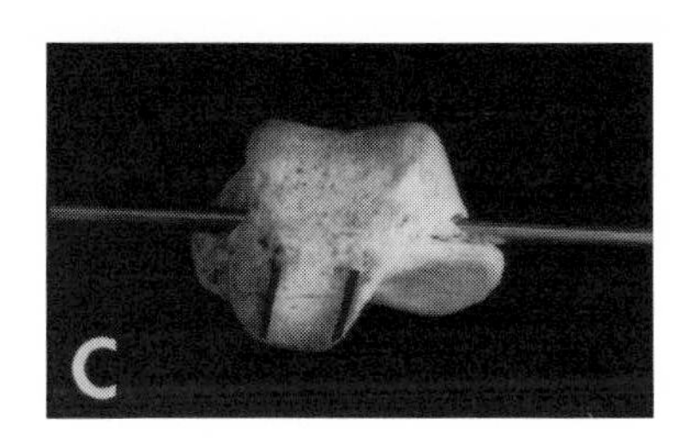

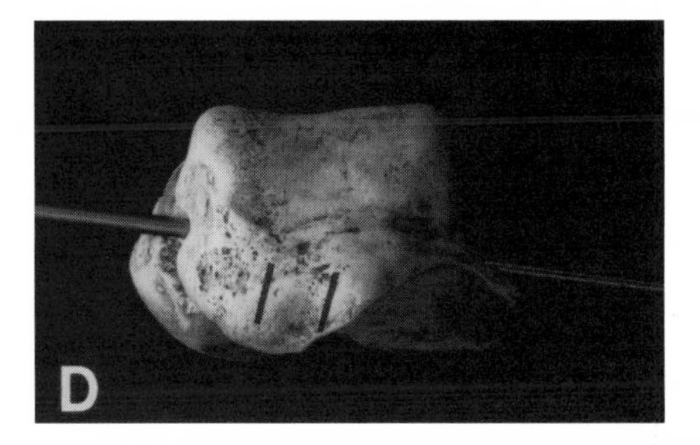

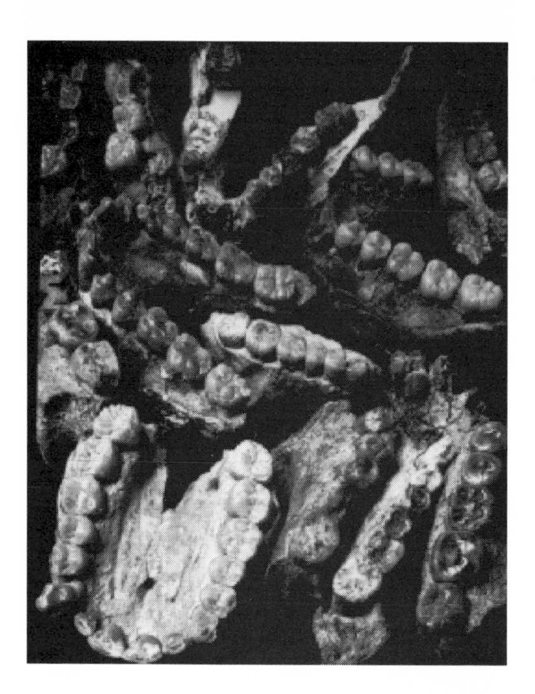

PLATE 8. The range of variation among the jaws of the First Family. You can see why Johanson assumed at first that they must have belonged to more than one species.

PLATE 9. Alan Walker's Black Skull, with its prominent cheekbones and sagittal crest.

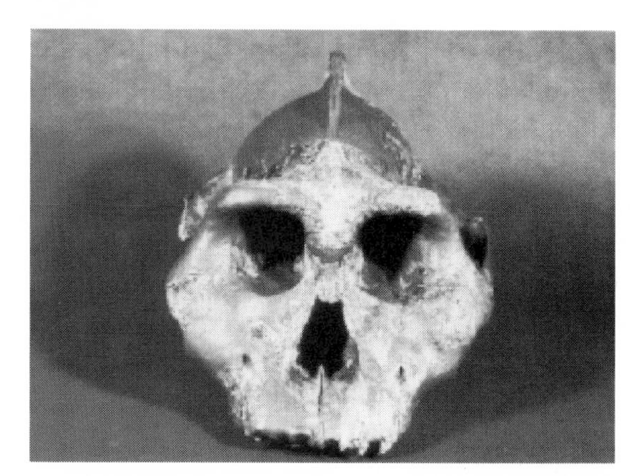

PLATE 10. Skulls of *(left)* a dachshund and *(right)* a Boston terrier. Every part of the terrier's skull has been foreshortened, some more than others.

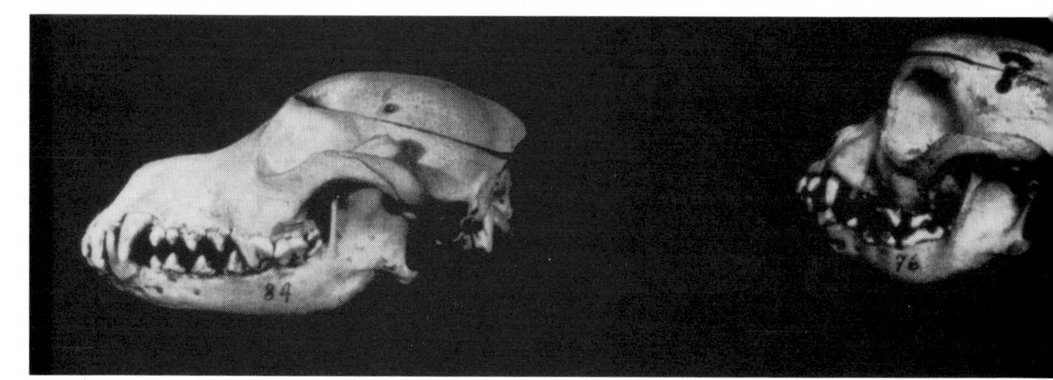

PLATE 11. Picture of an early Neanderthal skull from the Middle East. Note how the lower part of the face and its dentition are moved forward. No other hominid, ancient or modern, has so large a space between the back molars and the ascending ramus of the jaw—unlike in many modern humans, there was plenty of room for the wisdom teeth to erupt normally.

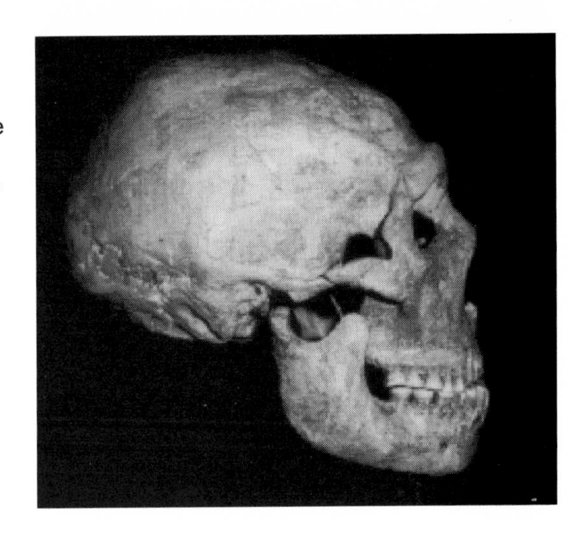

PLATE 12. A series of reconstructions of the Neanderthal face and head. *(Top center)* Marcellin Boule's reconstruction. In spite of the wide variation, attributable to the difficulty of adding flesh to bones, all the reconstructions show the forward-thrusting dentition and receding chin that give the Neanderthal mouth an almost beaklike appearance.

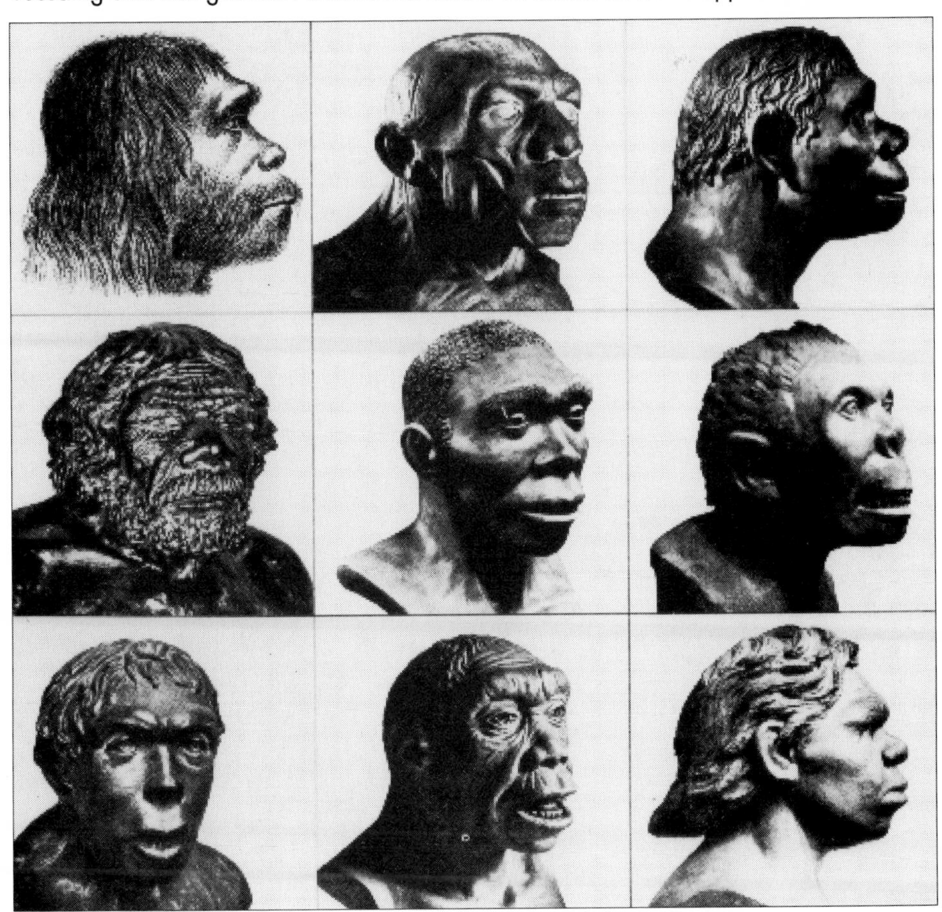

PLATE 13. A modern-appearing skull found at Skhul. Note the slightly primitive features, the projecting browridges, and the slightly prognathous jaw.

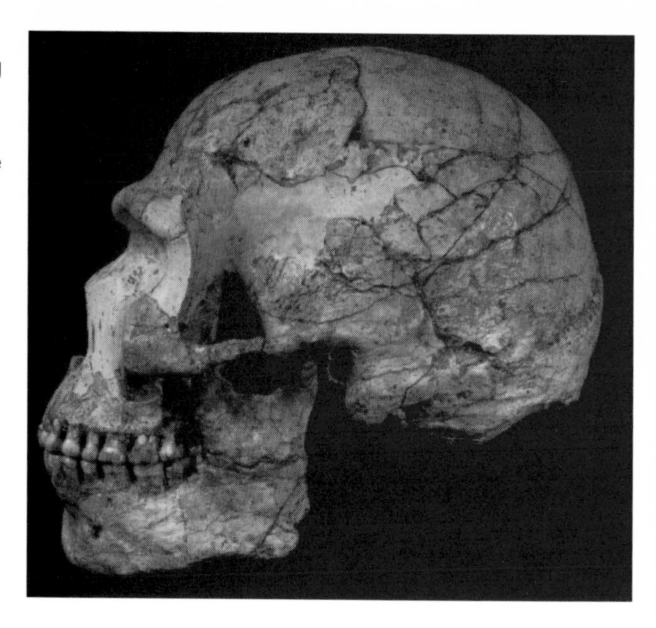

PLATE 14. The giant chromosomes of *Drosophila melanogaster*. In life, these chromosomes are curled up in the cell's nucleus, but in this preparation the nucleus has been broken open and the arms of the chromosomes have been allowed to spread. You can see the complex and highly specific banding pattern on each of the chromosomes.

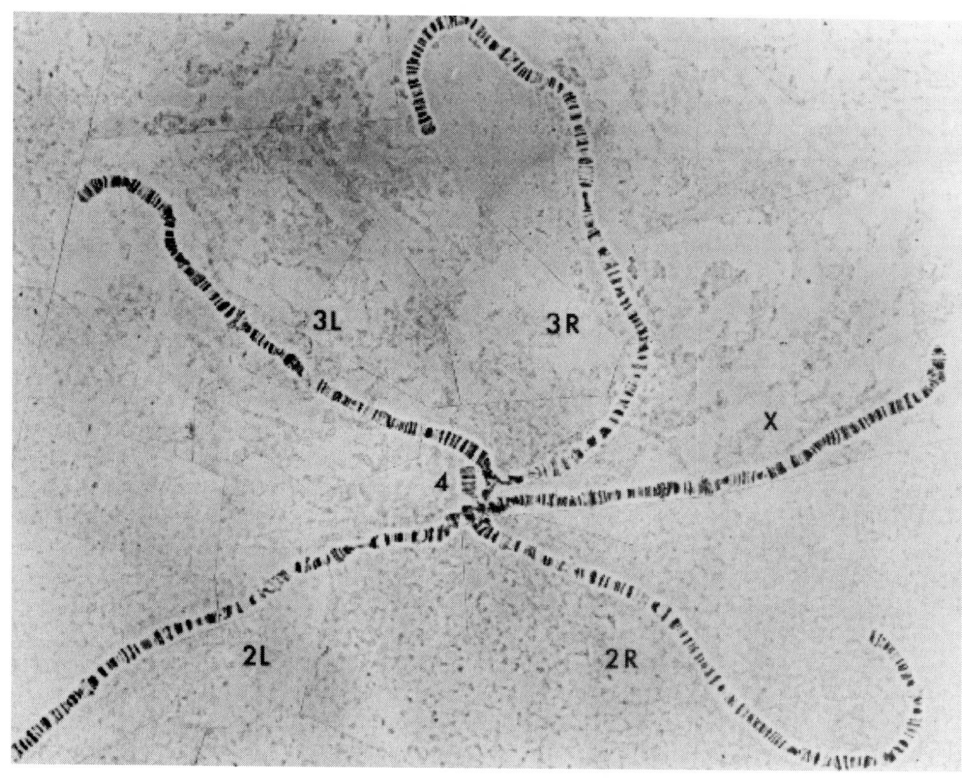

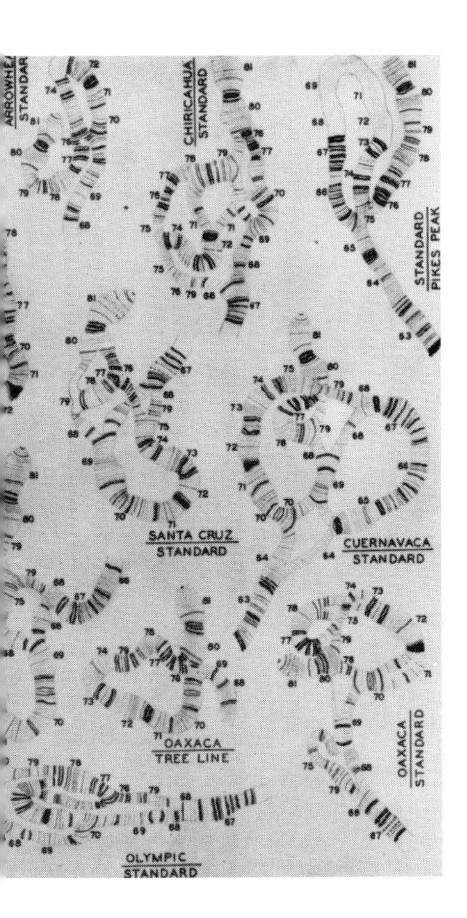

PLATE 15. Some of the complicated patterns that result when chromosomes carrying inversions pair up with each other. If the two chromosomes differ by a single inversion—for example, when a fly is heterozygous for the chromosomes called Arrowhead and Standard—a single loop results. If they differ by more than one, much more complicated contortions take place as the chromosomes try to pair with each other.

PLATE 16. The closely related Hawaiian *Drosophila, D. heteroneura (left*) and *D. silvestris (right*). In addition to the dramatic differences in head shape, the flies differ both in wing size and markings and in the pattern of markings on their bodies.

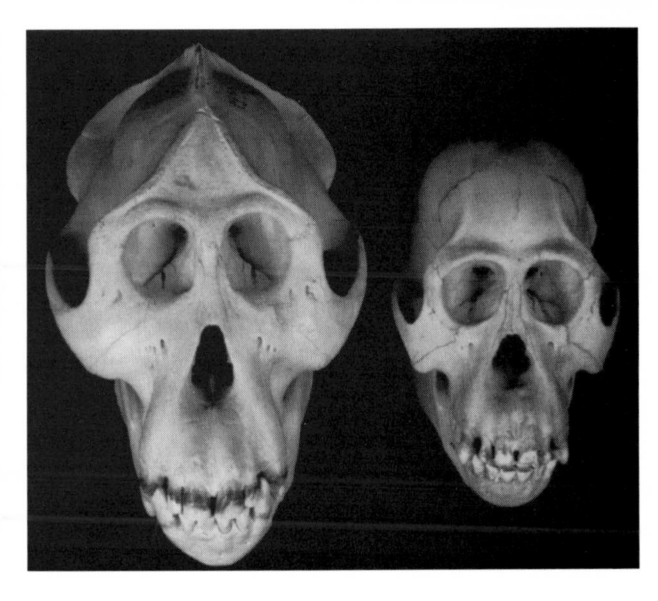

PLATE 17. Male and female orangutan skulls. The prominent and complex sagittal crest in the male is reduced to two slight superior temporal lines in the female. In consequence, her jaw muscles are far less powerful.

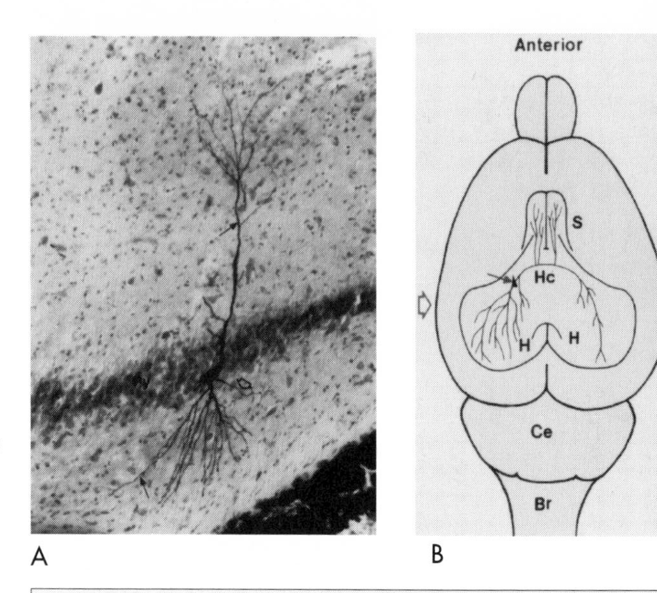

A B

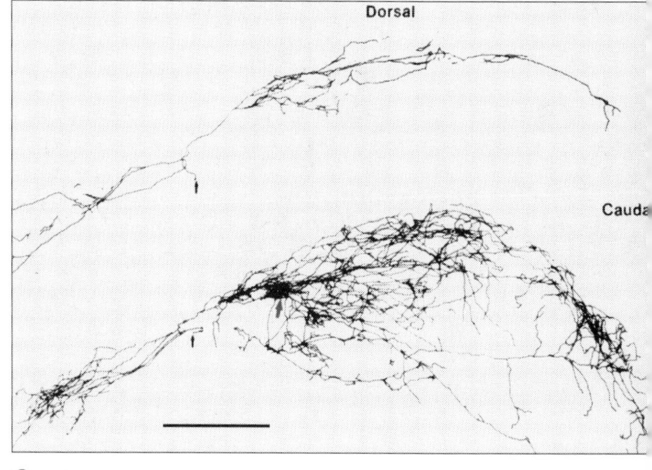

C

PLATE 18. A single neuron from the hippocampus of a rat's brain, carefully stained and followed in all its ramifications. The photograph *(A)* shows the cell body of the neuron, with two main branches leading to the dendrites *(small arrows)* and a small part of the axon (*open arrow*). But this picture tells only part of the story, for the axon branches and branches again, traversing the whole length of the hippocampus and even penetrating the other half of the brain *(B)*. Figure *C* shows *(gray arrow)* the location of the cell body surrounded by a dense cluster of dendrites. Seen from this vantage, the cell body is dwarfed by the enormous complexity of the axon. The bar below this part of the figure represents 1 millimeter.

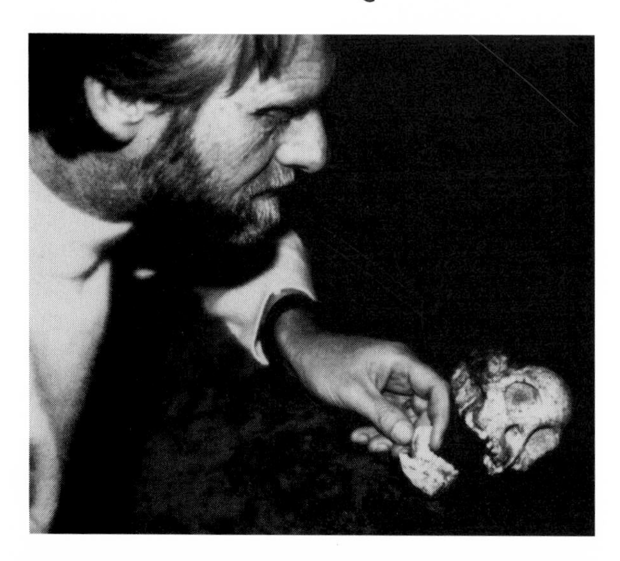

PLATE 19. The author examining the 2.5 million-year-old Taung skull found by Raymond Dart.

color of the pigment depends on which of a variety of different enzymes operate on it. People with a full complement of enzymes produce black melanin. If they lack tyrosinase, they produce no pigment and are albino. The genes for tyrosinase and for some of the allied enzymes have recently been cloned, and the ways by which they are turned on and off in the cell can now be studied in detail. Although tyrosinase seems to be made in large quantities in all melanocytes, the reason that all of us are not darkly pigmented is that the tyrosinase gene is subject to a complex pattern of regulation. Depending on the circumstances, the gene can be persuaded to make either active or inactive tyrosinase, and the ratio of active to inactive enzyme determines at least in part the amount of pigment that the melanocyte makes. This is done through a process called alternative splicing, in which the messenger ribonucleic acid (mRNA) is made in more than one form. To add yet another layer of complexity, the gene for an enzyme closely related to tyrosinase and also involved in melanin production has been examined in mice, and its structure indicates that the production of this second enzyme is under a very different system of control.

It is striking that the more these systems of regulation are investigated, the more highly structured they are discovered to be. The production of melanin is governed, not by some system that simply regulates the amount of tyrosinase that is made, but by a more elaborate system that controls the switch from inactive to active tyrosinase. The number of melanocytes and their activity is controlled not by the action of a single hormone, but by many different growth factors, within and outside the cell, that act together in complex ways. The black and white gene story is being supplanted by the realization that the amount and kind of pigment is determined by the effect of hormones on cell development and by the interaction of regulatory genes with genes that produce the enzymes involved. No such elaborate set of regulatory pathways could have evolved unless the melanocytes of our ancestors had sometimes been selected to produce and export large amounts of pigment and at other times been selected to do the opposite.

I expect that, as the story of skin pigmentation unfolds, there will be many examples of potential-altering and potential-realizing mutations that are found to affect this complex and ancient regulatory system. Of course, this skin color story is a bit of a cop-out because I am not yet able to point to such specific potential-altering and potential-realizing mutations in this system or indeed in other polymorphic genetic systems in humans. No marvelous stories like the one about the period gene have yet emerged from the study of human genes. The genes that really

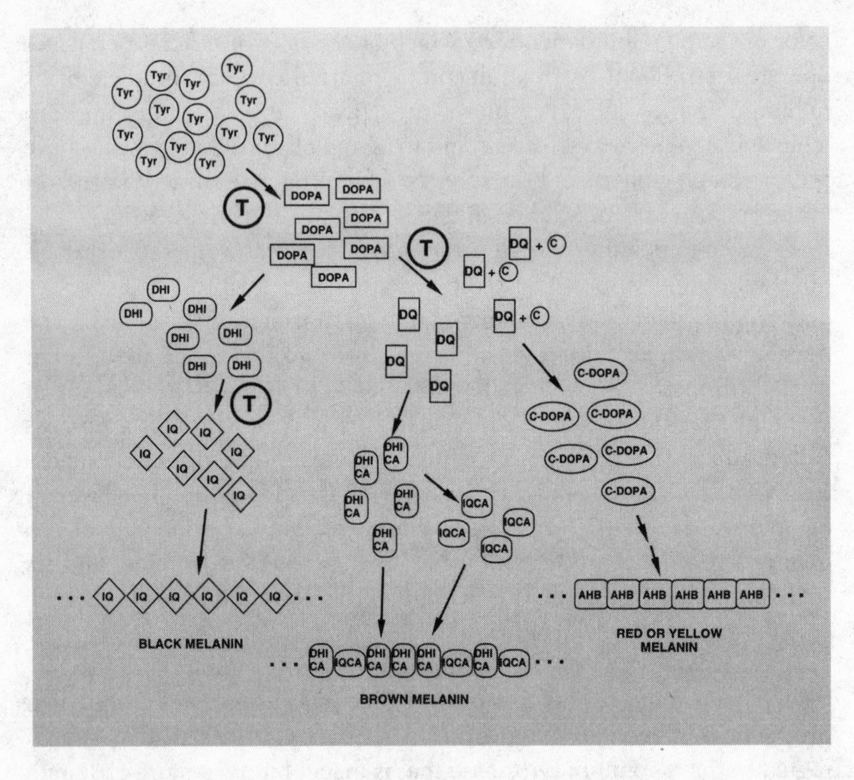

FIGURE 9.2. The complex series of biochemical events that produce the three major kinds of melanin. All start with the amino acid tyrosine, which is transformed by various enzymatic and nonenzymatic processes into a variety of building blocks. Some of these are in turn linked up into the long chains that constitute melanin. Many of these reactions can be inhibited to various degrees by a variety of mechanisms. The three reactions that are catalyzed by the enzyme tyrosinase are marked by the letter T.

distinguish us from our great ape relatives have yet to be cloned. So far, we have not discovered the genes that influence brain size, the distribution of body hair, or the shape of the pelvis or the foot. We have yet to clone genes that control the position and flexibility of the thumb or that regulate the sensitivity of the fingertips. And certainly no genes like the period gene have been found that explain the preference of humans to mate with humans, and chimpanzees to mate with chimpanzees. However, enough genes from ourselves and our near relatives have been cloned to make one thing very clear. Whatever these differences may turn out to be, most of them are not due to the evolution of new genes. Instead, they must be the result of molecular tweakings of genes that already existed in our ancestors. Most of those few hundred thousand meaningful differences between ourselves and chimpanzees must be small changes, either within preexisting genes or close to them.

Figuring these differences out, however, will be very hard. When differences between ourselves and chimpanzees are investigated, they often raise more questions than they answer. One of the proteins in the plasma of our blood (the liquid in which the blood cells float) is called haptoglobin. It is found in copious amounts, yet normally it seems to do very little. However, if a disease such as malaria or autoimmune hemolytic anemia ruptures our red blood cells and releases hemoglobin into the plasma, the haptoglobin begins to function. In a masterful display of recycling, it binds to the hemoglobin and prevents it from being broken down too quickly. This gives time for the iron of the hemoglobin to be rescued and reused.

Nobuyo Maeda of the University of North Carolina has explored the genetics of haptoglobin. She finds that most humans have two haptoglobin genes on chromosome 16, both perfectly functional. One makes the haptoglobin that salvages the iron. The other seems not to make a protein, or at least not to make one in any human tissue that has yet been explored. In some populations of African blacks, individuals have been found with as many as four extra copies of this second, mysterious gene. Now, it is surely not a coincidence that these extra genes are found in people inhabiting a part of the world where hemolytic diseases are common—but then why do these people have extra copies of the apparently nonfunctional gene, rather than the gene that we know salvages the iron?

Our ignorance about the haptoglobins is glaringly apparent. We think we understand what one of our two haptoglobin genes does, and yet most of the evolutionary activity seems to have been concentrated in the other gene. Mother Nature seems to know something that we don't. Dozens of similar, puzzling stories are emerging as the genetic divergences between ourselves and our relatives among the great apes begin to be explored. None of them, I suspect, will be understood until we have a good idea of the evolutionary history of the genes involved, the ways in which the genes interact with each other, and the ways in which they interact with the environment. We are finally beginning to attack that complex, subtle, and fascinating part of our genetic heritage that Dobzhansky's intuition told him was there all along. In the process we are moving away from those simple divisions of our genes into good and bad, or neutral and selected.

Esau My Brother Is a Hairy Man and I Am a Smooth Man

When smooth Jacob donned goatskins and pretended to be his hirsute brother Esau in order to obtain the blessing of their blind father, Isaac, he was trading on the fact that brothers are often very different from each other. Such differences are not surprising. Two humans chosen at random differ by about a tenth of 1 percent of their DNA. Two brothers differ by about half this amount. Since there are three million differences between unrelated people, Jacob and Esau must have been separated by one and a half million differences. A few of them happened to dictate how much body hair they had, and were obvious enough to be enshrined in a story that has lasted three thousand years. Jacob and Esau might also have differed in the number of their haptoglobin genes, but such essentially invisible differences do not form the stuff of legend.

Geneticists who deal with quantitative characters such as skin color or degree of hairiness have found over decades of work that one generalization seems to be holding up. However numerous the genetic differences between any two individuals are, when any given, narrowly defined character is examined in detail, it seems that a relatively small number of these differences will be found to be involved.

In the second part of this book we agonized along with the paleontologists about certain variations in the size and shape of fossil skulls, variations that are used to decide whether a fossil is a Neanderthal or a *Homo sapiens* or whether it is a *Homo erectus* or a *Homo habilis*. Just how many genes are likely to be involved in these differences, so central to the whole field of paleontology? We cannot answer this question directly, because such studies have not yet been carried out in humans. But to get a feel for what the answer will turn out to be, let us turn once again to our old standby, the fruit fly.

The big island of Hawaii is the youngest of the chain of Hawaiian islands, a mere four hundred thousand years old. It is the tip of a vast shield volcano with three major vents, Mauna Loa, Mauna Kea, and Kilauea. Kilauea is the most recent, accounting for the majority of the recent major eruptions. Over the millennia, every part of the island has been covered repeatedly by lava flows that, as they chop up the landscape, create an ever-changing patchwork of jagged black bands of fresh lava interspersed with native vegetation in various stages of recovery. This continual fragmentation of the environment has proved to be ideal for rapid speciation among the plants, birds and insects of the island.

On the islands as a whole, there are about eight hundred species of *Drosophila,* all descended from a few accidental immigrants that were blown or carried to this remote archipelago in the distant past. Many of the *Drosophila* of Hawaii are much more spectacular insects than the discreet little fruit flies that buzz around the cantaloupes on your picnic table. They can be as large as bees, and they often have vivid patterns on their bodies and wings. Some of them carry out elaborate courtship rituals. Many of the differences in appearance among the various species have been traced to these rituals and are apparently the result of sexual selection.

The young island of Hawaii has a number of species that are endemic, unique to the island, which means that it is very likely that they evolved in the short time since it rose from the sea. Two of these, the closely related species *Drosophila heteroneura* and *Drosophila silvestris,* are particularly intriguing. These flies belong to a group of about a hundred species that are called the picture-wing flies because of their striking wing patterns (see plate 16). In many ways these two species look very similar. They are the same size and have roughly the same markings on their wings. But there is one astonishing difference. The head of *D. silvestris* is approximately round, like the head of any other fly, but the head of *D. heteroneura* (its very name means "different head") is grotesquely distorted, stretched out laterally so that its eyes are set far apart like those of a hammerhead shark.

Both males and females of *heteroneura* have these distorted heads, so one might think at first that sexual selection did not play a role in their evolution. If it had, then you might suppose that only one sex or the other, probably the males, would show these strangely shaped heads. Yet sexual selection has almost certainly been involved. Both species gather in leks, places where the males fight each other and vie for the attention of the females. Fights between two males are quite spectacular and consist of pushing matches in which they butt their foreheads and strain against each other, at the same time extending their wings forward and vibrating them furiously. Possibly the larger heads of *D. heteroneura* have evolved by sexual selection to give males of the same species a distinguishing target, so that they won't butt heads with the other species by mistake. If so, then why do both sexes of *D. heteroneura* show the unusual head shape? One possible answer is that there has simply not been enough time for mutations to arise and be selected for that would limit the head shape to the male sex only. After all, these species began to diverge only in the very recent past.

Hampton Carson and his co-workers at the University of Hawaii have shown that the two species' giant chromosomes are indistinguishable under the microscope. Further, if males of one species are mixed together with females of the other in the laboratory, they mate readily and produce offspring. These offspring, unlike the offspring of hybrids between Dobzhansky's species, *D. pseudoobscura* and *D. persimilis,* are all perfectly fertile. This means that there must be few differences between *D. heteroneura* and *D. silvestris* at the gene level as well.

Even though their ranges overlap, the flies of these two Hawaiian species rarely crossbreed in the wild. They do, however, crossbreed more often than Dobzhansky's flies, for about 1 percent of them show signs of being hybrids. All this means that they are only barely species yet. The high proportion of hybrids and the genetic similarity between the two species shows that Carson and his co-workers have caught them near the beginning of their speciation process, before many gene differences have had a chance to accumulate.

The fertility of the hybrids allowed F. C. Val from the University of São Paulo, working in Carson's lab, to make a good estimate of the number of genes that determine the differences in head shape in the two species. He showed that the heads of the hybrids look like an average between the normal head of *D. silvestris* and the distorted head of *D. heteroneura,* because any genetic differences between the two species tend to be averaged. However, the hybrids are heterozygous for *silvestris* and *heteroneura* alleles at several different head shape genes, which means that they can give a variety of combinations of *silvestris* and *heteroneura* genes to their progeny. When a male hybrid is crossed with a female hybrid, the progeny of this second-generation cross show a great variety of head shapes, ranging all the way from *silvestris*-like to *heteroneura*-like. By chance a few of these progeny get all *heteroneura* genes, and a few others get all *silvestris* genes. Most of the progeny carry some mixture of the two types.

Val was able to show that one major gene on the X chromosome has a large influence on the head shape. It interacts with perhaps two or three other genes on the other chromosomes to produce the *heteroneura* head, and it does so in a complicated *epistatic* way. You will remember that epistatic interactions between genes, particularly genes found in natural populations, are very common and can often lead to results that are confusing to geneticists. Genes that have survived for a long time in natural populations tend to be the ones that have complicated interactions, and the genes for head shape in these flies are no exception.

The striking thing is how few genes seem to be involved. Alterations in three or four genes are enough to produce this enormous change in the flies' head shape. Dare we extrapolate these findings to our own evolutionary history? Perhaps a similarly small number of genes might be involved in the appearance or disappearance of projecting browridges, or sagittal crests, or the subtle changes in skull shape, jaw shape, and dentition on which so much of our story of human evolution depends. There can certainly be plenty of variation in characteristics such as the sagittal crest even within a single species of hominoid. Orangutan females have no sagittal crest and the males have two! (See plate 17.)

Of course, even if each individual feature of our morphologies is governed by a relatively small number of genes, there are many such features, and therefore many genetic differences, separating ourselves and our nearest relatives. Even the closely related Hawaiian *Drosophila* species have acquired many genetic differences in addition to those that dictate their head shape—differences affecting body and wing markings, behavior, and presumably many other things that we are not yet perceptive enough to detect. But whenever we are able to concentrate on and analyze a particular difference, it is surprising how few genes seem to be involved. In genetic terms it seems that we may not be as big a deal as we thought.

Do the same genetic processes that we have discussed here, processes that have molded our physical and biochemical characteristics, also apply to our brains? At the genetic level the evolution of our brains is subject to exactly the same laws as the evolution of head shape in fruit flies. As we will see, the difference lies, not in some unique properties of the genes governing the brain's development, but in the unique nature of that remarkable organ itself.

How to Turbocharge Natural Selection

The reductionism of twentieth-century biology has been enormously successful, resulting in, among other things, the discovery of genes and later of their actual biochemical structure. Reductionism is a powerful tool, but it has tended for decades to blind many biologists to the possibility that something as complex as behaviors might be selected for or against, in part because of strong antipathy to the idea that there might be a genetic component to human behaviors. Reductionists also steered away from the problem because the study of behavior seemed much fuzzier and less respectable than the study of clear-cut cases of inheritance. Reductionist geneticists, working with obviously mutant genes in the laboratory, have tended to forget that it is not genes that are being selected for or against, but phenotypes. And behaviors are phenotypes like any other, although their genetic basis tends to be very complex.

All this, as we saw earlier, changed with the publication in 1975 of E. O. Wilson's important book *Sociobiology.* Wilson brought together into one volume hundreds of studies showing how the behaviors of animals have been shaped by selective forces. Much animal behavior, he concluded, has a large genetic component and is subject to exactly the same evolutionary forces as any other genetically controlled characteristics such as tooth morphology or hair color. At the end of the book, however, he made the political mistake of extending these observations to human beings without surrounding his statements with a sufficient number of caveats. He neglected to emphasize that humans are, unlike most other species, not necessarily the prisoners of their behavioral genes.

Alister Hardy, a zoologist who spent his life working on oceanic plankton, examined the history of the concept that behaviors might evolve in his flawed but fascinating book *The Living Stream* (1965). He traced the idea of behavioral evolution all the way back to James Hutton, the eighteenth-century geologist who founded the influential school of uniformitarianism. This school contended that most changes in the surface of the planet have taken place gradually over very long periods of time. Hutton's geological observations inspired Charles Lyell, who in turn inspired Charles Darwin to look at the world through uniformitarian eyes.

In a manuscript titled *Principles of Agriculture,* on which Hutton was working at the time of his death in 1797 and which lay for a century and a half in the archives of the Edinburgh Geological Society, he wrote:

> In the infinite variation of the breed, that form best adapted to the exercise of those instinctive arts, by which the species is to live, will be most certainly continued in the propagation of this animal, and will be always tending more and more to perfect itself by the natural variation which is continually taking place. Thus, for example, where dogs are to live by the swiftness of their feet and the sharpness of their sight, the form best adapted to that end will be the most certain of remaining, while those forms that are least adapted to this manner of chase will be the first to perish; and the same will hold with regard to all the other forms and facilities of the species, by which the instinctive arts of procuring its means of subsistence may be pursued.

Although Hutton meant this remarkably Darwinian statement to apply to selection within a species, not the transformation of one species into another—after all, the received wisdom of the eighteenth century was that this could not happen—it is striking that it was behaviors he thought of first when he considered which characteristics might lead to the survival of the fittest. And it was behaviors that the French zoologist Jean-Baptiste de Monet de Lamarck first thought of when he produced his much-derided theory of evolution in the early nineteenth century. The way an animal behaved, Lamarck thought, would dictate the way it would evolve in the future.

A hundred years later at the end of the nineteenth century, just before the rediscovery of Mendelian genetics and the temporary triumph of reductionism, two different biologists independently arrived at a modification of Darwinian theory that at first sight seemed to have Lamarckian overtones. This modification has now become known as the Baldwin effect, after the British biologist Mark Baldwin, but it seems to have been first proposed by his compatriot C. Lloyd Morgan. Lloyd Morgan pointed out that, if a group of organisms is placed in a new environment, modifications in their "innate plasticity" will permit some of them to survive. He meant by *innate plasticity* what the Danish geneticist Wilhelm Johannsen would later call phenotypic variation. Lloyd Morgan knew that such variation is not inherited. "There is no transmission," he said, "of the effects of modification to the germinal substance"—what we now call the genes. Genetic variability—which he simply called variation—would be passed on if it tended to enhance this innate plasticity and would not be passed on if it did not.

In the 1950s, the British developmental biologist C. H. Waddington took this argument a step further. He pointed out that organisms are often subject to environmental pressures during their development, pressures that affect their phenotype. It was already known that litters of mice kept on starvation diets during critical times in their development grow up to exhibit a greater range of sizes and other physical characteristics than litters given normal amounts of food. Suppose, Waddington suggested, that this increase in phenotypic variation in a stressful environment was due in part to a greater degree of expression of underlying genetic differences among the litter mates. Environmental stress would increase the expression of that part of the phenotypic variation that was due to the genes. This would multiply the effectiveness of either natural or artificial selection.

Waddington was able to use *Drosophila* to demonstrate this. When the eggs of flies were subject to a sudden heat shock, or exposed to ether vapor, the adults that developed from these eggs were more phenotypically variable than adults that came from unshocked eggs. By choosing flies with some of these extreme characteristics and shocking their eggs in turn, Waddington was able within a few generations to produce flies that showed these new characteristics even without the aid of a heat or ether shock. Was this the inheritance of acquired characteristics? No, it was simply the result of the recombination and increase in frequency of genes already present in small numbers in the original *Drosophila* population. The phenotypic effects of these genes, normally largely concealed, were revealed as a result of the stress treatment in those few flies that originally carried them. Without the stress, Waddington might have sorted through tens of thousands of normal-appearing flies and never found any abnormal ones that would have allowed him to begin his selection. The shocks he applied to the eggs allowed him to see a larger portion of the flies' underlying genotypes reflected in their phenotypes.

Later, in his book *The Nature of Life* (1961), Waddington explicitly applied these ideas to behavioral evolution:

> Even within a single species different individuals differ hereditarily in their behaviour. . . . Thus the animal's hereditary constitution influences the type of natural selective pressure to which it will be subjected. And then, of course, the natural selection influences the type of heredity which is passed on to the next generation. We are dealing with a feed-back or cybernetic system in which there is nothing that is simply cause or simply effect.

Now, genetic reductionists tend to arrange things in the laboratory so that a particular allele always has a particular effect because it makes the design of experiments much simpler. This is done by deliberately keeping the environment as constant as possible, so that the genes they are following in the course of the experiment influence the phenotypes of the organisms carrying them in the same way each generation. Otherwise, the results of any genetic experiment quickly become unclear and confusing, since the genes that are being followed are expressed in different ways each generation. Waddington's experimental design produced a very different result because, by deliberately varying the environment, he encouraged the flies' genes to affect their phenotypes in different ways. His vision of how evolution operates is a very powerful one because it frees evolutionary biologists from that straitjacket of dull reductionism.

The simple view of evolution that we explored in chapter 7, that genes can be divided into good and bad, is a reductionist one. Basic to that view is the assumption that the environment in which organisms live is usually constant, although of course it does occasionally change. During the long periods of constancy, populations of even the best-adapted organisms accumulate, like lint under a bed, a collection of neutral mutations and harmful mutations that are largely recessive. If the environment changes, a tiny fraction of these accumulated mutations now become advantageous and spread through the population. If any new mutations arise that also happen to be advantageous in that new environment, they are selected as well.

Waddington's view of evolution is more complicated, and I find it much more satisfying. According to Waddington, populations of organisms do indeed accumulate genetic variability like lint under a bed. Just as in the reductionist view, much of this variability is not normally expressed if the organisms happen to grow up in environments to which they are well adapted. But if the organisms are exposed to the stresses of a new environment, these genetic differences are revealed as phenotypic differences on which natural selection can act. The Waddington effect magnifies the impact of the normally concealed genetic variability on the phenotypes of the organisms that carry it—and it does so precisely at the time that some of this variation might aid in the adaptation of the species to a new environment.

Does this really happen in humans as well as fruit flies? Certainly the phenotypic variability of human bodies increases under conditions of stress. Some years ago I stood for a while on the Galata Bridge that spans Istanbul's Golden Horn, where I watched tens of thousands of pedestrians and cyclists surge past. The appearance of the crowd made a stark contrast to the well-fed (indeed, altogether too well-fed) and pampered crowds that inhabit an American shopping mall. Many of the passersby had strabismus, clubfeet, hunched backs, or withered arms or hands. A few had pendulous goiters on their necks, and there was a great variety of dental problems ranging from undershot jaws to abscesses. Some had bodies distorted by years of toil, and it was not uncommon to see people, permanently stooped by those years of labor, carrying enormous burdens like refrigerators.

All this variation, you might think, was caused by the environment. Indeed, much of it undoubtedly was. Yet, it is highly likely that some people, pushed to their limits by the rigors of labor or the stress of disease, would be better able to survive under those extreme circumstances than others. Standing there, I reflected that I would not last very long among that crowd if I were required every day to carry refrigerators across the Galata Bridge while my rotting teeth pumped my body full of toxins.

So, the question we are now driven to ask is whether Waddington's view of evolution might apply to the brain as well as the body. We must begin by asking what happens to the brains of humans under stressful conditions. Does their phenotypic variability increase—that is, do the brains of different people react in markedly different ways to the same stress? Do stressful conditions for the brain include mental as well as physical stresses? Indeed, precisely what do we mean by mental stress?

It is possible that even the enriched environment experienced by those rats studied by Marian Diamond actually acted as a kind of stress, distinguishing among the rats on the basis of their ability to respond to it. If that is true, then there are two very different kinds of stress that can act on the brain during its development.

The first is the sort of stress with which we are becoming more familiar as starvation and disease become commoner in many parts of the world. Malnutrition and exposure to drugs can wreak havoc on the development of the brain, through denial of proper blood flow or the proper levels of nutrients during critical periods. It seems probable that people with certain genotypes are more likely to be able to withstand this kind of damage than others, but the brain's ability to survive such extreme conditions is not very likely to be correlated with its ability to do other things.

The second kind of stress, however, is much more subtle. It reveals, not the ability of the brain to survive some external insult, but the ability of the brain to handle an onslaught of *information*.

How good is the brain at doing this? If our ancestors spent at least part of their time reeling under a deluge of data, were some of them better at withstanding its effect than others? Perhaps this mental variability, if it exists, reveals underlying genetic variation that might otherwise be invisible, providing a handle for natural selection to grasp. If so, the Waddington effect may help to provide the evolutionary catapult that has shot the brains of our ancestors toward humanity.

IV

THE BRAIN

10

Escape from Stupidworld

Granted that natural selection is the only effective agency for producing change in biological evolution, a high degree of mental activity and mental organization could only have come into being if it was of biological advantage to its possessors. This at one stroke overthrows all theories of materialism, for they deny the effective reality of mind, or reduce it to a mere fly on the material wheel.

—Julian Huxley, *Evolution in Action* (1953)

Life Before the Runaway Brain

The late Allan Wilson, writing in *Scientific American* in 1985, remarked on two things that have happened in the course of the history of living things on our planet. The first is an accelerating, exponential increase in brain size among some groups of animals. The most striking examples of this have been among birds and mammals, particularly since the demise of the dinosaurs. The second is a tendency for species with large brains to last for a relatively short period of time (at least in evolutionary terms) before they are supplanted by other species.

Groups of closely related species are called genera (singular *genus*). Some years earlier, Wilson and his co-workers had calculated when various genera living at the present time had first appeared in the fossil record. The first representatives of present-day genera of lizards and amphibians made their appearance about twenty million years ago. The average first appearance for birds turns out to be less than four million years ago, and that for mammals a little over six million years. The bigger the brain, it seemed, the faster the turnover of species.

Wilson supposed that behavioral changes brought about by burgeoning brainpower must have increased the rate of morphological change. Increased brainpower increased the range of behaviors open to animals and birds, which in turn caused some of them to move into new ecological niches. The different selective pressures in these new niches, by bringing about rapid changes in appearance in their descendants, gave rise to new species.

At about that time, the existence of such strong selective pressures was being brilliantly demonstrated by Peter Grant and his colleagues as they investigated the finches of the Galapagos Islands. A century and a half earlier this collection of dull-plumaged bird species had provided Charles Darwin with some of the clues that led to the concept of natural selection. Grant's group showed that during periods of drought on the islands it was the birds with deep strong beaks, which enabled them to feed on thick-walled seeds, that tended to survive. This was a real genetic change, for these strong-beaked birds bred true. During periods of ample rainfall, when other kinds of foods became plentiful and thick-walled seeds were scarcer, the strong-beaked birds diminished in numbers. The strength and swiftness of this selection for and against beak size was remarkable, and it is likely that other characteristics, including behavioral ones, were being selected at the same time.

Unfortunately for Wilson's idea, though, evidence gathered over the next few years—evidence that emerged in part from his own laboratory—provided plenty of counterexamples. It seems that you do not have to be smart to evolve quickly.

The cichlid fish that are native to the lakes of Africa's Great Rift Valley are extremely diverse in size, morphology, color, mating rituals, and feeding habits. Some are carnivorous, some are vegetarian, and some actually make a living by plucking the eyes from other fish. Other unusual eating habits have evolved—very different groups of cichlids living in three different lakes have independently evolved a feeding habit in which they pull scales from other cichlids and eat them. More than two hundred of these morphologically diverse yet closely related species live in Lake Victoria and nearby lakes. Wilson's students showed that this diversity has evolved very recently. Using mitochondrial DNA, they found that the mitochondria of all these species could be traced back to a common ancestor (an Eve with fins, so to speak) that probably lived about two hundred thousand years ago. However, it was not brainpower that drove this rapid evolution, for the brains of cichlid fish are not noticeably larger than those of more slowly evolving species of fish.

The Hawaiian *Drosophila,* too, as we saw earlier, have undergone dramatic recent morphological evolution without the aid of significant

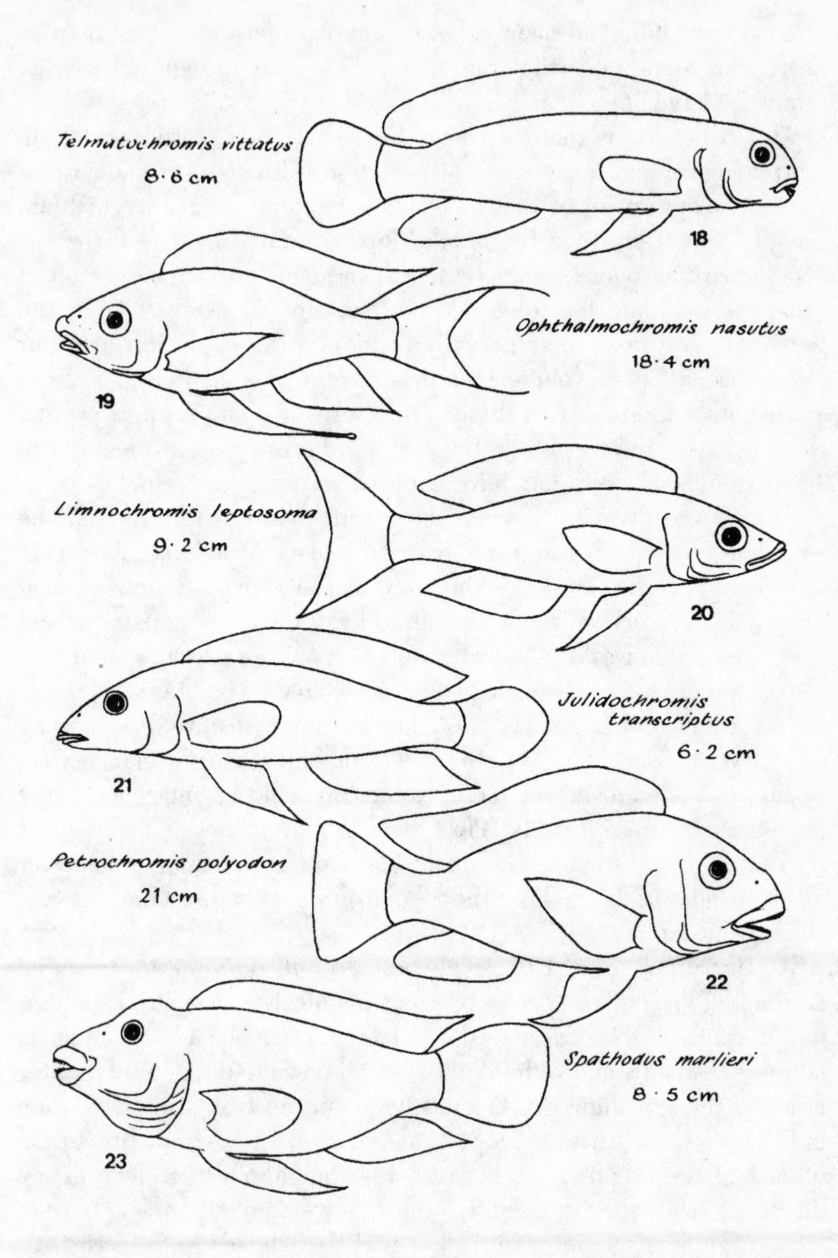

FIGURE 10.1 Drawings of some of the cichlid fish of Lake Victoria, showing their diverse body forms and mouthparts. The picture does not show the great diversity of coloration and pattern among these fish, ranging from red through silver to blue and green, with different species exhibiting a wide variety of spots and of horizontal and vertical stripes.

brainpower. These and many other counterexamples show that there is no reason to suppose that rapid morphological change necessarily requires a large brain.

This is not to say that increasing brainpower has played no role in morphological and other kinds of evolution. Although we know of no species except ourselves that has been catapulted toward larger brain size by a gene-culture feedback loop, animals in many different evolutionary lineages have increased their brain capacity as a result of other, slower, and less obvious feedback loops. For example, the behavior of animals may actually influence the evolution of other organisms, and that evolution can in turn influence the behavior of the animals that interact with them. The thorns of many bushes protect them against browsing by herbivores, and herbivores have learned to avoid them. Many insects have evolved warning (aposematic) color patterns that advertise they are dangerous or toxic—throughout the world, alternating black and yellow stripes means, Watch out, I sting! In *The Wisdom of the Genes,* I pointed out that this approach only works if potential predators are smart enough to learn the warning patterns and avoid them. In a world full of predators that are not very bright—let us call it Stupidworld—aposematic coloration would be pointless.

In such a world camouflage would be far more effective than warning coloration at ensuring survival. But in a world of predators with increasing brainpower, aposematic coloration would be highly advantageous to the potential prey, since it would allow them to move about unhindered. At the same time, it would actually contribute to the increasing brainpower of the predators by making their environment a more complex and challenging one.

As Harry Jerison of UCLA has pointed out, if you are a mammal surrounded by various species of smart mammals, it helps to be smart, too. Recently, as the African dusk was falling, I watched a herd of impala advance hesitantly and with infinite caution down to the edge of a water hole for their evening drink. Eyes wide and nostrils distended, they crept forward with tiny steps that ensured they would never be offbalance but would always be ready to spring into instant flight in any direction. When they reached the water they did not all drink at once—some members of the herd always had their heads high in the air, turning them continually to catch the slightest sound or scent.

The impala is an alert and cautious herbivore, and its complex behavior poses a continual mental challenge to the carnivores that prey on it. The presence of carnivores, in turn, dictates much of the impala's behavior. One wonders whether the great intelligence of predators and prey in Africa, powered by this feedback loop, might have helped to set the scene for our own runaway brain evolution. To what extent were the developing brains of our remote relatives challenged by those numbers and that diversity, and by the sheer cunning of the predators that stalked them? It is possible that on no other continent in the world of a few million years ago were there so many smart animals. If so, then our ancestors' achievement was not to leapfrog far ahead of other animals but simply to become the smartest of the smart.

Now, of course, we see no more than a trace of that feedback loop, for the animals of Africa are disappearing at a frightful pace because of human activity. A visitor today to the great dry plains of eastern and central Africa sees only a remnant of the vast herds that used to migrate across them only a generation ago. The piles of fossil bones heaped up in the washes and gullies of the region speak mutely of even greater numbers and much more diversity in the past.

In addition to the challenges offered by the interactions of predators and prey, the social interactions and other complex behaviors of many animals provide their own feedback loops. Robert Fagen, in his book *Animal Play Behavior,* has summarized some of the many ways animals of the same species play together. Play behavior among juveniles can be quite specific and often (though not always) clearly reflects in a kind of miniature the skills that they need to survive when they reach adulthood. Obviously, the more often juveniles practice through play, the more this helps them to acquire and refine those adult skills. As a result play behavior has been subtly modified in many ways. For example, an older juvenile often exhibits quite different, gentler, and less demanding play behavior with a younger animal than when it plays with one of its own age. There must be many times when by accident there is only one member of a particular age cohort in a group of animals. If there were no flexibility in play behavior among the older members of the group, that individual would be at a great disadvantage. The ability to adjust responses appropriately in this way must have constituted an enormous advance in mental powers.

The connections between the play behaviors of animals and the wondrous play behaviors of juvenile and adult humans are numerous and striking and have long been noted by human observers. Fagen points out that Plato, in his *Laws,* had remarked on how dance appeared to have originated in "the habitual tendency of every living creature to leap." Fagen also emphasizes the cooperative aspect of play, and castigates, quite properly, I think, the narrow views of many sociobiologists who assume that every cultural development in humans and animals has arisen because of some narrow and selfish benefit to the individual. Perhaps that was true in Stupidworld—now that we have escaped from it, it no longer is.

A Brief History of the Brain

To understand why it took so long for our ancestors to escape from Stupidworld, we must understand something of how our complex brains came to be. It goes without saying that the brains of our remote ancestors were far simpler than ours, but it is not generally appreciated that they were far simpler than those of comparable organisms alive today.

Our ancestors' brains consisted of a small number of cells of a relatively few types, connected to each other in a small number of ways. The oldest vertebrate brains of which we have a fossil record have been reconstructed (with the help of some imagination) from some small fossilized bony plates and the impressions left in them. Originally, these plates were armor covering the heads of jawless fishes that swam in the warm shallow Silurian seas over four hundred million years ago. As nearly as can be determined, the brains of these primitive fishes consisted of a linearly arranged series of hollow chambers, the walls of which were made up of neural tissue. Even these simple brains, however, already showed the structures that would eventually give rise to the far larger brains of mammals. At the rear, and continuous with the spinal cord, was a swelling making up the hindbrain. Immediately in front were the tiny bulges that made up the midbrain and the pineal organ. In front of these in turn, and directly connected to the olfactory capsules, was the smallest swelling of all, the rudimentary forebrain. These brains were far smaller and less complicated in their structure than those of the present-day jawless fishes, the lampreys and hagfish. Silurian jawless fishes were, it seems, a good deal less intelligent than lampreys. Even so, their tiny brains were among the largest of their time, and made them the intellectual giants of the Silurian world.

In these ancient fossils, traces can also be seen of the channels through which the various cranial nerves passed. These nerves correspond in their position to our own cranial nerves, and the little grooves that show where they used to lie provide a vivid link between our own brains and those simple brains of the past. In much the same way that our own cranial nerves do today, they brought information to the brain from the eyes and other sense organs and transmitted

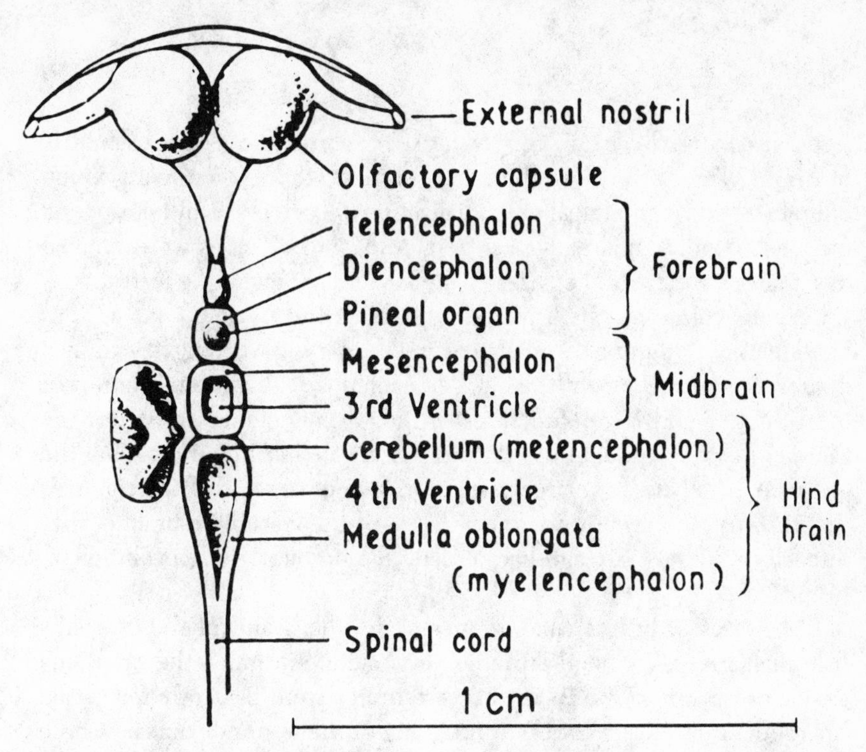

FIGURE 10.2. *Above*, a diagram of the brain of a Silurian fish, little more than a series of enlargements of the anterior spinal cord. *Below*, a cutaway view of a human brain, showing how these simple swellings have become enormously enlarged and transformed.

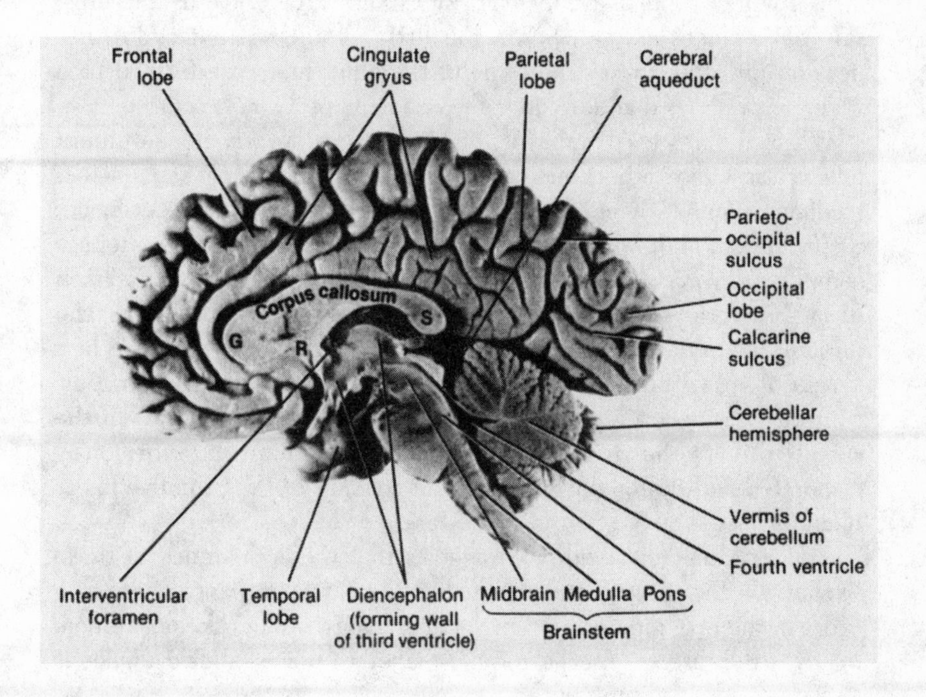

impulses from the brain to the muscles controlling various bodily functions. The existence of these nerves shows that even as early as the Silurian period a good deal of specialization of both brain and peripheral nervous system had already occurred. Already the nerves were capable of sending a variety of messages to the muscles. Some of the instructions sent were voluntary, such as the control of the muscle contractions needed for swimming. Some were involuntary, governing the almost automatic contractions of the gut that accompany digestion. The mix of these higher and lower functions in the various cranial nerves of the Silurian fish was undoubtedly different from the mix in our own, but the nerves themselves were largely in place. When we see the tiny channels that they used to occupy, we know that even the vertebrate brain of four hundred million years ago had already undergone a long evolutionary history.

The series of little chambers making up the brains of these jawless fish must have developed during embryonic life in much the same way as our own brain does. In all the vertebrates from fish to humans, the brain begins as a little crease or infolding of the upper or dorsal surface of the embryo. The cells lining this crease, which might under other circumstances have developed into skin cells, begin to differentiate instead into the neurons and glial cells of the nervous system. As this neural groove grows deeper its dorsal walls grow together and zip up to form a neural tube. The tube then sinks into the embryo as a layer of skin cells closes over it. Soon, the little chambers that eventually develop into the three main parts of the adult brain are formed by a series of swellings that develop at the head end of the neural tube.

When we are tiny embryos, our own brain's early development follows the same path as that of a fish, but soon the simple sequence of swellings of our early neural tube is followed by more complex events as the forebrain, midbrain, and hindbrain grow enormously in size, fold in complicated ways, and set up cellular connections that were undreamed of in the world of the Silurian. Still, in the heart of our brains, the remote descendant of the neural tube can be found in the form of a series of interconnected fluid-filled chambers called ventricles, continuous with a fluid-filled cavity running down the center of the spinal cord. In the final analysis, our brain is simply an enormously elaborated and thoroughly disguised descendant of the primitive linear brain possessed by our Silurian ancestors.

Recent molecular evidence suggests that the appearance of these swellings at the anterior end of the neural tube was perhaps *the* defining event in the evolution of the vertebrates. Invertebrates, like our remote relatives the insects, also have quite elaborate brains—but evidence is

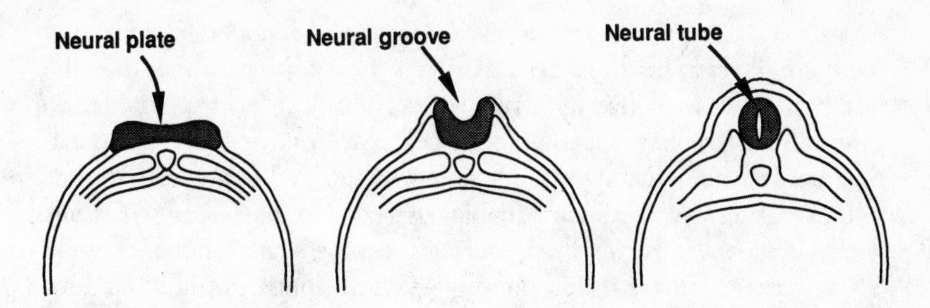

FIGURE 10.3. How the neural tube forms in an early embryo, starting with a plate of modified epidermal cells and finishing with a tube of neural tissue that sinks within the embryo itself, protected now by a layer of epidermal cells that has closed over it.

now accumulating that some parts of their heads and brains evolved independently from our own. Just as with many other structures such as wings, legs, and eyes, brains seem to have evolved more than once in the course of animal evolution.

This is not to say that our brains and those of insects have nothing in common, that their evolution was completely independent. Recent advances in molecular biology have led to the discovery of one large and diverse family of genes that appears to be directly involved in the evolution of the head region. We share these genes with the insects and many other groups of animals, which means that they can all be traced back to a common ancestral gene that must have arisen very early in animal evolution. By the present time, the genes of this family have become very different from each other. They do, however, have one thing in common, a little stretch of DNA called a *homeobox*. Diverse as these various homeobox genes are, they all carry the homeobox as a sign of their common ancestry, much as a phenotypically heterogeneous collection of Scottish Highlanders at a family gathering might indicate their relatedness by displaying their tartan.

The homeobox is about 180 bases long, and it codes for a part of the protein that binds to the DNA of other genes. Although it is not yet known precisely how homeoboxes operate, many experiments have demonstrated that their effects on gene regulation enable them to play a central role in the complex web of gene interactions that characterizes the development of multicellular animals. Because of their influence on many different developmental pathways, I suspect that they will eventually be found to be involved in some of those epistatic interactions that have made it so difficult for geneticists like Dobzhansky to understand how inheritance works in natural populations.

In mice, there are at latest count some thirty-eight different genes making up one particularly important group of homeobox genes, and more are likely to be discovered. Humans probably have about the same number, for in evolutionary terms not a great deal has happened since the time we last shared a common ancestor with mice. Confronted with this fact, we immediately wonder how we managed to acquire so many of these genes. As we all know, copies of genes are passed on to the next generation, but normally the offspring have the same number of genes as the parent. How is it that these genes have somehow multiplied like rabbits?

Homeobox genes have indeed multiplied, although nowhere near as quickly as rabbits. Like the haptoglobin genes we met in the last chapter, they have repeatedly gone through the process of gene duplication. Duplication results in a copy of the gene being inserted elsewhere in the genome, often close to the original gene on the same chromosome. Because gene duplication actually increases the number of genes that the organism possesses, it is perhaps the most powerful of all the mutational processes that have shaped our genome.

When a gene duplicates, the result is an organism that has two genes where there was only one before. If the duplication happens to be precise (sometimes it isn't), then immediately after the duplication both genes have exactly the same function. We have, for example, two essentially identical genes that each specify one of the components of our hemoglobin molecule. This is the result of a duplication event that occurred in the common ancestor of humans, apes, and monkeys, perhaps twenty million years ago.

During the twenty million years since that duplication, slight changes have accumulated in the regulatory regions of these genes, changes that are necessary in order to reduce the amount of hemoglobin being made by each gene. In the future, more mutational changes will accumulate both in the duplicated hemoglobin genes and elsewhere in the genome. These changes, by slowly altering the functions of the genes, will make them more and more unlike each other. Eventually, indeed, it might be very difficult for some geneticist of the remote future to determine, by inspecting the genes, whether they actually had a common ancestor.

Among the homeobox genes, this process of repeated duplication and subsequent divergence has gone on for more than a billion years. By now, the relationships among the most divergent of these genes might never have been detected if it were not for the happy circumstance that their homeoboxes, like the family tartan, have not changed as quickly as the other parts of the genes. The function of the homeoboxes is apparently so important that their sequence has been largely conserved, even though the other parts of the genes surrounding them have been much freer to change.

Different as they have become by this time, our thirty-eight homeobox genes can be grouped into six families, based on the various degrees to which they still resemble each other (see figure 10.4). Genes that fall into the same family have arisen relatively recently by gene duplication, so that they still show a good deal of resemblance. Those that fall into different families, on the other hand, have diverged so much from each other that their closest relatives have actually been found, not among other homeobox families in our own genomes, but

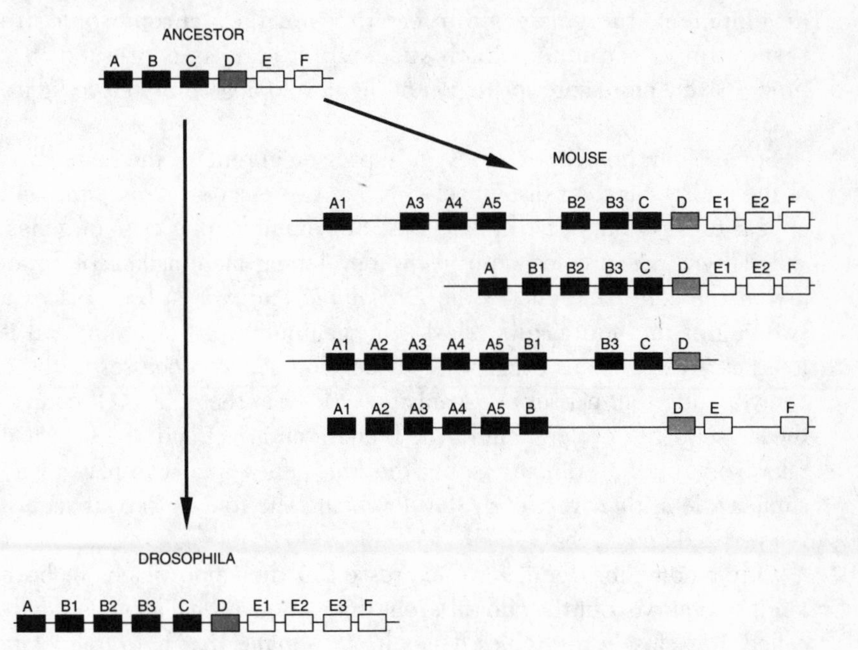

FIGURE 10.4. The history of one family of homeobox genes, known as the Antennapedia family, in the course of the evolution of the insects and the mammals. (The names of the genes have been altered for clarity.) The six genes in this family that were carried by the common ancestor of insects and mammals have undergone a few changes in the insect lineage, and many more in the mammal lineage. In the lineage leading to the mouse, the whole cluster of genes has been duplicated four times, each cluster ending up on a different chromosome. And, within each cluster, individual genes have also become lost or duplicated. The diagram does not show the many tiny changes that have taken place in each gene, but in spite of all these changes genes of type A in the mouse still resemble genes of type A in *Drosophila* more closely than they do genes of type B in the mouse.

among the homeobox genes of the fruit fly *Drosophila.* This means that genes that would eventually give rise to these six different vertebrate families must have arisen by gene duplication very early in animal evolution, long before our own evolutionary lineage parted company with that of the fruit flies.

In *Drosophila* the homeobox genes are grouped into two main clusters. In mammals, more gene duplications have taken place, and now various members of these six families have become grouped into four clusters on different chromosomes. Remarkably, in both flies and mammals, the positions of a homeobox gene on a chromosome actually reflects the places in the body at which the gene acts. During development, the genes within each cluster on a chromosome are turned on in sequence. Each succeeding gene is expressed in a progressively more anterior region of the developing embryo (see figure 10.5).

Some of our homeobox genes are expressed in roughly the same parts of the embryo as their distant relatives are expressed in *Drosophila* and appear to carry out very similar tasks. For example, one class of genes, called *Emx,* is expressed both in the developing mammalian forebrain and at the very head end of the *Drosophila* embryo. In both places a swelling of the neural tube marks the beginning of the brain, and it looks as if these genes trigger similar developmental processes in these two very different classes of organism—at least at the outset. Of course, the subsequent development of the brain of an insect and of a mammal takes very different directions, but the *Emx* genes appear to play a very similar role in their very early development. The role of *Emx,* it seems, is a very old one.

Other homeobox genes are expressed in the same places in both kinds of embryo, but they do different things. A mammalian gene family called *Msx* closely resembles genes in *Drosophila* that help to specify muscle development in the segments of the embryo. The *Msx* genes in vertebrates have become much more versatile, in part because they have undergone more duplications than the corresponding fly genes. As a result they have become involved in the development of a wider variety of organs—the heart, the little buds that develop into limbs, and various structures of the head such as the teeth and eyes.

Finally, some of the homeobox genes have taken on very different roles in mammals and insects. Although our forebrain may have something in common with the brain of insects, our midbrain and hindbrain seem to have had a very different origin. Genes of the *Dlx* family are expressed in our developing midbrain and hindbrain, but in

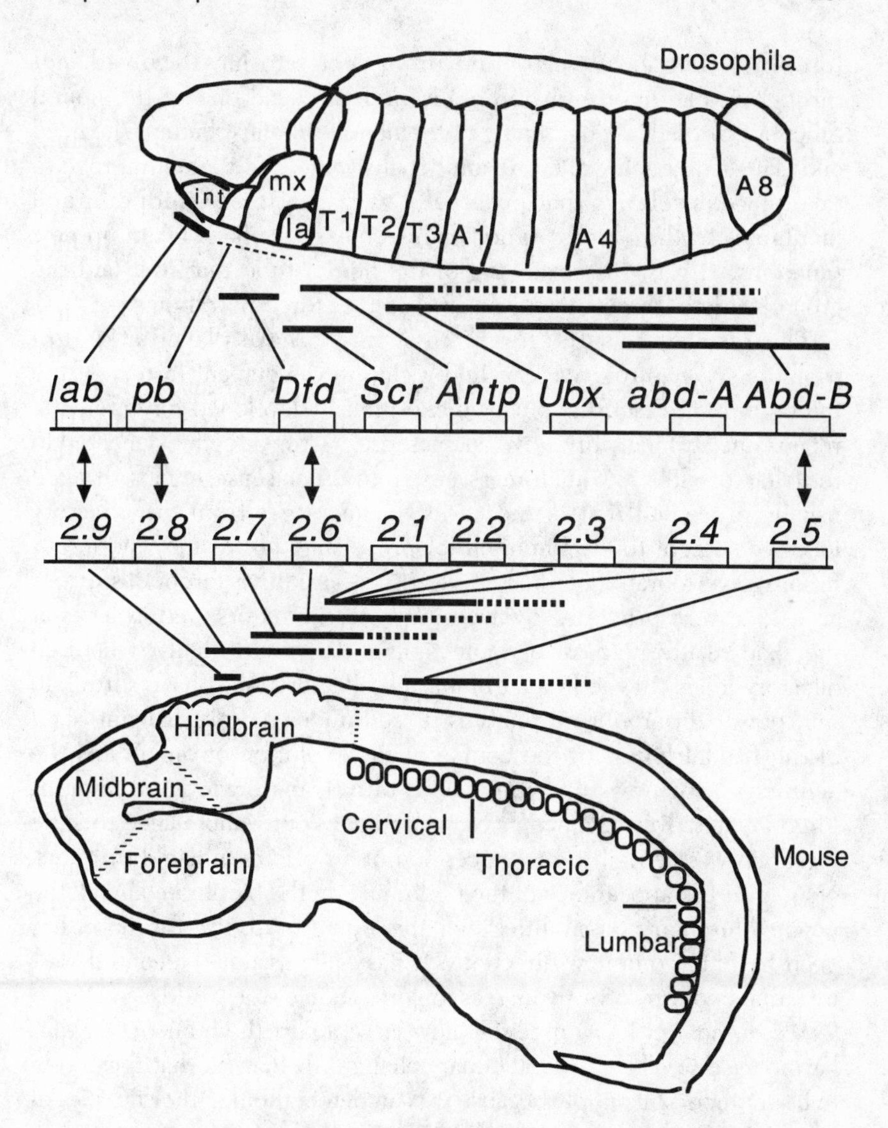

FIGURE 10.5. Diagram showing how the various homeobox genes are turned on in the *Drosophila* embryo (*top*) and the mouse embryo (*bottom*) in approximately the same order as they are arranged on the chromosomes.

Drosophila embryos these genes are expressed much farther back in the neural tube.

We tend to think of the forebrain as being the most recent part of our brain because it is so highly developed and has so many complex functions. But it now seems, ironically, that the midbrain and the hindbrain may actually have had a more recent origin than the forebrain. The early vertebrate brain seems to have evolved, not through the addition of a forebrain onto the head end of the neural tube, but through a lengthening of the neural tube just behind the head end. This lengthening allowed additional swellings of the neural tube to take place, swellings that eventually gave rise to the midbrain and hindbrain. In the process various homeobox genes have taken up new functions. It was this stretching of the neural tube that first laid the groundwork for our own enormously complex three-part brains.

The result is that the forebrain seems to have played a kind of evolutionary leapfrog. We can find a clue to the ancient history of the forebrain in the fact that our own enormously developed forebrain still retains one strikingly primitive characteristic—it is directly connected to the olfactory lobes. Unlike messages from other sense organs that are usually processed first in the midbrain, messages from the olfactory lobes go straight to the forebrain for processing, just as they did in our primitive vertebrate ancestors. Indeed, the primitive forebrains of our ancestors were primarily involved in processing odors, and as a result they had relatively massive connections to large and highly structured olfactory lobes. In the human brain, the olfactory lobes have shrunk to tiny, poorly differentiated nubbins tucked under the overhanging shelf of the frontal lobes. In the course of the evolution of our brains the acquisition of new functions has totally transformed the ancient forebrain and overwhelmed its original role. In our remote ancestors the forebrain was central to the perception of a world dominated by odors. Now, after the appearance of the midbrain and the hindbrain and all the new evolutionary possibilities that this opened up, the forebrain has been freed to go in new directions. It has again assumed a central role, this time in our own vastly increased perception of the world.

We do not yet know precisely how large a part the homeobox genes have played in all these evolutionary changes. It may be that they seem to be so important simply because we can detect them. Other families of genes, yet to be discovered, may turn out to be much more critical. Nonetheless, the pattern of evolution exhibited by the homeobox genes, through gene duplication and subsequent differentiation, is similar to the kinds of evolution we talked about in the last chapter. Whenever homeobox genes increase in number through gene duplication, the neural tube reflects these changes by lengthening and becoming more complex. In the terminology of the previous section of the book, homeobox-gene duplication changed the evolutionary potential of the developing embryo, and the subsequent differentiation of these duplicated genes helped to realize that altered potential. Selective pressures on the brains of vertebrates have been unrelenting, as the world has become more complex and their competitors smarter. We have many more homeobox genes regulating our brains' development than *Drosophila* does, and it seems that we need them all.

The Building Blocks of the Brain

The psychologist R. L. Gregory once remarked that the brain is hard to study because it resembles nothing so much as a lump of porridge. All lumps of porridge look alike as they sit on a plate. If we are to fully understand the differences between ourselves and other organisms, we must penetrate below the slight superficial differences between our lump of neural porridge and those of other animals and look at the detailed differences in the structures and biochemistry of the cells that make it up. At the embryological level, our brains are simply enormously developed swellings of that primitive neural tube. At the cellular level things are much more complex. The most important cells in our brains are the neurons.

As we saw in the last chapter, neurons appear to be related to the pigment-producing melanocytes of the skin, but the evolutionary split between these two classes of cell must have occurred in the very distant past. Tapeworms and other flatworms, which are among the simplest multicellular animals, already have melanocytes and neurons that look very much like ours, and their neurons respond to many of the small peptides and other chemicals that our own neurons use to communicate with each other.

Neurons are truly remarkable cells, the most specialized in the entire body. In the brain they are concentrated in the gray matter, a 2-millimeter-thick layer on the outer surface of the cortex. They are also found in the white matter, or medulla, that lies underneath the cortex, but not in such immense numbers.

When nineteenth-century microscopists first examined the crowded confusion of the cortex, they were able to see little more than a great many rather ordinary-looking cells, each with a nucleus. It was not until late in the century that the true complexity of each neuron was revealed, through a happy discovery by the Italian anatomist Camillo Golgi. Golgi found that he could stain not only the main bodies of the neurons but also some of the heretofore-invisible fine processes that extended out from them. He soaked thin slices of brain tissue in potassium bichromate and silver nitrate. The silver diffusing into a neuron would suddenly precipitate, revealing the cell in far more detail. Of course, if all the neurons in the slice had been stained by Golgi's method, they would have formed an impenetrable tangle under the microscope, but serendipitously his stain reacted with only one in every few hundred neurons. (Many other kinds of stain have this odd property, by the way—stains that are specific to a certain kind of protein often penetrate only a subset, and sometimes only a small subset, of the neurons in a slice of brain tissue. Although they all have the same general appearance under the microscope, neurons are an extremely diverse collection of cells at the biochemical level.)

Golgi, and later the Spanish physician Santiago Ramón y Cajal, were able to dissect the structure of neurons and determine how they grew in the course of the brain's development. Each neuron sends information to other neurons or to muscle fibers through a fairly thick process called an axon, while it receives information from other neurons or from the cells of sense organs through thinner, more complexly branched processes called dendrites. The term *axon* comes from the same root as *axis*, and *dendrite* is from the Greek for *tree*—though these terms are a bit misleading because axons are often almost as multiply branched as dendrites. Sometimes an axon grows straight to another neuron or muscle cell, and sometimes—especially in the cortex—it branches repeatedly and comes in contact with the dendrites or the cell bodies of many other cells.

Impulses travel along the length of an axon or a dendrite in the form of waves of altered membrane potential. Although the membrane of the axon or the dendrite normally excludes sodium ions, whenever it is stimulated a small region at the site of stimulation reverses its polarity and allows sodium ions to rush in. The region of reversed polarity then moves like a wave down the length of the axon or the dendrite. If it were to be left on its own the signal would soon tend to die away, but as it travels down the various branches it may be either boosted or suppressed by signals coming in from other neurons that have attached their own axonal processes at many different points along its path.

The places at which axon and dendrite come in contact, known as synapses, are essential to the function of the nervous system. When a wave of reverse membrane polarization reaches the tip of an axon, it stimulates the release of little packages of chemicals called neurotransmitters. Each little package is wrapped in its own layer of membrane, and it drifts rapidly across the tiny gap, only twenty-millionths of a centimeter across, that separates the axon from the dendrite.

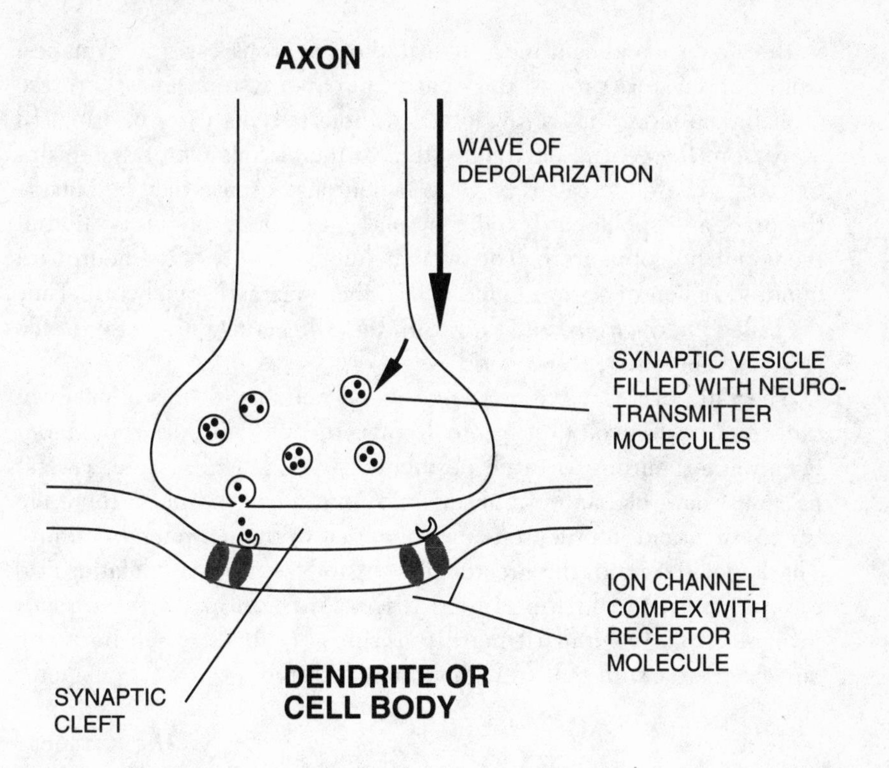

FIGURE 10.6. Diagram of a synapse. When a wave of depolarization arrives at the tip of an axon, it stimulates the release of tiny synaptic vesicles filled with neurotransmitter molecules. These molecules drift rapidly across the gap separating the axon from the cell body, or dendrite, of the postsynaptic cell on the other side of the gap, where they change the properties of the postsynaptic cell's membrane and can either stimulate or prevent a new wave of depolarization. Because many kinds of axons secreting different kinds of neurotransmitters can be connected to various parts of each postsynaptic cell, the flexibility and potential complexity of the system is enormous.

Unlike a contact between two transistors on a computer's circuit board, synapses do not simply transfer information unchanged from one part of the neural circuit to another. Each synapse is far more complicated and flexible than that. There are at least fifty different neurotransmitters, each playing its own special role in the transfer of information and each with its own set of receptor molecules waiting on the other side of the synapse to recognize it. The flexibility of such a system of connections, and its ability to change over time, is immense. Throughout the life of the organism each neuron undergoes subtle alterations that change the numbers and kinds of its many different synapses. Intense efforts are being made to fathom how these changes result in short-term and long-term learning.

In our own bodies a mere five million neurons carry information from our sense organs to the central nervous system, and there are probably no more than a few hundred thousand that carry information away from the central nervous system to the various motor systems of the body. These are called *peripheral* neurons because they lie outside the brain and spinal cord, and they make up only a tiny proportion of the population of neurons. The overwhelming majority of the neurons, a hundred billion or so, are found within the brain and spinal cord. They are called *interneurons,* and they form the connecting links between the world we perceive and the world we act on.

The ratio of interneurons to peripheral neurons tells us a lot about the remarkable evolution of our brains. In rats there are only about twenty interneurons for each peripheral neuron. The Silurian jawless fish must have had an even lower ratio. In humans, however, there are twenty thousand interneurons for every one of the peripheral neurons. The larger the ratio, the greater the possibilities for manipulating and comparing information. Indeed, this burgeoning collection of interneurons is what ultimately defines the difference between ourselves and our distant and dim-witted ancestors.

Is Brain Size the Way to Escape from Stupidworld?

Obviously, although our brains are much bigger and more complicated than those of Silurian fish, animals like elephants and whales have brains much bigger than ours, so sheer size alone is not enough to explain the unique mental properties of humans. At the beginning of this book I suggested that one important difference between our brains and those of any other animal is that our brains do a great deal of their development after birth. During that critical development period we are exposed to far more complex and varied stimuli than any other animal, for much of their brain development takes place when they are sequestered in the womb. Are there any other things that particularly mark our brains out? Yes, there are, although often they are not unique to humans. Consider the microscopical organization and structure of the brain itself.

You might suppose that our brains would be more densely packed with neurons than those of animals that have simpler neural equipment. Generally, however, the reverse seems to be true. The cortex of a rat's brain is packed with one hundred thousand neurons per cubic millimeter, while that of a human may have as few as ten thousand. (The reason we have so many more neurons than a rat is because our brains are much larger and have a much thicker cortex.) This paucity of neurons per cubic millimeter does not mean that our brains are simpler than those of rats, however—quite the reverse. We have fewer neurons because each neuron has far more dendrites and a much more complexly branched axon. Under the microscope the central cell bodies of human and rat neurons seem to be about the same size. This does not reflect their true sizes, for if you could take a neuron from a human and one from a rat, including all their branched dendrites and axonal ramifications, and squeeze them down to a sphere, the squeezed human neuron would have up to ten times the volume of that of the rat. Each human neuron can make contact with between ten and a hundred times as many other neurons as a rat neuron can. Couple this with the fact that we have ten thousand times as many neurons as a rat, and it is hardly surprising that our brains are far cleverer. Plate 18 shows the astounding ramifications of a single neuron in the hippocampus of a rat—human neurons are even more complex!

Further, the effect of the environment, even on the simple neurons of a rat brain, can be enormous. When Marian Diamond studied the brains of rats raised in deprived and enriched environments, she found that the more detailed her examination, the more pronounced were the differences she could detect. Neurons from rats raised in an enriched environment showed the effects of that environment in the number and type of connections that they made with other neurons. Still, not even the most complicated neuron from such an enriched-environment brain approached the complexity of human neurons. That difference has been the result of a long period of evolution.

Even though there are great differences between rat and human neurons, they still lie more in the direction of quantity than quality. The fundamental organization of the neurons is the same—we have not evolved neurons that are completely different from those in the brains of rats. Rats and humans also make the same variety of neurotransmitters, although the mix is a little different. It may be that relatively minor changes in regulatory genes have been enough to produce the increased numbers and complexity of neurons in the human brain.

However, although there are pronounced differences between the neurons of various animals, the farther we back away from the cellular level and approach the "porridge" level, the smaller the differences become. If a slice through the cerebral cortex is examined in great detail, it can be seen that the neurons in the slice are organized into tiny columns of cells that extend from the lowest to the highest layer of the cortex. The cells of these columns are in turn connected in complicated ways to the medulla underneath. Many neurophysiologists believe that it is these columns and not the individual neurons that are the fundamental unit of the brain's structure. Reinforcing this idea is the fact that the number of these columns per unit surface area of the cortex turns out to be about the same in brains from a wide variety of animals. At this level of organization the cortex is a tessellated pavement of neural columns, with more and more columns simply being packed in as it expanded in the course of evolution.

The differences become even smaller at higher levels of organization. In his 1977 book *The Dragons of Eden,* Carl Sagan popularized a view of the brain as a progressively more sophisticated series of layers, each added on to the previous ones and with the forebrain making up the latest and most sophisticated addition. However, as we saw, new molecular biological evidence indicates that the forebrain may in fact be the oldest part of our brain—and, although the forebrain has undergone enormous changes, the midbrain and the hindbrain have not stood still during our evolution. Popular discussions notwithstanding, our brains are not the result of an immensely expanded, overgrown cerebral cortex that has come to dominate the midbrain and the hindbrain through a wild mushroomlike growth.

Harry Jerison of the University of California at Los Angeles has carefully examined the relative sizes of the various parts of the brain, and finds that to a surprising extent they increase in size in concert. The brains of anthropoids, including the monkeys, apes, and man, can vary nearly twentyfold in size from one species to another. Yet the forebrain remains quite constant at about 75 percent of the total brain. Over a tenfold range of brain sizes among the prosimians, including our more remote relatives such as lorises and lemurs, the forebrain makes up about two-thirds of the total brain mass. The human brain is not out of line with the brains of our near relatives. The cerebellum, the hippocampus, and the pons have all increased in synchrony with the cortex. At this level the only unusual feature of our brains, which we share with those of the other anthropoids, is a dramatic shrinkage of the olfactory lobes compared with those of other mammals.

We are also not out of line compared with our near relatives when it comes to other measures of the brain. One of these measures is the surface area of the cortex compared with the volume of the brain, and another is the number and complexity of the cortical folds. By both measures the human brain is a typical anthropoid brain. It seems that the farther we back away from the neurons, the more ordinary our brains become.

Where we are out of line, however, is in our degree of *encephalization.* This is a term used to describe the relative sizes of brain and body. Over the years, and after much agonizing, it has been generally agreed that a reasonable measure of encephalization can be obtained by determining the ratio of brain size to body size and then comparing it to the ratio you would find in an average animal of the same body size. This is called the encephalization quotient.

By this definition, of course, an average animal would have an encephalization quotient of 1. If we search out mammals with the same body size as ourselves and then compare ourselves to them, we find that our encephalization quotient is much greater than 1, meaning that our brain is larger than it "ought" to be. Of course, how much larger depends on the group of animals with which we are being compared.

If we match ourselves with other primates, for example, our encephalization quotient is about 3—we have three times as big a brain as we would have if we were an average primate of the same size. If we compare ourselves to our more remote relatives the prosimians, our encephalization quotient rises to 10. (There are no living prosimians anywhere near as large as we are, although some very large lemurs inhabited Madagascar as recently as historical times. We can, however, determine how big a human-sized prosimian's brain would be by extrapolation.) Or, if we compare ourselves to yet more remote relatives such as the shrews and hedgehogs, our quotient rises to 25.

So our brain size is well out of line compared with that of most other mammals, but not as far out of line when we compare ourselves with our closest relatives. Although our runaway brain clearly marks us off from other animals, we must ask whether such atypically large brains are really that unusual, or have similar discrepancies appeared at earlier stages in the evolution of life?

As the age of dinosaurs approached its dramatic end, a number of strikingly advanced dinosaurs appeared. One of these was *Stenonychosaurus inequalus,* remains of which have been found in Alberta. This dinosaur looked rather like an ostrich, although the similarity was more in its appearance than its habits. It had forelimbs instead of wings, and the digits of its forefeet were remarkably flexible—it has even been suggested that its "thumb" had a degree of opposability.

Enough skulls of various dinosaurs have been found to enable estimates to be made of the size of an average dinosaur's brain, and *Stenonychosaurus* turns out to have an encephalization quotient almost ten times as great as the dinosaur average. Its quotient actually overlaps those of living birds and mammals that have the lowest quotients for their group. This made *Stenonychosaurus* an intellectual giant among dinosaurs—but of course it didn't have much competition! Now the world is filled with much smarter animals. Even ostriches' brains are larger than the brain of *Stenonychosaurus,* but these birds are still not terribly bright—an instance has been recorded of an ostrich accidentally getting itself tangled in the fork of a tree and, in a panic, pulling off its own head.

It has been seriously proposed that *Stenonychosaurus,* with its opposable thumbs and binocular vision, might—if not for that infamous asteroid—have eventually given rise to descendants as intelligent as the mammals. Perhaps, although it seems it had a long way to go! We will never know whether or not *Stenonychosaurus* had a rosy evolutionary future that the asteroid nipped in the bud. Still, the fact remains that the brain of this creature from eighty million years ago was as huge compared with those of the rest of the dinosaurs as our own brain is compared with those of our distant prosimian relatives such as the potto and the loris. This phenomenon is not unique to the dinosaurs. In every group of vertebrates that has been examined, including extinct groups, the encephalization quotients of the members of the group do not fall along a neat line when they are plotted against body size. Instead, they tend to fan out. Some species have high quotients and others have low quotients compared with the average of the group.

In the world of the dinosaurs, it seems, *Stenonychosaurus* was a mental giant. It is difficult for us to imagine an entire planet in which the smartest creature had roughly the intellectual stature of an ostrich, even though that world existed a mere sixty-five million years ago. How much more difficult it is to imagine a world with such poverty of intellect as must have been exhibited by the creatures of the Silurian period. Remember that, to judge by their brain sizes, even the parasitic jawless fishes of today have a brainpower that far exceeds that of the jawless fish of four hundred million years ago.

The evolution of the brain of *Stenonychosaurus* was certainly not driven by the kind of runaway cultural change that has catapulted our own brains into the upper end of the encephalization quotient. Nonetheless, it seems that a slower and less dramatic brain-environment feedback loop was at work during the age of dinosaurs, just as it was during the age of mammals before the accretion of culture began to accelerate the process in our immediate ancestors.

We think of ourselves as being alone on an intellectual pinnacle, but in fact we are surrounded by creatures that are, in absolute terms and in contrast to what has gone before, almost as smart as ourselves. Our brains are the result of runaway evolution, but, as Harry Jerison has pointed out, the brains of many of the animals and birds that inhabit our current world have become larger as the result of a less powerful but fundamentally similar process of accelerating evolutionary change.

So, it is plain that our brains occupy the most recent stage of a long evolutionary story. How do they differ, both genetically and functionally, from the brains of Stupidworld? And how do they differ from the brains of the animals that have accompanied us part of the way in our escape from Stupidworld? What really separates our brains from those of other animals?

11

The Master Juggler

Perhaps no mortal has ever thought as hard and continuously as Newton did in composing his Philosophiae Naturalis Principia Mathematica. . . . *Never careful of his bodily health, Newton seems to have forgotten that he had a body which required food and sleep when he gave himself up to the composition of his masterpiece. Meals were ignored or forgotten, and on arising from a snatch of sleep he would sit on the edge of his bed half-clothed for hours, threading the mazes of his mathematics.*

—Eric Temple Bell, *Men of Mathematics* (1937)

Some years ago I attended a performance by the Royal Shakespeare Company of a play by Arthur Wing Pinero. In the course of the performance, and with remarkable cunning, one of the actresses performed a magic trick.

The trick was the very reverse of the usual actor's craft. The actress played an old lady who was largely ignored by the other characters. At one point, in the course of a complicated and emotion-filled scene, she sank into a high-backed Victorian chair in full view of the audience. The scene was played out, and at the end of it the other characters made their exits in assorted states of dudgeon and despair, leaving the stage empty.

Until the actress suddenly moved. Then the entire audience of several thousand uttered a single gasp of astonishment, for she had sat so quietly in her chair that we had all completely forgotten her existence. Our attention had been totally engaged by the other actors, so that, although her image had undoubtedly impinged on all our retinas, it had been ignored by all our minds.

Magicians, of course, use similar kinds of misdirection all the time, but in most cases the audience never knows that it has been fooled. Magicians, like the actress, depend on the fact that we all are subject during our waking moments to an enormous influx of information from our senses. This influx is so vast that we are forced to prioritize and concentrate only on the parts that previous experience has shown us to be relevant. For most of us, these priorities can be swiftly and easily changed, but even so there is no way that we can pay attention to everything at once.

One thing our brains have evolved to do very expertly is to make decisions about which parts of this deluge of information are likely to be important. We are also, without apparent effort, able to juxtapose two or more inputs, actual or remembered, and to draw inferences about the ways they are related. This enables us to act accordingly and to communicate these inferences to others. Perhaps most astonishingly, we are able to take fragments of remembered and half-remembered information and create something startlingly new from them.

When Walt Disney created Mickey Mouse, nothing like Mickey had ever been seen or imagined before, although certainly Disney's invention drew on various aspects of both people and mice that he had known. Thanks to filmed animation, Mickey's many attributes, particularly his somewhat raffish early persona, have been preserved for everyone to see. This would never have been possible if Disney had been confined to drawing still pictures of his creation. Given access to animation, Lucas Cranach or Hieronymus Bosch would undoubtedly have left us even more vivid (and terrifying) images than those that populate their canvases. And the Neolithic artists of the Perigord could certainly have presented us, not with static images of bison, but rather with whole thundering herds.

Is the human brain unique in this remarkable ability? Of course not, for other animals must juggle sensory inputs as well, and juggle them expertly. We differ from other animals not in the ability to juggle but in the number of balls that we can keep in the air simultaneously.

Using the Whole Brain

The rate at which complex electrophysiological pulses travel along an axon or a dendrite is about 100 meters per second, only a millionth of the speed that an electrical impulse can travel through the circuits of a computer. Because of the computer's great advantage in speed, its designers have been able to concentrate all of its activity at a single point called the processor. The processor performs arithmetical operations one after the other, in an order that is specified by the programmer. The British cryptographer Alan Turing pointed out that such a processor could eventually carry out even the most complex arithmetical calculations. His insight opened up tremendous possibilities for computer design. By now, however, this single processor has become a bottleneck preventing further advances.

So-called massively parallel computers, designed to try to circumvent the Turing bottleneck, are really nothing but a great many processors operating simultaneously. However, it turns out to be extremely difficult to design programs that can utilize such a collection of Turing bottlenecks without making some of them too busy and leaving others idle. Each program must be carefully and individually tailored to ensure that each Turing bottleneck gets a fair crack at the data. (New computer chips are now being designed, however, that are more neuronlike in their properties.)

The brain, with its sluggardly impulses traveling down innumerable pathways simultaneously, has no Turing bottlenecks but has found other, and far more subtle, ways to process this surging maelstrom of impulses. Because it is so immensely flexible in the way it can juggle inputs, the brain does not resemble an ordinary computer in the least. The programming of the brain is so flexible that there are always many alternate pathways for data to be processed—or nearly always.

Because I have a "lazy eye," when I am tired, particularly when I am reading, my left eye tends to wander off the printed page on an errand of its own, leaving the right eye to do the work. This does not present a problem because, when it happens, I simply ignore the input from my left visual field. At first this suppression of one of my two visual fields must have been quite unconscious, but now that I have noticed it I find that I can easily replicate it. When I deliberately cross my eyes and separate the two visual images, the image from my right eye immediately predominates in my perception, and that from my left eye largely disappears.

With some effort, I can reverse these priorities, by concentrating on the left-eye image and ignoring that of the right. This is not as satisfactory, however, for the left-eye image tends to dance about, and the right-eye image keeps popping back into my attention if I do not keep it, with a deliberate effort, at bay.

Recently, in the laboratory of Vilayanur Ramachandran at the University of California at San Diego, Ramachandran and I tried some simple tests to see whether we could localize where in the brain this prioritizing was taking place. Rama gave me a pair of special spectacles that are normally used for playing three-dimensional computer games but that he has adapted ingeniously for his experiments. These spectacles have liquid crystal lenses that can flick instantaneously from dark to transparent and back again. They are adjusted so that when one lens is dark the other is transparent and vice versa, and they flick back and forth several times a second.

The glasses were attached to a computer, on the screen of which two different images could be displayed. The computer, in synch with the glasses, displayed first one image and then the other. When I was not wearing the glasses, the images both occupied the center of the screen, flickering rapidly from one to the other. When I put on the glasses, one eye saw one image, and the other eye saw the other. Because of persistence of vision, each image appeared to be steady. When I crossed my eyes, I could separate the two images.

Normally my tired left eye, when it moves, takes with it an image that is essentially identical to the one the right eye sees. The unusual feature of Rama's setup was that the computer could be programmed so that each eye saw a different image. This trick enabled him to ask what properties of the image might be important when I suppressed the image from my left eye. Could I manage to suppress images in my left visual field that clashed with the image my right eye saw?

The answer was that I could. We started with two identical white spots. I could easily move the left-visual-field image away from the right and then suppress it. Then, changing the color of the spots, Rama made the spot I saw with the left eye red and the one I saw with the right eye green. Again I had little problem. My ability was not a function of the colors involved, for I was able to suppress the left-eye image even when the colors were switched. I even had little difficulty in suppressing the left-visual-field image when it consisted of two spots rather than one and when the pair of spots occupied different positions relative to the spot seen by the right eye.

All this suggested to Rama that the suppression process was taking place quite early in the hierarchy of perceptions that constitute the brain's visual pathway. It was possible, he thought, that this juggling of priorities between the visual fields was occurring somewhere in the lateral geniculate nucleus near the base of my brain, where images from my eyes are first processed. If it were farther up, at a stage in processing where colors and shapes are distinguished, it should have been harder for me to suppress left-visual-field images when they were very different from those of the right eye.

But things turned out not to be so straightforward. Now that I was confronted with this computer-generated set of images that simplified what was normally a very complicated visual field, I was able to detect an interesting phenomenon. Whenever I blinked during the period when I was deliberately trying to suppress the left-eye image, the suppressed image popped back into my attention momentarily. Then I could make it disappear again. Rama checked the existence of this phenomenon by hitting a key on the computer's keyboard in front of me that made the right eye's image blink off for a moment. When it disappeared, the suppressed images in my left visual field suddenly blossomed into my attention.

"Let's try a control," he said. His hand darted toward the keyboard but did not depress the key. His trick was so unexpected that I was not sure what I had seen. "Try it again," I suggested. This time, the left-eye image suddenly leapt into my consciousness as his hand was starting toward the keyboard! I had anticipated what he was going to do, and the way my brain juggled the inputs from the two visual fields had reflected that anticipation. This, as Rama pointed out, was not a phenomenon connected with the brain's base. It was taking place somewhere in my cortex. It was apparent that several parts of my brain, at widely different levels of organization, were taking part in the suppression process. It was not a case of the more advanced cortex dominating the activities of the less advanced midbrain—all the parts of the brain contributed to the suppression.

Rama has conducted similar experiments with people with involuntary strabismus caused by weakness of the muscles of one eye and finds similar phenomena. My advantage as an experimental subject was that I could deliberately suppress one of the two fields, while people with involuntary strabismus have been suppressing one field for so long that it has become largely automatic.

Thinking Like a Fish

The brains of our ancestors were able to juggle sensory inputs quite well, although without a doubt the human brain currently holds the title of master juggler of them all. This image of the brain as master juggler gains force when we consider brains that, for a variety of reasons, are unable to carry out this function normally. By comparing the normally functioning human brain with such brains, we can begin to understand more clearly what separates our brains from those of other animals.

Burst blood vessels, disease, or external injury can destroy part of the brain or deprive a specific region of needed oxygen. In 1839 the French doctor Jean Baptiste Bouillaud examined a man who had attempted suicide and found that injuries to the frontal lobes had deprived him of the ability to speak. Bouillaud showed that the man could still understand speech perfectly normally. Paul Broca, building on Bouillaud's observations, demonstrated that this inability to speak could be localized quite precisely to a region now known as Broca's area, at the base of the frontal lobe. Broca's area is found in only one of the two lobes, usually the left. This was the first clear indication that specific parts of the brain are involved in specific functions.

Different functions were soon localized to other parts of the brain. Experiments on monkeys showed that much of the processing of visual impulses is carried out in a very different region of the brain, the occipital lobe at the brain's very rear. Not all of it, however, takes place there. There was a raging argument at the end of the last century about whether removal of a monkey's occipital lobes made it blind—some investigators found that it seemed to, others found that monkeys behaved as if they were blind immediately after the operation and later recovered, and one experimenter claimed that they were hardly affected at all.

Why was there such a wide discrepancy? The visual pathway, like all the other pathways that have been investigated in the brain, is a highly complicated one. There are at least six different routes taken by retinal nerves to different parts of the brain. The main one goes to the ventral part of the lateral geniculate nucleus, a knee-shaped cluster of cells in a small structure called the thalamus that is buried deep in the middle of the brain. Another major branch leads to the superior colliculus, a structure tucked behind the thalamus. Fascinatingly, among nonprimate animals, this second branch is more substantial than the first—at some point in the early evolution of the primates, the collicular pathway became less important than the path leading to the lateral geniculate nucleus.

Still other branches lead to other parts of the midbrain. Arising from all these different structures, a bewildering neural network leads to various regions of the cortex. The best studied of these interneuronal pathways is the one that leads from the lateral geniculate nucleus to the very rear of the brain. There it fans out into the center of the brain's occipital lobe, an area known as the striate cortex. In the striate cortex, the different parts of the visual field can actually be mapped onto the surface of the brain, and it was this area that was removed in the monkey experiments.

The striate cortex is the most studied of the regions involved in vision, partly because of its accessibility. Much less is known about the other pathways, although other seemingly less-detailed maps of the visual field have been located in dozens of different parts of monkey brains. How much do these other paths contribute to vision, and what kind of contribution is it?

Some idea of the extent and nature of this contribution comes from studies that were carried out by Lawrence Weiskrantz of Oxford University on a very unusual patient. The patient, named D. B. in the literature to protect his identity, was a young man who, by his early thirties, was suffering almost continuously from migrainelike headaches. Frighteningly, these headaches were accompanied by the appearance of a huge blind spot in his left visual field. His problem was traced to a cluster of malformed and engorged arteries at the very rear of his right occipital lobe. An operation to remove this arterial mass was successful, but at the cost of removing all his right striate cortex. After the operation, he was functionally blind in his left visual field, a condition called hemianopia. This was a relatively small price to pay, for the migraines disappeared and he was able to resume a normal and very productive life.

A surgeon at the hospital, Michael Sanders, noticed that after the operation D. B. was able to do something quite unexpected. When Sanders held out his hand in the region of D. B.'s blind left visual field and asked D. B. to reach for it, he could do so almost as accurately as if he could actually see the hand. Yet D. B. insisted that he was indeed blind in that half of his visual field, and tests showed that he was normally unresponsive to events happening there unless his attention was somehow drawn to them. When it was, then in some indefinable fashion he could see, or at least perceive, them.

Sanders told Weiskrantz about the remarkable abilities of D. B., and D. B. himself turned out to be eager to learn more about his strange situation. Weiskrantz subjected him to detailed tests during a period of over a decade. He found that when a spot of light or a line drawn on a piece of paper were presented to D. B. in his blind visual field, he could correctly guess the position of the spot or the orientation of the line more often than would be expected by chance. When told that he had guessed correctly, he was astonished, for he claimed to have seen nothing. Weiskrantz called this phenomenon blindsight. Blindsight is not unique to D. B. A year earlier, Ernst Pöppel had described four brain-injured patients with similar abilities, and many others have been studied since.

One striking feature of blindsight was recently pointed out by Daniel Dennett in his book *Consciousness Explained.* In most cases, Dennett noted, the subject had to be prodded. It did no good for the investigator simply to position a spot of light in the blind visual field and await a response, for none would be forthcoming. The subject had to be *asked* to guess where in the visual field the spot might be.

There was one exception to this, and it had to do with motion. When a spot of light was adjusted to move up and down cyclically on an oscilloscope screen, D. B. would not respond until the speed of the spot's motion exceeded a certain threshold. Then he would claim to see or feel motion. The visual image that he claimed to see was not a moving spot but rather a curved gridwork of lines, sometimes moving ever upward, sometimes radiating from a central point.

During all these experiments the visual input from D. B.'s left eye was normal, and presumably the input proceeded normally through the branching pathways of the thalamus and the superior colliculus. It was how this input was perceived by the higher visual centers that had altered. It is something of an oversimplification, but one way to describe D. B.'s condition is that he was thinking like a fish. A fish, presented with a stationary spot of light, does not respond. It does respond to a rapidly moving spot, for such an object often spells danger or food. A fish's brain can juggle priorities and respond appropriately (for a fish) but over a much narrower range than the responses of a human brain. Destruction of D. B.'s right striate cortex destroyed, not his ability to perceive stimuli, but his ability to juggle them in a humanlike fashion. In that half of his visual field, he had lost his ability to pay attention to changes in stimuli that would interest a human. Instead, he had been reduced to the abilities of a fish.

Experiments, probably impossible to do, suggest themselves. Suppose D. B. had volunteered to have his functioning right visual field covered for some period of time, rendering him—at first blush—functionally blind. Under those circumstances, forced to pay attention to the crippled inputs coming from his left field, would he have begun to prioritize them in a more human fashion? How would his perception of the world have changed over time—would he, like those monkeys that had seemed unaffected after their striate cortex had been removed, learn to move about his world with some assurance without actually seeing it in the way that normal people do? Would he finally, actually, come to see it, and how would his sight differ from that of normal people?

Perhaps he would learn to overcome his blind spot, because the resilience of the brain is legendary. Monkeys with very large brain lesions have often, over a period of months, been able to recover much of their cognitive function. Occasionally, people with essentially normal brain function have been found on autopsy to have one entire half of the brain missing. When such anomalies arise during early development, they give ample time for the circuits of the growing brain to readjust and compensate for the brain's missing parts. Indeed, because of the brain's redundancy, many extra circuits are already in place and ready to begin functioning if an injury takes place elsewhere in the brain or elsewhere in the body. Ramachandran and his co-workers have recently reported on a young man whose left arm had been amputated above the elbow just four weeks earlier—far too short a time for new neural pathways to have been established. They found that this young man had, on the lower left part of his face, a kind of map of the missing arm. If a part of his jaw was touched lightly, he would report sensation both from his jaw and from a specific part of his missing hand. It seems likely that the map of the hand had always been there—the region of the cortex that processes sensation from the face lies adjacent to the region that processes sensation from the hand, and there could easily be some sharing of neurons between the two regions. Until the hand itself had been removed, however, the map of sensations on his face had been repressed, presumably at some fairly low level. After the amputation, his brain began to juggle the inputs inappropriately, confused by the disappearance of the flood of real sensations that had formerly come from his arm.

It is very likely that, given time, this odd effect will fade as the young man's brain learns to juggle the inputs properly. Yet Ramachandran's experiments on the young amputee, performed during what was probably a brief window of opportunity, show just how redundant the brain's systems really are. Our ancestors did not simply add functions one at a time to their brains. Many parts of the brain are involved in each function, and, although the various maps of the world that are found in different parts of the brain have different properties, one map may be able to substitute for another to a surprising extent.

Juggling a Single Ball

Thomas Fuller, a black slave living on a Virginia plantation at the end of the eighteenth century (he died in 1790), was a calculating prodigy. With no education, and after only the sketchiest instruction in arithmetic, he was able to calculate in his head the number of grains of wheat needed to sow a particular field or the number of shingles needed to cover the roof of a barn. He was also fascinated by surveying and could accurately remember the lengths and directions of lines that had been surveyed months or years earlier.

Yet, despite his prodigious memory for numbers, he had no memory for faces. He could not recall people who had spent hours talking to him the day before. It is possible but unlikely that this was actually a clinical condition called prosopagnosia, in which damage to the brain renders its victim quite unable to recognize faces (although able to recognize other things). Of course, accurate diagnosis is impossible after a lapse of two centuries, but it seems more probable that Fuller suffered instead from a mild case of the pathological self-centeredness called autism. He was simply not interested in the people who presented him with arithmetical problems, only in the problems themselves.

Calculating prodigies come in a bewildering variety of types and defy categorization. Some, like the autistic man portrayed so brilliantly by Dustin Hoffman in *Rain Man,* are or appear to be mentally retarded but have what Michael Howe has termed a "fragment of genius." Others have perfectly normal and sometimes even extraordinary intelligence. It is striking, however, how often a childhood injury to the brain seems to uncover unusual calculating abilities.

Further, the range and kind of calculations that prodigies can perform, and the ways in which they carry them out, vary widely. Bertold Stokvis tells of two brothers of normal intelligence, Wim and Leo Klein. The two could carry out highly complex calculations and remarkable feats of memory involving numbers. Wim, for example, who actually worked as a kind of human computer at Geneva's Center for Nuclear Research, could multiply two nine-digit numbers together in less than a minute, a feat that can also be performed by some mentally retarded calculating prodigies. He did it not by painfully multiplying each digit of the first number by each digit of the second, but by multiplying two-digit chunks of the first number by two-digit chunks of the second and summing the intermediate products as he went. Incredible as this procedure may seem to ordinary mortals, it is at least possible to imagine mastering it eventually. Of course it would be helpful if, as Wim had, you could memorize the multiplication table up to ninety-nine times ninety-nine!

The two brothers performed their calculating feats in very different ways. Wim had an auditory memory and, when memorizing long lists of numbers, would move his legs, arms, and whole body, meanwhile exhibiting a variety of facial contortions. Leo, almost equally proficient, would sit motionlessly and stare at the paper before him. He was memorizing a vivid picture of the numbers rather than their sounds.

My own abilities as a calculator are sadly limited, but on analyzing how I do it I think I am closer to Wim than to Leo. For simple calculations, I sound the numbers out in my head, usually making slight (and I hope invisible) movements of my mouth as I do so. For more complex calculations, I can draw a shape of the numbers in front of me, simultaneously sounding the words out in my mind and making slight (and again I hope invisible) movements of my fingers that help me to form the shape. The shape that results, such as it is, is not a true image but rather a kind of memory traced on my visual field of the movements that my fingers have made. Although readers who are capable of conjuring up vivid pictures of numbers may be totally puzzled by this odd description of what I cheerfully admit is a very odd procedure, I hope that it strikes a chord among others who are as deficient at visualizing images as I am. Perhaps what I am using is a kind of blindsight, a different way of utilizing the visual field. Whatever it is, it is remarkably ineffective—I can perform only limited manipulations before my memory of the shapes that I have drawn disappears.

People with mental retardation who exhibit extraordinary calculating, musical, or artistic ability were until recently commonly known as idiots savant. The term has now been replaced by the less brutal *savant syndrome*. Many authors have concluded that the abilities of people with savant syndrome have resulted from some loss of balance in the brain. This is not an unreasonable assessment because what is chiefly remarkable about such people is the immense amount of time and effort that they put into developing their talents. It is a myth that people with savant syndrome are somehow born that way. They single-mindedly, indeed obsessively, develop their skills, often over a period of many years.

People with savant syndrome who show extraordinary musical ability are often blind, which may in part explain their concentration on the auditory world. They are, as well, largely solitary in their efforts, and they usually turn to the piano rather than some other instrument that requires ensemble playing. A group of Canadian psychologists recently reported on a severely retarded, blind, and partially paralyzed young man, J. L., who has, as nearly as can be estimated, a mental age of about two. He is able to use only his left hand and foot. He grew up in a musical household and was drawn to the piano at age three. Institutionalized at fifteen, he is now in his mid-thirties. For many years he has been able to play, accurately and skillfully, a wide variety of pieces ranging from classical to show tunes and jazz, and to do so on the piano, organ, or guitar. He can improvise pieces in a given style and can transpose effortlessly to accommodate another player or singer. Remarkably, being one-handed has posed few limits—by a clever combination of digital dexterity and the use of the pedal, he can play both melody and chords. He is able to reproduce sequences of notes as well as any thoroughly trained musician. None of these abilities appeared suddenly, however, but instead they were developed over long hours of practice, often extending far into the night.

It is this singleness of purpose that marks out people with savant syndrome. Because of injury, disease, genetics, or some combination of all three, they cannot shift their attention to that great range of inputs from the surrounding world that occupy much of the attention of normal people. They concentrate on only one thing, and the result, although often extraordinary, is often flawed. Mentally retarded musical prodigies, for example, while they may play with a great degree of expertise, generally do not seem to be able to communicate much emotion through the music. Thier inability to communicate normally with others seems to extend to the musical level as well.

Such isolation is perhaps at its extreme in autism. This condition, luckily rare, is often marked by a complete lack of ability to relate to others. Clara Claiborne Park, in her book *The Siege,* emphasizes that overcoming this inability to relate was by far the most difficult task that faced her as she tried to reach her child "Elly" (her real name is Jessy). Jessy exhibited very early many of the classical symptoms of autism. She was slow to speak, and then, when she did learn, she had great deficiencies in grammar and articulation. Abstract concepts such as metaphors remained completely beyond her. Easily upset by slight changes in routine that would not disturb a normal child, she would sometimes throw violent and terrifying tantrums. She was interested only in certain narrow and highly specialized aspects of her environment, chiefly those that had to do with pattern and position. She was not, at first, the least bit interested in other people and refused to interact with them in normal ways.

Instead, Jessy was fascinated by numbers and rules and obsessively made long lists of things that seemed to be arbitrarily chosen. When she was young, she had some of the abilities of a calculating prodigy. Her mother had given her a little instruction in arithmetic, which fascinated her. On her own, she discovered that some numbers were prime, and she quickly found and memorized all the prime numbers less than a thousand. But these calculating abilities gradually slipped away as she grew older.

In part they slipped away because Jessy's mother, after years of besieging her daughter's autistic shell, finally discovered a way to penetrate it by using Jessy's own obsession with numbers and order. The family learned about another autistic child who had been helped by a trick suggested by a clinical psychologist. The trick was to put a golf counter on the child's wrist. This counter is a device that is normally used to add strokes and subtract penalty points during a golf game. The child's parents awarded positive behavior by various numbers of points and penalized negative behaviors by subtracting points. After a certain number of points were accumulated a more tangible reward, such as a popsicle, was forthcoming.

Jessy, who by this time was fifteen, was immediately fascinated by the system, which was tailor-made for her obsession with rules and numbers. This clever reinforcement was soon used by her family to teach her skills that ranged from clearer speech articulation to swimming. Most importantly, it was possible to teach her to interact appropriately with others, to pay attention to their feelings and the ways that they reacted to her, and even to execute some of the essential social grace notes, such as saying hello, that are an important part of normal daily life. Jessy, now in her twenties, holds a part-time job with some responsibility—sorting and forwarding mail, a task that she loves—and is leading a far richer life. At the same time that her obsessive behavior patterns have slipped away, so have her limited savant capabilities. She is no longer capable of—or indeed interested in—memorizing the immense lists of things that dominated her childhood years. Her brain has now learned to juggle more than a single ball.

Jessy's mother wonders what might have happened had this reinforcing behavior been started when Jessy was much younger. How much of the damage, whatever it was, could have been undone? How much could Jessy's brain's own flexibility have aided in its recovery? The full value of such reinforcement therapy, tailored to the problems of the individual child, has yet to be explored.

As we can see from these examples, our ability to be master jugglers of information stands out vividly when we compare ourselves to people who, for a variety of reasons, cannot carry out this process properly. Yet our prehuman ancestors were obviously not braindamaged, or they would not have survived. Their brains, limited as they were, functioned superbly in the environments to which they were exposed. At every stage, that environment played an essential role. Once the environment became sufficiently complex that it began to accelerate the brain's evolution in its current runaway course, it also began to influence in fundamental ways the master-juggler capabilities of the brain itself.

The Zoom Lens of Attention

The impala that we met in the last chapter, cautiously approaching their water hole, were not of course giving equal attention to every aspect of their environment. Rather, they were attuned to anything abnormal or unusual that might signal danger. Their brains were master jugglers, capable of bringing any unusual sight, sound, or smell instantly into the foreground of their attention. An impala that cannot do this, or that cannot do it very well, soon falls victim to predators.

No obsessive-compulsive animal or bird lasts very long in the wild. A monkey that is obsessively worried about snakes, and that peers shortsightedly at each branch before it ventures along it, can easily fall prey to a passing cheetah. In zoos, however, particularly in the Victorian prisonlike zoos that are slowly being phased out in favor of more natural settings, animals often show abnormalities that bear a strong resemblance to human mental diseases. They endlessly pace to and fro and perform repetitive activities, as some schizophrenics do, or they withdraw into the animal equivalent of autism. They have, at least temporarily, lost some of the flexibility of behavior that they showed in the wild—although it is striking to see how the behavior of such animals often becomes more normal when they are placed in more congenial surroundings.

Our brains, too, have been selected to be master jugglers of information, and this selection has been going on for a very long time. Yet, our perception of the world is different from that of other animals, and we must ask how it differs and how that difference evolved.

To begin with, it is obvious that what separates our brains from those of other animals is not simply the amount of information coming in from our sense organs. It is true that while we see better than many other animals, we hear and smell far less acutely. Nor are we unique in our ability to pick out and isolate some of this information, for other animals can be far better than we are at noticing the unusual. Although humans are capable of learning from their experience, other mammals, and even insects, can do this as well.*

We differ from other animals in two ways: The first is the sheer amount of juggling that our brains can do. We are able to choose and extract more pieces of information from a given set of sensory inputs than other animals can, and we are able to add these pieces of information to our long-term memories. Because the store of memories that results is so large and detailed, we can make subtle and complex

*The sense of smell usually has priority over the sense of vision in many animals, but a sufficiently remarkable visual stimulus can overcome it. Recently I took my dog for a walk and we passed a house nearby where there was a garage sale in progress. Standing on the sidewalk was a carved wooden carousel horse, about the size of a Great Dane. As we neared, my dog's jaw dropped, and she circled the horse cautiously. Then she approached and sniffed the horse's rear end! Immediately she lost interest and we proceeded on our walk. Half an hour later on our return we passed the horse again, and my dog showed no interest in it.

comparisons between present-day inputs and our previous experience. As we add to our stored memories, this can in turn change the pattern of juggling that our brains do with new incoming information. The brain is always reorienting its priorities so that certain kinds of information receive more attention than others.

Although most of this shaping of the way that we perceive the world takes place in the first few years of life, it took the autistic child Jessy two decades, and much help from her supportive family and friends, to accomplish even a part of this process. Her brain was not a master juggler. For reasons that we are not even close to understanding, she ignored or deemphasized some inputs and put too much stress on others. Because of this defect, she behaved in the same way that an obsessive monkey or a less-than-alert impala might. Like them, it seems certain that she would not have survived had she been born at any other point in human history or prehistory.

The second way in which we differ from other animals is the degree to which we can voluntarily override this continual sorting and prioritizing of new information. This enhanced ability to ignore the immediate world has itself been a result of our evolution. An impala thinks about one thing too long at its peril. We, however, because we are insulated from the dangers of our environment by the shield of our culture, have the luxury of being able to concentrate. Most of us are able to concentrate on a single thing, either a new input or something remembered, by the expedient of deliberately ignoring the continued stream of inputs from our surrounding environment. Some of us can do this for longer periods and to greater effect than can others—see this chapter's epigraph about Isaac Newton. The human juggler can leave some balls suspended in the air for a while. We feel, as we regard Rodin's statue of *The Thinker,* that the sculptor has captured an important essence of humanity—and we are right.

Were our ancestors always able to perform this trick? Certainly they were not. If they had spent long periods mooning about, furrowing their receding foreheads in thought, they would have been eaten. Let me emphasize this point because it is essential as we try to understand the factors that converted the evolution of our brains from the relatively leisurely pace enjoyed by our mammalian ancestors to the frantic pace of the last few million years. It *must be* that this capability for concentrated thought has been selected for during a period when it did not carry great risks. This means that our ability to lengthen our attention span voluntarily *must* have arisen very recently, during the period when we were also modifying our environment in order to reduce its dangers. It must have happened during the time of that accelerated feedback between the genes that control our brain's development and the ever-more-complicated (and safer) environment created by our ancestors. This, of course, was the very feedback loop that led to our runaway brains.

Our current way of thinking, however, has not completely diverged from that of the impala. Our ability to concentrate is still, for most of us, easily overridden by a sufficiently powerful or specific stimulus, either internal or external. A smell of burning can rouse us from deep thought. In my case, even the anticipation that Rama's hand would dart toward the keyboard of the computer was enough to make me stop suppressing my left visual field.

Yet is this ability to pause in our ceaseless juggling of environmental inputs, to leave the balls suspended in the air for a little while, so remarkable? Does it really set us so clearly apart from other animals? If the sheer time that can be spent concentrating is partly under genetic control, and if the ability to lengthen and shorten that time has been important in the past, then it may not be beyond the realm of possibility to suppose that a few genes, perhaps structured like the period gene in *Drosophila,* might be involved. One can go a step further and suppose that, like those different alleles of the period gene that have been found in northern and southern *Drosophila* populations, different alleles at these cogitation loci exist in the human population. This would explain, in part, why some people are better at cogitating than others.

Even among present-day humans people who can concentrate for very long periods, shutting out the rest of the world, are unusual. Often they tend, like Newton, to be rather unsuccessful at other aspects of their lives. Although the products of their ruminations may sometimes benefit society, society tolerates rather than embraces them. You will recall George Wallace's memorable gibe about "pointy-headed perfessors who cain't even park their bicycles straight!"

It is not unreasonable to expect that many of the properties of the human brain that we are examining in this part of the book will eventually be traceable to a relatively small number of genes. These genes surely did not spring up de novo during our recent evolution, although slight modifications of them must certainly have occurred. Even these modifications, as I suggested in chapter 10, might have an increased likelihood of taking place because of the way genes have been shaped by their long history. Brains have increased in size many times during the course of vertebrate evolution. Our ability to cogitate, and to exert control over our cogitation, has also grown. The incredible speed with which our runaway brains have evolved over the past few million years was made possible largely because we have been able to draw on this genetic capital.

In spite of all that has happened to our brains lately, this capital is not yet exhausted. The cogitation alleles that came together so fortuitously in the genome of Isaac Newton still exist scattered throughout the human population. If there was one Newton in the seventeenth century when the world's population was a tenth the size it is now, probability dictates that there must be about ten Newtons today, presumably bringing their formidable powers of concentration to bear on quite different problems. It is just such unusual minds that, having done much to alter the environmental part of the gene-environment feedback loop, bring the rest of us along in their wake.

The genetic end of the loop is amply taken care of by the enormous genetic diversity of our species, and by the fact that each of our brains has been shaped by runaway evolution to take advantage of environmental enrichment. Even so, we cannot yet predict just when or where a fortuitous combination of genes and environment will give rise to the next Newton. It is impossible for us to seize control of runaway brain evolution and decide which parts of our genetic capital are more important than others. We do, however, have far more control over the environmental end of the feedback loop. Rather than trying to modify our genes, we would do well to concentrate on enriching the environment of every individual, for in the process we will be drawing more fully on our remarkable genetic capital. As we will see in the next chapter, such an approach is likely to have an enormous payoff.

12

A Sponge for Knowledge

Meanwhile the desire to express myself grew. The few signs I used became less and less adequate, and my failures to make myself understood were invariably followed by outbursts of passion. I felt as if invisible hands were holding me, and I made frantic efforts to free myself. . . . After awhile the need of some means of communication became so urgent that these outbursts occurred daily, sometimes hourly.

—Helen Keller, *The Story of My Life* (1902)

In Extremis

Helen Keller, born in 1880, was the first child of a middle-class Alabama family. Very precocious both physically and mentally, she claimed that she was able by six months to say a few words, and she was walking at the end of her first year. Then, at nineteen months, she was stricken by a severe fever. When she recovered, she was profoundly deaf and completely blind.

During the first few years after her illness, she herself managed to overcome to a surprising extent the immense handicaps of her new existence. She was able to communicate her desires to others through gestures that she invented herself, and she received positive feedback whenever family members or servants did what she wanted. Intensely imitative and sensitive to all the aspects of her environment that she could touch, taste, or smell, she quickly learned to do household and kitchen chores and delighted in them. For some years she spoke one word, *water,* which gradually degenerated into a barely recognizable *wah-wah.* But she learned, slowly and with mounting frustration, that there were things going on around her that she could not take part in.

She touched the faces of family members as they talked and realized that they were somehow passing information back and forth by moving their lips. Frustration at not knowing what was going on sent her into repeated rages.

The story of what happened when Helen was seven and her dedicated teacher, Anne Mansfield Sullivan, came to live with the family has been told repeatedly. Miss Sullivan quickly taught Helen a number of words for nouns and verbs, showing her how to make the signs for the series of letters that made up each word. But Helen, who readily imitated the signs, did not at first make the connection between the thing and the sign for it. Finally, Miss Sullivan managed to breach Helen's incomprehension by running water from the well-house pump over her hand and repeatedly spelling the word for water into Helen's other hand.

Once the barrier to her understanding was broken, indeed on the very same day, Helen charged delightedly into the world of words. By the end of that day she had learned dozens of words. The sudden realization that had come to her in the well house, that everything in the world had a name, unleashed a remarkable mind that until that moment had been fully formed but held in abeyance.

Her teacher was firm with Helen from the beginning, insisting that she should talk in complete sentences even though it slowed down communication. Helen quickly learned—or rediscovered—the rules of grammar and from the beginning showed a delight in acquiring all the tools of language. Her brain was a sponge for knowledge, and with the aid of her dedicated teacher she absorbed it with amazing speed.

Helen Keller's subsequent career would have been dazzling even for a person without handicaps. After graduating cum laude from Radcliffe, she devoted her life to helping others in her situation, writing numerous books and magazine articles and traveling around the world. She became accomplished in several languages. All those who met her, including the famous literary figures of the day, were impressed with her knowledge and boundless enthusiasm.

Helen Keller's explosion of comprehension could never have taken place if she had not, during the five and a half years between her illness and the arrival of her teacher, been exposed to a great variety of stimuli. During that dark period she had spent much time in the garden, with its profusion of plants and numerous barnyard animals. She played for hours with the child of one of the servants and quickly learned to order her about by using the gestures that she had invented. Always intensely aware and curious about everything that was going on around her, she extracted every bit of information that she could. There is no doubt that her mind was unusual. There is also no doubt that, deprived of so much sensory input during those critical years of development, it would have been crippled were it not for her family and her teachers.

"Genie," a child with normal sight and hearing, was deprived in exactly that way. Born in Los Angeles in 1957 to a woman whose husband repeatedly beat her, Genie's early life was a nightmare of terrifying proportions. Apparently normal at birth, she was slow to develop, in part because her mother, whose own health was worsening, paid little attention to her. Her father took a violent and irrational dislike to her. Her paternal grandmother, the only person who stood between Genie and her father, died when Genie was twenty months old.

After that, the real nightmare began. During the day, her father harnessed her to an infant potty seat, and at night (when he didn't forget) she was placed in a sleeping bag that had been sewn into a kind of straitjacket. The family was forbidden to speak to her—indeed, her father only growled and barked at her like a dog. If she tried to make a noise, she was mercilessly beaten. The father terrified the rest of the household into silence, so that Genie could not even hear others speaking. There was no radio or television, and Genie stared for so long at a blank wall 10 feet away that by the time she was rescued at age thirteen and a half she was unable to focus at closer or longer distances. The father, about to be put on trial for child abuse, killed himself. He left a suicide note stating: "The world will never understand." He was right.

After her rescue, Genie was kept in a hospital for almost a year before being moved to a foster home. Never having worn clothing, she was indifferent to heat and cold. She was unable to chew solid food, and indeed it took years for her to learn to do so. She was incontinent and spat on everybody and everything that came near her. Her vocalization consisted only of whimpers, and she understood only a few words. Observers compared her behavior to that of a newly captive animal.

Yet her curiosity was intact, and after a few months she began to learn the names of objects. Her rate of learning soon began to accelerate, and within half a year she had learned and could recognize hundreds of words for objects. Her speaking vocabulary increased rapidly at the same time, and although at first it consisted only of nouns, it gradually began to incorporate other parts of speech. She exhibited intense interest in the world around her, developing at the rate of about one year in mental age for each year she had passed in freedom.

Was Genie retarded? This has been argued, but it is striking that only four years after her rescue she was able to accomplish some of the nonverbal parts of various IQ tests at her chronological age level or above. She had an outstanding ability to understand shapes and to perceive and remember features in pictures she was presented. Such is the resilience of the human brain that in spite of the appalling abuse she had suffered she quickly acquired many of the cognitive skills of a normal child.

But not all of them. Although Genie's vocabulary increased rapidly, her grammar skills did not. She could couple nouns and verbs appropriately, but the result was strings of words that were formed in a kind of pidgin English. All her sentences employed transitive verbs, requiring an object, and she never grasped the idea of an understood object. She had great difficulty in formulating questions—when she wanted to know the name of something, she gestured or pointed at it but did not learn to ask its name. Often, if she wanted information or desired something, she would simply repeat one of her primitive sentence strings until she got a response.

One wonders whether Helen Keller, too, might have fallen into some of the same developmental traps if it had not been for her teacher's firm and patient insistence that she phrase her thoughts using the appropriate grammatical constructions. Genie was a far more difficult subject, and nobody spent the endless hours with her that Helen Keller received from her teacher. Still, one gets the impression that Genie's incredibly deprived brain was blocked from achieving high levels of certain verbal skills and perhaps would never have been able to achieve them even with the most concentrated therapy. The very different story of Helen Keller's development suggests that the sensory inputs that she received from her environment before the arrival of her teacher, limited as they were, were enough to allow her brain to develop normally.

Sensory deprivation of the kind that Genie suffered during critical stages of development can indeed stunt the development of certain functions in a normal brain, but the deprivation has to be severe and prolonged. The brain *wants* to learn, and it will do so in spite of almost every obstacle that is put in its path. This is hardly surprising, for learning is exactly what the brain's recent evolution has shaped it to do. Nothing demonstrates this more clearly than the evolution of our ability to speak and to understand spoken language.

Where Does Language Lie in the Brain?

Wittgenstein once said, "If a lion could talk, we could not understand him." I think, on the contrary, that if a lion could talk, that lion would have a mind so different from the general run of lion minds, that although we could understand him just fine, we would learn little about ordinary lions from him.

—Daniel Dennett, *Consciousness Explained* (1991)

Most of the marsupial mammals of Australia are unnervingly silent. This fact could not be imagined by Europeans, which is why Robert Hughes in *The Fatal Shore* is able to quote mockingly the poet Robert Southey's "Ode to Botany Bay," written soon after the penal colony was founded:

> *Welcome ye wild plains*
> *Unbroken by the plough, undelv'd by hand*
> *Of patient rustic; where for lowing herds*
> *And for the music of the bleating flocks,*
> *Alone is heard the kangaroo's sad note,*
> *Deepening in distance.*

It seems that mammalian vocal communication systems began to evolve some time after the split, some two hundred million years ago, between the ancestors of marsupial animals such as the kangaroo and opossum and those that like ourselves nourish their young in the womb through a placenta. By now many placental mammals have vocal communication systems, sometimes of considerable complexity. So do birds, although it appears from both neuroanatomical and evolutionary evidence that their communication systems arose independently of those of mammals.

Thus, while nothing marks us off more clearly from other animals than our ability to speak, and to understand grammatically complex spoken language, these abilities did not appear suddenly and without any forerunner in the course of our recent evolution. They have evolved from neural systems that were already largely in place. Allan Wilson was certainly wrong when he suggested that the ability to speak arose suddenly in our recent ancestors and that it formed the watershed event that defined the appearance of true humans. The acquisition of language was more complicated than that, and was spread over a long period of evolutionary time. As we will see, it depended more on the evolution of the brain's capacity as a sponge for knowledge than on the acquisition of specific and clearly defined abilities.

The physical requirements for speech probably began to appear fairly early in our evolution. Upright posture freed the larynx from its constricted position at the top of the trachea and resulted in more room in the back of the throat for the tongue to move. This provided the flexibility needed to produce a variety of vowels and consonants, a flexibility denied to the apes. These changes probably did not take place immediately after upright posture was achieved four or more million years ago. Like those equally important changes in the hand and the pelvis, they were spread over a span of millions of years. Unfortunately we have no fossil record of all this, except for the single Neanderthal hyoid bone from Kebara Cave in Israel, a fossil that must come from somewhere near the end of that long evolutionary process.

Parallel changes must have occurred in the brain, but they are even more difficult to trace. There is a hint from fragments of fossil skull that the cortical temporal lobes of *Homo habilis* were larger, and overgrew the underlying medulla farther, than the corresponding regions of the brain of Australopithecines. This is exciting because, as we saw earlier, the temporal lobes, particularly the region of the left lobe known as Wernicke's area, are important in the comprehension of language.

The temporal lobes are important, but not central. The more deeply the language capabilities of the brain are investigated, the more one astonishing fact emerges. It is not just Broca's area in the frontal lobe and Wernicke's area in the temporal lobe that take part in language production and comprehension—virtually every part of the brain is involved! Any explanation for the evolution of our ability to speak and understand language must take this immensely significant fact into account.

Indeed, it appears that the pathways involved in language production and comprehension are not all just simply recent elaborations of connections within the cortex but often run through the very deepest and, according to some authors, the most primitive parts of the brain. Language ability is not an add-on like fancy hubcaps on a car. The evolution of this ability has involved long-term and deep-rooted changes in the very organization of the brain itself.

A combination of new and old techniques has shown just how extensively the fabric of every part of our brain is involved in language. Over a period of decades, thousands of patients with lesions in many different parts of the brain have been tested extensively to determine how those lesions affect the way they speak and understand. Paul Broca had to wait until his subjects died before he could examine their brains to determine what might have been wrong with them. Now, noninvasive techniques such as computerized axial tomography (CAT) scans and nuclear magnetic resonance can be used to locate a lesion precisely in a living patient. Real-time motion pictures of brain activity can now be made using the very new technique of positron emission tomography, which can follow the rate at which radioactive dyes are washed away by increases in blood flow to various parts of the brain. This enables the experimenter to follow, although still crudely and at low resolution, the very process of thought itself.

The classic aphasias are Broca's and Wernicke's. A patient with Broca's aphasia can understand speech quite well but finds it difficult or even impossible to speak. Memory, too, is impaired, and often there is partial paralysis of the limbs on the right side of the body. Broca's is an example of a *motor* aphasia. A patient with Wernicke's aphasia, on the other hand, shows no paralysis, speaks easily and fluently, but speaks nonsense. One word is substituted for another, meaningless words are invented on the spot, and the order of the words is scrambled. These aphasics also have great difficulty comprehending the speech of others. Wernicke's and similar aphasias have been called *sensory* aphasias.

So, it might appear at first blush that there is a clear compartmental separation between language production and language comprehension. Unfortunately for this simple model of language, many patients have been found with lesions of the more anterior parts of Wernicke's area who have problems with language production, and many patients with lesions of the more posterior parts of Broca's area are found to have difficulty with comprehension. It appears that the anterior parts of both areas are involved in the motor functions of language and the posterior parts are involved in the sensory functions. It is as if one region were a much-altered but still recognizable copy of the other.

In their property of apparent duplication, these areas resemble the way the brain handles visual input. Many different visual maps are scattered through the posterior part of the brain, and we saw in chapter 11 that different parts of the brain interpret the visual world in somewhat different but nonetheless related ways. A surprising amount of the brain is involved in vision, and a surprising amount, parts of it actually overlapping with the visual areas, is involved in language. As more is learned about the genes that control development of these regions of the brain, there is a good chance that it will be discovered that the genes have multiplied through the process of duplication and subsequent divergence and have given rise to structures in the brain that have themselves duplicated and diverged. As I pointed out in the last section, it is much more likely that genes duplicate and take up new functions in the course of evolution than that completely new genes appear.

Even now, signs of such processes are visible in the structure of the brain itself, although of course the genes underlying these structures have yet to be discovered. You will recall the shift in emphasis that has taken place in the primate visual pathway, away from the paths leading to the superior colliculus that are so prominent in other mammals and toward a new emphasis on the paths leading to the lateral geniculate nucleus. Totally new pathways have not arisen during primate evolution, but the emphasis placed on them during development has shifted. This can be seen both at the level of brain structure and at the level of language use.

Positron emission tomography studies of people involved in various language-intensive tasks show that many parts of the cortex, extending all the way from the frontal to the occipital lobes, show patterns of increased blood flow. Language is not simply a cortical function, for regions located far beneath the cortex are involved in language production and comprehension. The caudate nucleus, which lies just above the brain stem, receives input from many parts of the cortex and then, through an exceedingly complex set of connections, influences motor activity and sends signals back through the nearby thalamus to other parts of the cortex. Lesions in the anterior part of the left caudate nucleus produce an aphasia rather like Wernicke's, but it is accompanied by muscle weakness on the right side of the body. Damage to the thalamus, as well, can produce similar symptoms.

Indeed, parts of the brain that until recently were thought to have nothing whatever to do with higher brain function are now beginning to be implicated in language. One of these is the cerebellum, the "little brain" tucked behind the brain stem (see the bottom section of figure 10.2 on page 251). People with cerebellar lesions tend to have problems with posture and balance, and as a result for many years this part of the brain was thought to be primarily concerned with very primitive functions. More and more information is now emerging that the cerebellum, particularly a more recently evolved part of it called the neocerebellum, interacts in complex ways with the cortex, just as the caudate nucleus and the thalamus do. Here the connection with genes that control the brain's development suddenly becomes strikingly clear.

A rare condition called Williams syndrome is caused by mutations in one or more genes that are involved in calcium metabolism. People with this syndrome are retarded—adolescents and adults manage to achieve an IQ score averaging about 50, roughly the same as is achieved by adults with Down syndrome. Like Down syndrome, Williams syndrome is often connected with heart defects. The physiognomy, too, is unusual and unique to the syndrome—people with Williams syndrome have pixielike faces, with pointed chins and noses and unusually broad foreheads. The irises of their eyes have an unusual lacy appearance.

Unlike those with Down syndrome, however, people with Williams syndrome show an abnormal sensitivity to sounds and an intense interest in words. Their language acquisition is retarded at first, but eventually they begin to pick up words and phrases with remarkable speed. Their conversation is peppered with unusual and adult-sounding turns of phrase. Indeed, their spoken language, in both vocabulary and grammar, is surprisingly normal. In tests of vocabulary, they are often able to come up with unusual and appropriate words—asked to list as many animals as they can, they might include *brontosaurus*, *ibex*, and *vulture*. They can tell complex and imaginative stories. Yet they are able to read and write only poorly and show many other defects in comprehension.

These children have some of the characteristics of those people with savant syndrome whom we looked at in chapter 11. They single-mindedly attack spoken language in much the way that somebody with savant syndrome might attack multiplication or music. But, while each person with savant syndrome is unique, usually the result of some developmental accident, Williams syndrome can be traced to single gene defects. Because dozens of children and adults with the syndrome are available for study, this opens up the possibility of tracking down some feature or features of the structure of their brains that they all share.

Magnetic resonance imaging shows that the cerebral cortex of people with Williams syndrome is only about 80 percent as massive as that of normal people. The cerebellum, however, is as large as it is in people without the syndrome, and in particular the neocerebellum is actually larger than it is in a normal brain. This raises the exciting possibility that this cerebellar hypertrophy has much to do with the remarkable sparing of their language ability.

As we have seen, the whole brain has evolved in concert in hominid evolution, all its parts becoming larger and more complex. It is therefore not surprising that language capabilities seem to be spread through much of the brain—although it must be remembered that certain parts of the brain are more important in language than others, and small lesions can sometimes produce surprisingly specific effects. Nonetheless, the impression is growing among neurobiologists that the evolution of language ability has been a massive process that has involved an increase in the brain's capabilities at every level.

Patients with either Broca's or Wernicke's aphasia have a very interesting feature in common. Often, they can perform simple tasks such as counting out loud almost unimpaired. It is when stimuli become more complicated that difficulties ensue—Broca's patients begin to trip over all but the simplest words and sentences, and Wernicke's patients begin producing a stream of nonsense words or performing inappropriate actions. It is as if a landslide has blocked part of a river and turned a deep, calm current of information into a choked and turbulent fraction of the former flow. When you watch such patients, you gain the impression that the problem is not so much damage to their comprehension or motor centers (although that must play some role) as it is an impediment to the flow of information.

If there is so much information flow from one part of the brain to another during the comprehension and production of language, and if so many parts of the brain are involved, then this strongly suggests that the ability to communicate verbally has had a long evolutionary history. Then why do we seem to be unique among animals in our ability to utilize language? Runaway brain evolution certainly provides part of the answer, for only during the last few million or perhaps the last few hundred thousand years has our physical and social environment become so complex that a really complex language is required to deal with it. Still, if our brains are simply building on capabilities that in large part already existed before runaway brain evolution began, then why are other animals not cleverer at language than they seem to be?

The Uniqueness of Human Language—a Persistent Myth?

In 1966 two psychologists at the University of Nevada at Reno, Beatrix and Allen Gardner, were given an opportunity to adopt a ten-month-old chimpanzee straight from the wild. They called her Washoe, and began to teach her the signs and gestures of American Sign Language (ASL). Up until that time, scientists and others who had adopted chimpanzees had attempted to communicate with them verbally, and although those chimpanzees could learn the meaning of a number of spoken words and respond appropriately, they were unable to produce words of their own. Such early studies were often cited to show the so-called unbridgeable gap between the world of humans and that of chimpanzees. The Gardners' project, which has continued down to the present and now includes several other chimpanzees, was the first in which unequivocal lexical communication between humans and chimps could be shown to be taking place.

Over many years, the chimpanzees learned a great variety of ASL signs, which they usually used appropriately. In a series of careful experiments, people familiar with ASL watched the signs made by chimpanzees who were presented with various objects that the human observers could not see. They were easily able to guess the names of the objects from the chimps' signs. The Gardners also collected much anecdotal evidence to the effect that older chimps could teach younger ones some signs, although this was certainly more a matter of imitation than of conscious teaching. The chimpanzees often signed to each other and would respond appropriately to other chimps' signs. Intriguingly, they would sometimes sign to inanimate objects, toys, or household pets, in the same way that a human child might talk to a favorite toy or to a dog. While leafing through picture books or magazines, they would make signs appropriate to the people or objects they saw in the pictures. A chimpanzee all by itself would often spontaneously make signs, much as an isolated young child might say random words or phrases to itself.

An acrimonious debate soon broke out over whether this was simply an elaborate case of monkey see, monkey do, a debate that has persisted to the present day. Many attempts have been made to decide the matter one way or the other, one of the most interesting of which was carried out by Sue Savage-Rumbaugh, Duane Rumbaugh, and their colleagues at Emory University's Yerkes Regional Primate Research Center. They examined an eight-year-old bonobo (pygmy chimpanzee) that had been raised by humans. It had been exposed to many different situations and to much spoken language exchange among the people caring for it, but there had been no deliberate attempt to teach it language. They compared its ability to comprehend language with that of a two-and-a-half-year-old human child. Without any prior training, chimp and child were each exposed to 660 different sentences, spoken by a person invisible to them, that directed them to do a variety of simple tasks. In order to confront them with a variety of grammatical structures, not simply the nature of the words but the word order and grouping were varied.

Remarkably, chimp and child did about equally well! The child did a little better than the chimpanzee with sentences like, "Take the pen and the pencil and put them on the piece of paper," the chimp tending to forget the second half of the task once the first half was completed. Otherwise there was little difference between them.

We have seen that our brains have evolved to become, to a greater degree than those of any other animals, sponges for knowledge. Helen Keller sucked all the information she could out of her very limited world, and, when new worlds opened up to her, her brain was ready for them. Genie's brain was sadly crippled by her ghastly early environment, but even so it recovered to a remarkable degree. Acquiring knowledge is what our brains are good at, and they acquire whatever knowledge is available.

The chimpanzee's brain, too, seems to be able to comprehend more than we might give it credit for, though for a variety of reasons it is unable to express this comprehension in words. No chimpanzee in the wild is exposed to such a set of elaborate verbal cues as the bonobo in the Rumbaughs' experiment, but the nonverbal cues it must comprehend and act on are probably even more challenging. The chimp's brain, like ours, is a sponge for knowledge, including new kinds of knowledge such as words and sentence order. It has evolved into such a sponge even though chimpanzees have presumably not begun to undergo runaway brain evolution. But chimpanzees are not as good as we are at concentrating on certain tasks, excluding other external inputs as they do so. Most tellingly, perhaps, chimpanzees do not draw as much information from the world around them as we do. They can see the stars, but they do not regard them.

Limited as their abilities are, the brains of chimpanzees are still ahead of their bodies. They can comprehend language much better than they can produce it, in part because their vocal apparatus is not up to the task. This is exactly what we might expect to have happened in the early stages of our own runaway brain evolution—like that of the chimpanzees of today, the mental reach of our distant ancestors must have quite literally exceeded their grasp. Runaway brain evolution has by now driven the evolution of our bodies as well, so that they can now, however imperfectly, respond to the demands put on them by our brains. Some of us can fly helicopters, paint pictures, sing Verdi, play the piano—activities that would have been impossible if runaway brain evolution had not in turn driven the evolution of our postures, our vocal cords, our hands, to the point where our bodies can finally do what our brains can imagine.

One cannot help but wonder what would happen if a chimpanzee were provided with a noninvasive electronic device designed to extend the capability of its brain, perhaps a kind of electronic larynx that it could learn to control readily and easily with slight movements of its throat and face in order to produce the range of sounds of human speech. Of course it is unlikely that, provided with such a device, a chimpanzee could begin to converse learnedly on Wittgenstein—but how far *could* it go? The result of such an experiment might be exceedingly surprising—and exceedingly humbling. Our nearest relatives might be farther along the path of runaway brain evolution than we currently imagine.

Brain Duplication

The evolution of language is inseparable from the evolution of the many other capabilities of our minds that have made us uniquely human. These capabilities in turn are intimately connected with the capacities of our brains to absorb, and to be master jugglers of, incoming and stored information. Part of this evolutionary process, although certainly not all, has been a direct result of the increased mass of the brain itself. Because the brain's circuitry is so highly parallel and the rate at which information is processed in each set of circuits is so slow, it seems that it has not been possible simply to add to the complexity of a single map in only one part of the brain and to make it more and more detailed in the process. Instead, the brain distributes information into many different, less detailed maps in different regions.

Once regions of the brain have appeared that are capable of producing several such visual or auditory maps of the environment, differentiation among them becomes possible. Many people have pointed out that this process of differentiation is rather like the process of gene duplication that we discussed in chapter 10. You will recall that, once genes have been duplicated, accumulating mutations, chance, and selection will gradually cause them to diverge. Eventually they end up performing quite different functions. There may be, as I pointed out earlier, a process of brain duplication going on that is the result of gene duplication. Indeed, one of the most exciting aspects of brain research over the next few decades will be to discover whether the duplication of maps in the brain really is the result of underlying gene duplication, just as the appearance of grosser structures such as the midbrain and the hindbrain seems to have been in part the result of duplication of the various homeobox genes.

I explored in chapter 12 some of the numerous ways in which visual images are processed in the brain. They can be extremely divergent, and indeed some of them, like blindsight perceptions, are not processed in a particularly visual way. The organizational resemblances among these areas of the brain make them prime candidates for this process of brain duplication. Language information can be processed in similarly divergent ways. Ursula Bellugi of the Salk Institute has shown that lexical information can be processed through both auditory and visual pathways.

She and her co-workers have found that, in spite of the large superficial differences between sign language and spoken language, there are remarkable similarities in the ways that children learn them. Neither deaf nor hearing children master their language until between the ages of seven and ten, and up until that time both go through remarkably similar stages in development. Deaf children learning sign language tend to make grammatical mistakes that are very like those made by hearing children of the same age who are learning to speak English or other spoken languages. Further, if a young child is trying to tell a complicated story that includes a number of different characters, then whether the child is communicating in English or in sign he or she has trouble keeping the characters straight and keeping them clearly differentiated. It is not the actual mechanics of the gestural language that are the problem, for the child who is learning sign language quickly grasps the idea of reserving different regions of the space in front of it for different characters in the story. Rather, the problem facing the deaf child seems to be the same as that facing the hearing child, simply the difficulty of remembering and keeping track of a number of characters at once.

Although different pathways are certainly involved in these two kinds of language processing, there is some overlap. Both sets of pathways seem to involve Broca's and Wernicke's areas. Deaf people who suffer strokes affecting these areas can lose their ability to sign intelligibly or to understand signs. Yet it is certain that the overlap is not complete. Bellugi's co-workers Antonio and Hanna Damasio of the University of Iowa have found that some hearing people who have speech and comprehension defects resulting from strokes in these areas can learn to communicate more readily through sign language. These and many similar results show how enormously adaptable our brains can be to new circumstances. It is as if our stomachs could learn to digest cellulose or tin cans.

Our ability to switch our attention from one set of these maps to another has led to all kinds of new possibilities. The details of all these switchings escape us as yet, although I suspect that, when we do understand them, we will have traveled a good way toward understanding human consciousness itself. Some aspects of these switchings, however, already give us glimpses of the evolutionary pressures that drove the runaway brain and allowed our ancestors to acquire language.

First, sign language has many of the characteristics of mime, and indeed in order to communicate nuances of meaning the speaker of sign moves his or her whole body and accompanies the signs with a variety of facial expressions. Some people are far better at mime than others, far better at communicating their thoughts or desires through body language. As Merlin Donald has pointed out, there is an obvious evolutionary advantage to being an effective mime. The great mimes of the silent screen made the whole world fall in love with them. Can we not suppose that the great mimes of the more distant past cut swaths through the opposite sex that were limited in size only by the fact that the mimes could not be everywhere at once?

Second, at some time in our past, our ancestors attained the ability to use their stored memories to construct a world that never existed—to tell stories that never happened or to invent deities in an attempt to explain the origin of the tribe or of prominent features of the landscape. Some members of the group were undoubtedly better at this than others, just as some people today have a talent for telling ghost stories around a campfire. (It is possible to scare the more suggestible members of such a group into near catatonia—I know because I have succeeded in doing so once or twice.) The ability to manipulate others by supplying their imaginations with ready-made tales and beliefs has been the source of most power, both secular and religious, through all the human history that we know. There is no doubt that it played a powerful role in our prehistory as well. The selective advantage that resulted from this ability to construct vivid and persuasive imaginary worlds must have been enormous.

Third, we have seen that the ability to speak may be older than we think. This is not to say that complex, grammatically rich languages of the kind that all peoples of the world speak today are themselves very old. Some historians of language, such as William Noble and Iain Davidson of the University of New England, argue that all human languages can be traced back only about thirty-two thousand years, because it was only then that iconic objects—such things as sculptures and cave art—appeared. These iconic objects were not simply utilitarian tools, of the kind that have been found in the fossil record up to that time. Rather, they exemplified a new way of looking at the world in which representations of objects rather than the objects themselves had become important. Language as we know it, these historians feel, must have evolved at the same time because it also deals with iconic representations of the world.

They may be right. But the genetic alterations that were able to provide the physical and mental machinery needed to invent such languages must by that time have been largely in place—too many parts of the brain are involved in language for all these new pathways to have appeared at once, and we know that we share a surprising number of these abilities with chimpanzees. If Noble and Davidson are right, the development of complex languages might have provided the impetus for the latest dramatic spurt of our runaway brain evolution. Nonetheless, it could never have happened if the genetic, physiological, and social groundwork had not been laid long before, if the brain had not become a superbly adaptable sponge for knowledge of any kind.

The Dilemmas Revisited

Now that we have explored the kinds of genetic changes and the kinds of changes in our brains that have taken place in the course of our recent evolution, we can finally confront the dilemmas with which I began this book. The apparent directionality of our evolution becomes much less puzzling if we think of it as being driven by the runaway brain. The dilemma posed by possible multiple origins of our species also becomes far less puzzling.

If the mitochondrial Eve lived as much as a million years ago, then different groups in various parts of the Old World must have evolved from *Homo erectus* to *Homo sapiens* at different places, and perhaps at different times and at different speeds. Carleton Coon and later Milford Wolpoff suggested that this process was probably aided by gene flow, and certainly this seems likely. However, let us imagine the most politically incorrect of all the possible scenarios, one in which various human groups like the Khoisan and the Australian aboriginals have remained isolated from the rest of the human gene pool so that no gene flow could take place. Might these groups have evolved less far than other human groups?

No, for most of our evolution had already taken place by the time of the appearance of *Homo erectus*. Both before and since that time, the imperative of runaway brain evolution has been the primary factor responsible for driving the evolution of all human groups, and it has done so in remarkably similar ways in all the parts of the world which our species has penetrated. As I have repeatedly emphasized, unique genes for humanness did not arise de novo in our immediate ancestors, as Coon's simple diagram of human evolution (p. 65) suggested. Allele frequencies at many genes have shifted in the course of our recent evolution, and perhaps gene duplications and potential-realizing mutations have taken place at certain gene loci. These things can happen easily, far more easily than the appearance of hypothetical new genes for humanness. It is such small genetic events as these that separate us from our *Homo erectus* ancestors. We are not the result of some dramatic change in the direction of evolution but of a gradual and continuing process that would still have taken place among the *Homo erectus* of Zhoukoudian or the *Homo erectus* of Java or the people of southernmost Africa even if all the hominids living elsewhere had disappeared.

13

Reining in the Runaway?

We cannot escape the conclusion that [nearly one hundred thousand years ago] man's evolution towards manness suddenly came to a halt.

—Ernst Mayr, *Animal Species and Evolution* (1963)

The paleontologist Francis Thackeray of Pretoria's Transvaal Museum has recently examined the dentition of a number of human remains found in southern Africa that have been dated from between fifty thousand and one hundred thousand years ago. The jaws were found at sites scattered over a wide area: Springbok Flats in the Transvaal, Border Cave in the Lebombo Mountains near the eastern coast, and Klasies River Mouth on the southern coast.

Without exception, the shapes of the dental arcades of these Late Pleistocene jaws were clearly different from those of modern jaws. Viewed from above, the jaws were wider from side to side, and the teeth as a consequence formed a flatter arch with the molars further from the midline. All the modern human jaws Thackeray looked at, from people as different from each other as Khoisan and Caucasians, have narrower jaws. So these ancient peoples were clearly different morphologically from the present-day aboriginal inhabitants of the region, the Khoisan. Were they replaced by the Khoisan, or did they pass their genes down to them?

The Khoisan, living now in very restricted areas, used to range far more widely throughout southern Africa. Khoisan cave paintings, found in many places south of the Limpopo River, extend from very recent times all the way back to eighteen thousand years ago, and possibly as long ago as twenty-six thousand years. Between one and two thousand years ago, the Khoisan began to be displaced from their ancient hunting-gathering areas by Bantu-speaking agriculturalists from the north. The Khoi tribes adopted the invaders' way of life, while others, the San, retreated into the Kalahari desert and retained their hunter-gatherer culture.

Cultural continuity can be clearly traced among these peoples over the last few tens of thousands of years. Hilary Deacon of Stellenbosch University thinks that in fact cultural continuity extends even further back, providing a direct connection between the San hunter gatherers of the present time and the peoples who lived fifty thousand years ago in Border Cave and nearly a hundred thousand years ago in the caves overlooking the sea at Klasies River Mouth. His argument is that advanced stone artifacts that have been dated to between fifty and seventy thousand years ago correspond in their geographic distribution to the known distribution of Khoisan hunter gatherers before the recent invasions of tribes from the north. There are also, he points out, many other points of resemblance between the culture of the Khoisan and that ancient culture.

If Deacon is right, then Ernst Mayr is wrong, and there has been substantial morphological evolution in our species, even as recently as over the last hundred thousand years. Our evolution, then, has not come to a halt. If morphological evolution has taken place, it seems not improbable that brain evolution has taken place as well.

Still, the idea persists that at some point tens or hundreds of thousands of years ago our ancestors somehow managed to escape the evolutionary process. J. S. Jones of the University of London recently wrote an op-ed piece for the *New York Times* entitled "Is Evolution Over?" in which he contended that human evolution has indeed come to an end for us. *Discover* magazine asked me to respond, and I jumped in with both feet.

Jones argued, among other things, that we are running out of our capacity for genetic change. He cited the undoubted fact that the average human life span, which has increased greatly in First World countries over the last century as a result of improved sanitation, better nutrition, and immunization against disease, is now increasing more and more slowly. We have reached, he suggested, a limit to further evolutionary change.

There are two obvious things wrong with his argument. The first is the assumption that long lives are necessarily advantageous and are currently being selected for. The things that we enjoy are not necessarily the things that lead to the survival of our species—if short lives were advantageous, they would surely be selected for. We have at the moment absolutely no idea what, if any, selection is going on in the human population for increased or decreased life spans. There has certainly been no deliberate effort on the part of governments or society at large to encourage long-lived people to have children.

The second error is the assumption that we are necessarily running out of the genetic variation that affects the lengths of our life spans. We simply do not know the answer to this, although I suspect that we are not. What we are doing is improving the *environment* for a lucky fraction of the human race and allowing them to achieve life spans something close to their maximum. Improving the environment allows the expression of our genetic potential, but does not itself change the genetic makeup of our species.

Different laboratory strains of mice can have very different life spans, a characteristic for which they breed true. We know that there has been selection for increased life span in our own past because our maximum life span is almost twice as long as that of chimpanzees. Have we run out of genetic variability for life span? Probably we have not, for there is much data to show that such genetic variability exists in human as well as in mouse populations at the present time. Where there is genetic variation, it is usually possible to select for it.

Of course, the question of whether there has been recent selection for life span, or for the shape of the dental arcade, is simple compared with the much thornier problem of whether brain evolution, runaway or otherwise, is still going on. How, if it is, might our brains have changed in the very recent past, during the latest stages of runaway brain evolution? It is impossible to come up with a clear-cut answer to this question, for a variety of reasons:

To begin with, there is the matter of the role that genes play. The genetic components of such characteristics as longevity and the shape of the jaw are complex and little understood. How much more complex, then, must be the genetic components of variation in how the brain works. All that can really be said at the moment is that there are such components. Thomas Bouchard and his colleagues at the University of Minnesota recently studied fifty-six pairs of identical twins raised apart. These twins tended to resemble each other in more ways than would be expected if the environment were the sole influence on their development. Although Bouchard found that there are detectable genetic influences on various kinds of test-taking ability and even on various aspects of personality, he and his group are still investigating the question of whether some of these pairs of twins are more influenced by their genotypes than others. They are not able to address at all the question of whether these genetic similarities would still be detectable if the twins raised apart were reared in extremely *different* environments—if one twin were raised in an American suburb and the other among a tribe of Eskimos, for example.

Second, the selective pressures that have shaped the brain are probably very different from those that have shaped our bodies. As I pointed out earlier, mental stresses and challenges must be very different from physical ones. We really know nothing about such challenges and how they might have affected our ancestors' survival.

Third, most of us have never experienced the full effect of the natural forces that our ancestors were exposed to. We are cocooned in the snug security of our present postindustrial society, a world very different from that of even a century or two ago. We can, of course, get some idea of what conditions might have been like then by examining the situation in Third World countries at the present time. High infant mortality in Third World societies must be an indicator of severe selective pressures—unless, of course, such mortality strikes children entirely at random, which seems unlikely.

Fourth, it is usually difficult or impossible to connect selection pressures with particular genotypes, and this is particularly the case when the connection between genotype and phenotype is ill understood and tenuous at best. There is for example a very strong correlation between illiteracy and infant mortality among Third World countries at the present time—the children of illiterate parents are much more likely to die in infancy than those of literate parents. We cannot conclude from this, however, that there is selection for the ability to be literate in these countries, because most illiterates in Third World countries are unable to read through no disability of their own.

We, on the other hand, live in a society completely new to human experience, a society of virtually universal literacy. Most of you who are reading this book spend the greater part of your waking hours reading and writing, yet you need trace your ancestry back only a few generations to find that the majority of your ancestors were illiterate. The brains of those ancestors were essentially the same as yours, and they were exposed to the same density of information from the outside world. Yet they lived in a world essentially free of the written word. I am sure that you, like me, have as much trouble imagining such a world as eighteenth-century Europeans must have had in imagining a world in which animals were noiseless. My life is totally dominated by the written word—I read so much that I often *dream* I am reading! What would I have dreamed if I were illiterate? How different would the neural hookups of my brain be? There is no more dramatic illustration of the immense flexibility of our brains than their ability to encompass a totally new skill *without any genetic change whatsoever.*

This does not mean, however, that selection is not operating. Illiteracy is due to a whole complex of factors, different for each individual. Now that the shift to literacy has taken place in our society, many people who are illiterate in spite of being exposed to universal education are at a great disadvantage. In a recent study of thirty-five adult illiterates in the United States, many were found to have various perception difficulties, and some had problems with short-term memory. Most of the group had trouble adjusting to society—twenty-seven of them were unemployed. They might have had even more problems, one imagines, in a less forgiving society.

What, if anything, does illiteracy have to do with genes? Although the data are sketchy, it is strongly suspected that many learning difficulties can be traced to factors such as poor nutrition and exposure to environmental pollutants such as lead. There is at least one contributing cause of illiteracy, however, that because of its specificity is unlikely to be due entirely to the environment. Dyslexia, or reading dysfunction, has been shown to have a genetic component, though it is a complex one. (It should be noted that, in spite of recent publicity in the media, there is no "reading gene.") And dyslexia is not alone. Although genetic studies are in their infancy, there are strong indications of genetic components to autism, stuttering, and other kinds of language impairment. The associations are weak, complex, and often partially masked by environmental effects, but they seem to be real.

So the ingredients for selection for and against different kinds of brain function seem all to be present in today's human population. There is a great deal of phenotypic variation for mental abilities in the population, with indications that at least some of this variability can be traced to genetic differences. There are also strong culture-based selection pressures for and against particular aptitudes of the brain. Let me emphasize three points before I go further:

First, this range of abilities is found in all human groups. Dyslexia and stuttering know no racial boundaries. As I have emphasized again and again in the course of this book, the pool of genetic variation in our species is so vast, and racial differences so relatively slight, that all human groups have been driven by runaway brain selection to essentially the same levels. All groups have grammatically sophisticated language, which means that all groups are capable of mastering high levels of cultural complexity. As Phillip Tobias recently pointed out to me, the Khoisan of southern Africa have the most tonally complex languages in the world. Cultural chauvinists might profitably compare their elegant language skills with the slurred excuse for speech, brutalized by television, that passes for language in much of the United States.*

Second, just like the skills for language, the skills needed for literacy must have predated the invention of written language itself. The genes that contribute to literacy are genes that have many other functions. Recently, it has been shown that many dyslexics have trouble discriminating between pairs of slightly different sounds when they are separated by a short interval of silence. Dyslexia, it seems, is only the most obvious manifestation of other perceptual difficulties, but it happens to have been thrown into high relief by our culture's emphasis on reading. The skills for reading, like the skills for language, are not an evolutionary add-on—they build on abilities that, to varying degrees, our brains already possess.

Third, the skills or indeed the brain organizations that are at a premium in some societies may not be in others. In a thought-provoking book, *The Origin of Consciousness in the Breakdown of the Bicameral Mind* (1976), Julian Jaynes argues that the shift to emphasis of left-hemisphere skills occurred at about the time of human urbanization. He suggests further that it was this shift that resulted in the emergence of the idea of self. My guess is that the concept of self probably arose a good deal earlier—Jaynes ignores the fact that people from nonurbanized societies at the present time seem to have a perfectly good concept of self. Suppose that Jaynes does turn out to be partly right and that there was such a shift to emphasis of left-hemisphere activity at some point or points in the recent past for most of the human race. Selection for genes that would aid that shift would certainly take place. The genetic shift would probably be only a slight one, however, for individuals carrying those genes would already have existed in the earlier population. There would be no necessity for new genes or for new wiring capabilities of the brain, but the people who survived the urban transformation of society would have had a slightly different genetic mix from their immediate ancestors.

Runaway brain evolution is undoubtedly continuing in our species. Can it continue indefinitely? What shape might future brain evolution take?

*William Safire, the language maven of the *New York Times*, has given some wonderful examples of this. My favorite is "ommina," as in "Ommina go to the store."

Why the Runaway Brain Is Truly Unique

In this book we have ranged more widely through time and space than in most books about human evolution. I have tried to present our evolutionary development as part of the whole sweep of life from the earliest multicellular animals to the present. As I have repeatedly emphasized, there is nothing unique about the evolutionary processes that have molded us. Our genes have been shaped by the same processes of mutation, selection, and chance that have shaped the genes of all the other organisms with which we share the planet. What is unique is not the process but the result. That is why the subtitle of this book is *The Evolution of Human Uniqueness*, not *The Uniqueness of Human Evolution*.

Innumerable theories have been proposed about the kinds of selection pressures that have driven human evolution. Darwin attributed much of our evolution to the power of sexual selection, which he was the first to examine in detail. He thought that the physical and mental differences between the sexes, which as a good Victorian he emphasized, were produced primarily by selection on the part of males for what they perceived to be beauty and desirability in females and on the part of females for such things as wealth and power in the males.

Robert Ardrey, following the lead of Raymond Dart, thought that the primary motivating force was the development of hunting and warfare skills, which in turn encouraged communication and cooperation among the hunters or warriors. In a reaction to his bloody scenario, the anthropologist Nancy Tanner has taken a feminist approach and proposed that the real push was provided by the women of the group, who kept things going by gathering plant foods and later deliberately growing them while the men were out being macho.

Clifford Jolly has suggested that bipedalism and manual dexterity evolved because a long period of drought required our ancestors to pick up and manipulate tiny seeds in order to survive. Elisabeth Vrba points out that around two and a half million years ago there was a dramatic change in the fauna of southern and eastern Africa, as a result of climatic changes that are not yet fully understood. She suggests that this may have presented the hominids of the time with new opportunities and driven the burst of hominid evolution that seems to have occurred at about that time. The late Glynn Isaac, on the other hand, emphasized the tremendous new capabilities provided by the invention of new types of tools. Many, many other perfectly reasonable suggestions have been made.

Who is right? Well, of course, they all are. Some of these influences may have been more important than others, but all of them contributed to the evolution of our remarkably flexible and multitalented species. All of them enriched our ancestors' environment and contributed to the runaway evolution of their brains. On the scale of the events dealt with in this book such processes, fascinating and important as they are, become details. Instead, we have backed away to look at our evolution on a broader canvas. Rather than examine in detail specific events that might have flowed from such things as our acquisition of the ability to walk upright, I have talked about the roles of gene duplication and of regulatory mutations. I have discussed the evolution of our brains as part of an ongoing process of gene duplication and divergence that has been at work since long before the Silurian period. We have seen that similar though far less dramatic periods of accelerated brain evolution took place during previous geological epochs and may be taking place among various groups of animals at the present time.

This runaway process, the result of a strong and increasingly swift gene-environment feedback loop acting on our species, is the reason why there seems to be that unnerving appearance of *progress* in human evolution—not because there is anything intrinsically progressive about the process of evolution itself but rather because our cultures and our brains have become more complex. This has led to what we define as progress, although a frog in a pond destroyed by our activities or a dolphin dying from the effects of our sewage might look at things rather differently.

The runaway process is also I suspect the primary reason why our evolution seems to have been so remarkably continuous. Different parts of our bodies have certainly evolved at different rates, sometimes in fits and starts, but from the evidence of the fossil record there has been no time during the last three and three-quarters million years when our evolution has stopped. The appearance of *Homo erectus* is the largest apparent discontinuity that can be seen in that record, and even here the postcranial skeleton shows no marked difference from those of the hominids that went before it. The mitochondrial Eve might or might not mark some important event in human evolution—the genetic evidence suggests that it does, but the fossil evidence is quite silent on the matter. However, it was not some unique defining event that marked the emergence of fully modern man. Fully modern man is a fiction. Whatever the event accompanying the mitochondrial Eve might have been, it was only one of dozens or hundreds of similarly important events scattered over the last few million years—and, in the larger sense, of thousands of evolutionary events scattered over the two billion years since mitochondria first came to live in the cells of our remote ancestors.

I hope I have convinced you that the evolution of the mind has not yet stopped for the human species. We are now, of course, using technology to extend the capabilities of our brains in ways that were unimaginable just a few years ago. This is in itself not biological evolution, but it may give a biological advantage to those individuals who are best able to utilize these new capabilities. We are also learning, through the Human Genome Project and many other allied research efforts, more and more about the genes that influence brain development. As I have argued elsewhere, we will not be able to use this new information to influence directly the genetic makeup of our species—gene therapy is unlikely to have much effect on the gene pool of a species that numbers five and a half billion individuals and counting. Perhaps we can use this knowledge instead to ensure that the environments of all children are rich and challenging ones and to allow them to develop their mental powers to the fullest degree possible. Each brain not fully utilized is two billion years of evolution wasted.

There is a caveat. At the beginning of this book, I pointed out that runaway evolution of any kind cannot continue indefinitely. Sooner or later, the tails of some peacocks will grow so large that the danger they are placed in as they try to haul them around will outweigh any advantage provided by sexual selection. Our runaway brains, by enabling us to multiply unreasonably in numbers, have put such pressures on the planet that ecological and nuclear catastrophes have become a real possibility. If there are any survivors of such disasters, selection will continue to operate on them, and it may not necessarily be selection in the direction we would like. If destruction were so widespread that the survivors could not even form social groups, then the complex and culture-driven capabilities of our brains might over many generations disappear.

Yet the runaway brain has an astounding property that makes it very different from a runaway peacock tail. The peacock with a tail that is too large is truly a prisoner of its genes, doomed to die because of its species' history of runaway sexual selection. The unique nature of runaway brain evolution is that it has the capability of *releasing* us from the prison of our genes. Surely we are now too smart to go on breeding ourselves into extinction, destroying most of the rest of the species on the planet in the process. Surely we are too smart to blow ourselves up with the nuclear weapons that have been provided by our runaway brains. The long saga of the human species has been one of selection for intelligence, not stupidity. It is time we woke up to that fact.

NOTES

Preface

p. xxii The first extended discussion of runaway sexual selection can be found in Ronald A. Fisher, *The genetical theory of natural selection* (London: Oxford University Press, 1930).

Introduction

p. 2 The story of the early development of studies on the genetics of learning in rats is in N. K. Innis, Tolman and Tryon: Early research on the inheritance of the ability to learn, *American Psychologist* 47 (1992): 190–97.

p. 3 Marian C. Diamond, *Enriching heredity: The impact of the environment on the anatomy of the brain* (London: Collier, 1988), traces the detailed studies of the brains of enriched and deprived rats.

p. 5 Early chimpanzee development was studied by A. H. Reisen and E. F. Kinder, *Postural development of infant chimpanzees* (New Haven, Conn.: Yale University Press, 1952) and Adolph H. Schultz, Growth and development of the chimpanzee, *Carnegie Institute of Washington publication no.* 518 (1940): 1–63.

p. 7 S. Leatherwood and R. R. Reeves, eds. *The bottlenose dolphin* (San Diego, Calif.: Academic Press, 1990), examines the differences between human and dolphin development.

p. 10 The quotation is from Richard E. Leakey and Roger Lewin, *People of the lake: man, his origins, nature, and future* (London: Collins, 1979).

p. 10 Culturgens are discussed in Charles J. Lumsden and Edward O. Wilson, *Genes, mind, and culture: The coevolutionary process* (Cambridge, Mass.: Harvard University Press, 1981) and the genetic effect on cultural evolution in E. O. Wilson, *Sociobiology: the new synthesis* (Cambridge, Mass.: Harvard University Press, Belknap, 1975).

p. 11 Luigi Luca Cavalli-Sforza and Marcus W. Feldman, *Cultural transmission and evolution: a quantitative approach* (Princeton, N.J.: Princeton University Press, 1981), examine mathematical models of cultural transmission but do not consider the effect of culture on large evolutionary changes.

Chapter 1: Eve's Companions

p. 18 L. Margulis and R. Fester, eds., *Symbiosis as a source of evolutionary innovation: speciation and morphogenesis* (Cambridge, Mass.: MIT Press, 1991), and R. Adam, The biology of Giardia spp., *Microbiological Reviews* 55 (1991): 706–32, are two of many references to early cellular evolution.

p. 26 A mitochondrial disease is discussed in N. J. Newman and D. C. Wallace, Mitochondria and Leber's hereditary optic neuropathy, *American Journal of Ophthalmology* 109 (1990): 726–30.

p. 28 W. M. Brown, Polymorphism in mitochondrial DNA of humans as revealed by restriction endonuclease analysis, *Proceedings of the National Academy of Sciences of the United States* 77 (1980): 3605–9, is the first paper detailing mitochondrial DNA variation in humans. The first complete sequence of human mitochondrial DNA was published by S. Anderson, A. T. Bankier, B. G. Barrell et al., Sequence and organization of the human mitochondrial genome, *Nature* 290 (1981): 457–64.

p. 28 A summary of Sanger's remarkable work is in Christopher Wills, *Exons, introns, and talking genes: The science behind the human genome project* (New York: Basic, 1991).

p. 33 A recent branch of the human and great ape lineages was proposed by V. M. Sarich and A. C. Wilson, Immunological time scale for hominid evolution, *Science* 158 (1967): 1200–1203. The controversy this engendered is recounted in J.

Gribbin and J. Cherfas, *The monkey puzzle* (New York: Pantheon, 1982).

p. 34 The great diversity of chimpanzee mitochondrial types was first explored by W. M. Brown, E. M. Prager, A. Wang et al., Mitochondrial DNA sequences of primates: Tempo and mode of evolution, *Journal of Molecular Evolution* 18 (1982): 225–39.

p. 36 J. C. Avise, J. E. Neigel, and J. Arnold, Demographic influence on mitochondrial DNA lineage survivorship in animal populations, *Journal of Molecular Evolution* 20 (1984): 99–105, criticized Brown's suggestion of a size bottleneck in our past.

p. 38 My own argument that there probably was a bottleneck is in C. Wills, Population size bottleneck, *Nature* 348 (1990): 398.

p. 42 Some of the many problems facing scientists who want to build molecular family trees are discussed in J. Felsenstein, Phylogenies from molecular sequences: inference and reliability, *Annual Reviews of Genetics* 22 (1988): 521–65.

p. 44 The famous paper that sparked the mitochondrial Eve controversy is R. L. Cann, M. Stoneking, and A. C. Wilson, Mitochondrial DNA and human evolution, *Nature* 325 (1987): 31–36.

p. 47 See Jared M. Diamond, *The third chimpanzee: The evolution and future of the human animal* (New York: HarperCollins, 1992).

p. 48 The two main competing models of human origins were first discussed in W. Howells, Explaining modern man: Evolutionists versus migrationists, *Journal of Human Evolution* 5 (1976): 477–95. More recent discussions of the two polar points of view are A. C. Wilson and R. L. Cann, The recent African genesis of humans, *Scientific American* 266, no. 4 (1992): 68–73; and A. G. Thorne and M. H. Wolpoff, The multiregional evolution of humans, *Scientific American* 266, no. 4 (1992): 76–80. I have also discussed them from the genetic standpoint in C. Wills, The role of molecular biology in human evolutionary studies: Current status and future prospects in *Molecular genetic medicine,* ed. T. Friedmann (San Diego, Calif.: Academic Press, 1991), 169–232.

p. 50 The first real criticisms of the African origin of the mitochondrial Eve were A. R. Templeton, Human origins and analysis of mitochondrial DNA sequences, *Science* 255 (1992): 737, and D. R. Maddison, African origins of human mitochondrial DNA reexamined, *Systematic Zoology* 40 (1991): 355–63. The justice of these criticisms was admitted by S. B. Hedges, S. Kumar, K. Tamura et al., Human origins and analysis of mitochondrial DNA sequences, *Science* 255 (1992): 737–39.

p. 51 My criticisms of the ways molecular family trees are presented are in C. Wills, Human origins, *Nature* 356 (1992): 389–90, and C. Wills, Phylogenetic analysis and molecular evolution, in *Biocomputing: Informatics and genome projects,* ed. D. W. Smith (Orlando, Fla.: Academic Press, in press).

p. 52 The new and very extensive mitochondrial DNA sequence data were collected by L. Vigilant, M. Stoneking, H. Harpending et al., African populations and the evolution of human mitochondrial DNA, *Science* 253 (1991): 1503–7. They were recently reanalyzed by M. Stoneking, S. T. Sherry, A. J. Redd et al., New approaches to dating suggest a recent age for the human mtDNA ancestor, *Philosophical Transactions of the Royal Society (London),* ser. B, 337 (1992): 167–75.

p. 54 My own entry into the Eve-dating sweepstakes, in which I present a number of lines of evidence suggesting she might have been as much as a million years old, is C. Wills, When did Eve live?—An evolutionary detective story, *Philosophical Transactions of the Royal Society (London),* ser. B, (submitted).

p. 56 Some of many reports on the Iceman include B. Rensberger, "Iceman" yields details of Stone Age transition: Portrait of life 5,300 years ago emerges (remains found in the Austrian Alps), in *Washington Post,* October 15, 1992, p. A1, and L. Jaroff, Iceman (study of the frozen remains of a man who lived 5,300 years ago), in *Time,* October 26, 1992, 62–66.

Chapter 2: An Obsession with Race

p. 58 Some of the fascinating books by and about Carleton Coon include Carleton S. Coon, *Adventures and discoveries: The autobiography of Carleton S. Coon* (Englewood Cliffs, N.J: Prentice-Hall, 1981); Carleton S. Coon, *Caravan: The story of the Middle East* (New York: Holt, 1958); Carleton S. Coon and Edward E. Hunt, Jr., *The living races of man* (New York: Knopf, 1965); Carleton S. Coon, *A north Africa story: The anthropologist as OSS agent, 1941–1943* (Ipswich, Mass.: Gambit, 1980); Carleton S. Coon, *A reader in general anthropology* (New York: Holt, 1948); and C. S. Coon, *Tribes of the Rif* (Cambridge, Mass.: Harvard University, Peabody Museum, 1970).

p. 61 The unpleasant early results of mixing human genetics and eugenics are found in Charles B. Davenport, *Heredity in relation to eugenics* (New York: Holt, 1911); and R. Ruggles Gates, *Heredity and eugenics* (London: Constable, 1923).

p. 62 Coon's magnum opus on the diversity of humans is Carleton Coon, *The origin of races* (New York: Knopf, 1962). Dobzhansky's review of it was T. Dobzhansky, A debatable account of the origin of races, *Scientific American* (1963): 169–72.

Chapter 3: An Obsession with God

p. 67 Many of the details of paleoanthropology in China can be found in W. Rukang and J. W. Olsen, eds., *Paleoanthropology and paleolithic archaeology in the People's Republic of China* (New York, Academic Press, 1985).

p. 69 Details of the excavations at Zhoukoudian are in Jia Lanpo and Huang Weiwen, *The story of Peking man* (Oxford: Oxford University Press, 1990); and L. Zechun, Sequence of sediments at locality 1 in Zhoukoudian and correlation with loess stratigraphy in northern China and with the chronology of deep-sea cores, *Quaternary Research* 23 (1985): 139–53.

p. 72 Surveys of the life of Pierre Teilhard de Chardin include Mary Lukas and Ellen Lukas, *Teilhard* (New York: Doubleday, 1977); J. Mortier and M.-L. Aboux, eds., *Teilhard de Chardin Album* (New York: Harper & Row, 1966); and Robert Speaight, *The life of Teilhard de Chardin* (New York: Harper & Row, 1967). The Piltdown forgery is examined in Joseph S. Weiner, *The Piltdown forgery* (London: Oxford University Press, 1955); Ronald Millar, *The Piltdown men* (New York: St. Martin's, 1972); Charles Blinderman, *The Piltdown inquest* (Buffalo, N.Y.: Prometheus, 1986); and Frank Spencer, *Piltdown: A scientific forgery* (New York: Oxford University Press, 1990).

p. 75 Pierre Teilhard de Chardin, *The future of man* (New York: Harper & Row, 1964); Pierre Teilhard de Chardin, *Toward the future* (New York: Harcourt Brace Jovanovich, 1975); and Pierre Teilhard de Chardin, *Christianity and evolution* (New York: Harcourt Brace Jovanovich, 1969), are representative of Teilhard's many philosophical writings.

p. 76 Syntheses of Teilhard's philosophical views can be found in H. James Birx, *Pierre Teilhard de Chardin's philosophy of evolution* (Springfield, Ill.: Thomas, 1972); and Paul Grenet, *Teilhard de Chardin; The man and his theories*, trans. R. A. Rudorff (New York: Eriksson, 1966).

p. 77 The Darwin-Wallace controversy can be found in Alfred Russel Wallace, *Letters and reminiscences* (New York: Arno, 1975),

Alfred Russel Wallace, *My life: A record of events and opinions* (New York: Dodd, Mead, 1905).

p. 78 The quotes are from Robert Broom, *The mammal-like reptiles of South Africa and the origin of mammals* (London: Wetherby, 1932); Henri Bergson, *Creative evolution*, trans. Arthur Mitchell (Westport, Conn.: Greenwood, 1975), and John C. Eccles, *Evolution of the brain: Creation of the self* (London: Routledge, 1989).

Chapter 4: The Crystal Brain

p. 85 Dart's discovery of *Australopithecus* is recounted in Raymond A. Dart, *Adventures with the missing link* (New York: Viking, 1959).

p. 86 The paper reporting the discovery is R. A. Dart, *Australopithecus africanus:* The ape-man of South Africa, *Nature* 115 (1925): 195–99.

p. 88 The remarkable life of Robert Broom is recounted in George H. Findlay, *Dr. Robert Broom, F.R.S.: A biography, appreciation, and bibliography* (Cape Town: Balkema, 1972). Broom's own story of his discoveries is Robert Broom, *Finding the missing link* (London: Watts, 1950).

p. 92 The controversial monograph detailing the discoveries is R. Broom and G. W. H. Schepers, *The South African fossil ape-men: The Australopithecinae* (Pretoria: Transvaal Museum memoir no. 2, 1946). Many details of the caves and their formation are given in Charles K. Brain, *The hunters or the hunted? An introduction to African cave taphonomy* (Chicago: University of Chicago Press, 1981).

p. 93 The story of Le Gros Clark's conversion is told in Wilfrid E. Le Gros Clark, *Man-apes or ape-men? The story of discoveries in Africa* (New York: Holt, Rinehart & Winston, 1967).

p. 94 Ardrey's violent view of human evolution is set forth in Robert Ardrey, *African genesis: A personal investigation into the animal origins and nature of man* (New York: Atheneum, 1961).

Chapter 5: The Birth Canal of Our Species

p. 97 The East African landscape and geology are depicted strikingly in Colin Wilcock, *Africa's Rift Valley* (Amsterdam: Time-Life, 1974).

p. 100 Some of the details of the history of the lakes are given in

Robert E. Hecky, Late Pleistocene-Holocene chemical stratigraphy and paleolimnology of the Rift Valley lakes of central Africa (Woods Hole, Mass.: Woods Hole Oceanographic Institution Reference no. 73–28, 1973).

p. 101 Some surveys of East Africa's complex past are given in E. S. Taieno Odhiambo, *A history of East Africa* (London: Longman, 1977); Thomas Spear, *Kenya's past* (London: Longman, 1981); and Norman R. Bennett, *Arab versus European: Diplomacy and war in nineteenth century east central Africa* (New York: Africana, 1986).

p. 102 Glynn L. Isaac, *The archaeology of human origins,* ed. Barbara Isaac (New York: Cambridge University Press, 1989); and Glynn L. Isaac, *Olorgesailie: Archaeological studies of a Middle Pleistocene lake basin in Kenya* (Chicago: University of Chicago Press, 1977), trace some of the history of tool use in the area. The oldest Acheulian tools are reported in B. Asfaw, Y. Beyene, G. Suwa et al., The earliest Acheulean from Konso-Gardula, *Nature,* 360 (1992):732–35.

p. 104 The oldest hafted ax yet found was reported in L. Groube, J. Chappell, J. Muke et al., A 40,000 year old human occupation site at Huon Peninsula, Papua New Guinea, *Nature,* 324 (1986): 453–55.

p. 105 Important books by and about Louis Leakey include Sonia M. Cole, *Leakey's luck: The life of Louis Seymour Bazett Leakey* (New York: Harcourt Brace Jovanovich, 1975); Louis S. B. Leakey, *By the evidence: Memoirs, 1932–1951* (New York: Harcourt Brace, 1974); Louis S. B. Leakey, *White African: An early autobiography* (Cambridge, Mass.: Schenkman, 1966); and Louis S. B. Leakey, *Kenya: Contrasts and problems* (London: Methuen, 1936).

p. 108 Mary Leakey's striking copies of ancient paintings are in Mary D. Leakey, *Africa's vanishing art: The rock paintings of Tanzania* (London: Hamilton, 1983). The quotation is from Mary D. Leakey, *Disclosing the past* (Garden City, N.Y.: Doubleday, 1984).

p. 109 A. Walker and M. Teaford, The hunt for *Proconsul*, *Scientific American* 260 (1989): 76–82; and A. Walker, Louis Leakey, John Napier and the history of *Proconsul, Journal of Human Evolution* 22 (1992): 245–54, examine various aspects of this distant Miocene ancestor of ours.

p. 115 Roger Lewin, *Bones of contention* (New York: Simon & Schuster, 1987), details some of the arguments that have surrounded fossil dating in East Africa.

p. 117 Some important books by and about Richard Leakey and his work include Richard E. Leakey and Roger Lewin, *Origins:*

What new discoveries reveal about the emergence of our species and its possible future (New York: Dutton, 1977); Richard E. Leakey, *The making of mankind* (New York: Dutton, 1981); Richard E. Leakey, *One life: an autobiography* (Salem, N.H.: Salem House, 1984); and Richard E. Leakey, *Origins Reconsidered* (New York: Doubleday, 1992).

p. 122 M. D. Leakey and J. M. Harris, eds., *Laetoli: A Pliocene site in northern Tanzania* (New York: Oxford University Press, 1987), describes the footprints in detail.

p. 125 Donald Johanson's discoveries are recounted in Donald C. Johanson and Maitland Edey, *Lucy: The beginnings of humankind* (New York: Warner, 1982); and Donald C. Johanson and James Shreeve, *Lucy's child: The discovery of a human ancestor* (New York: Morrow, 1989).

p. 131 Alan Walker, R. E. Leakey, J. M. Harris et al., 2.5 myr *Australopithecus boisei* from west of Lake Turkana, Kenya, *Nature* 322 (1986): 517–22, describes the finding of the Black Skull.

p. 132 The *H. habilis* skull fragment is described in A. Hill, S. Ward, A. Deino et al., Earliest *Homo, Nature* 355 (1992): 719–22.

p. 133 The remarkably complete and very old *H. erectus* skeleton is described in F. Brown, J. Harris, R. Leakey et al., Early *Homo erectus* skeleton from west Lake Turkana, *Nature* 316 (1985): 788–92.

p. 134 C. L. Brace and M. F. Ashley Montagu, *Man's evolution: An introduction to physical anthropology* (New York: Macmillan, 1965), details the one-species hypothesis.

p. 135 One of the best and most complete descriptions of comparative hominid anatomy is Leslie Aiello and Christopher Dean, *Human evolutionary anatomy* (San Diego, Calif.: Academic Press, 1990). Various aspects of the evolution of the hand are described in J. R. Napier, *Hands* (London: Allen & Unwin, 1980).

p.136 The gracile Australopithecine hand bones are described in D. E. Ricklan, *The precision grip in Australopithecus africanus* in *From apes to angels,* ed. G. H. Sperber (New York: Wiley-Liss, 1990), 171–83.

p. 136 The discovery of robust Australopithecine hand bones is reported in R. L. Susman, Hand of *Paranthropus, Science* 240 (1988): 781–84.

p. 137 R. R. Skelton and H. M. McHenry, Evolutionary relationships among early hominids, *Journal of Human Evolution* 23 (1992): 309–49, details the case for possible parallel evolution in early hominids.

Chapter 6: The Latest Steps

p. 141 A useful life of the discoverer of Java man is Bert Theunissen, *Eugène Dubois and the ape-man from Java: The history of the first missing link and its discoverer* (Boston: Kluwer Academic, 1989).

p. 142 The argument about gradual evolution versus stasis in *H. erectus* of East Asia is set out in M. Wolpoff, Evolution in *Homo erectus:* The question of stasis, *Paleobiology* 10 (1984): 389–406; and G. Philip Rightmire, *The evolution of Homo erectus: Comparative anatomical studies of an extinct human species* (Cambridge: Cambridge University Press, 1990).

p. 144 These new dating techniques are described in H. P. Schwarcz and R. Grun, Electron spin resonance (ESR) dating of the origin of modern man, *Philosophical Transactions of the Royal Society (London),* ser. B, 337 (1992): 145–48, and H. Valladas, G. Valladas, O. Baryosef et al., Thermoluminescence dating of Neanderthal and early modern humans in the Near East, *Endeavour* 15 (1991): 115–19.

p. 145 The arrival of ancestors of Australian aborigines is described in R. G. Roberts, R. Jones, and M. A. Smith, Thermoluminescence dating of a 50,000-year-old human occupation site in northern Australia, *Nature* 345 (1990): 153–56.

p. 146 Bruce Chatwin, *The Songlines* (New York: Viking, 1987), brilliantly describes the songs the aboriginals sing.

p. 147 The fascinating story of Australian prehistory is set out very readably in Josephine Flood, *Archaeology of the dreamtime,* 2d ed. (Sydney: Collins, 1989).

p. 149 The Kow Swamp measurements are in A. G. Thorne, Mungo and Kow Swamp: Morphological variation in Pleistocene Australians, *Mankind* 8 (1977): 85–89.

p. 151 The story of the discovery of the jaw is told in A. H. Brodrick, *Father of prehistory: The Abbe Henri Breuil, his life and times* (New York: Morrow, 1963). A discussion of the Ethiopian cave is in Gary D. Richards, *Porc-Epic Cave, Ethiopia: A re-evaluation of the artifactual component and published data,* (University of California, Berkeley, Ph.D. thesis, 1979).

p. 152 The story of what may turn out to be the earliest migration of our ancestors from Africa is dealt with in G. P. Rightmire, The dispersal of *Homo erectus* from Africa and the emergence of more modern humans, *Journal of Anthropological Research,* 47 (1991): 177–91.

p. 154 Useful references about the Neanderthals and the history of their discovery are found in F. Spencer, *The Neanderthals and their evolutionary significance: A brief historical survey* in *The origin of modern humans,* ed. F. H. Smith and F. Spencer (New York: Liss, 1984), 1–49; and E. Trinkaus, *The Shanidar Neanderthals* (New York: Academic Press, 1983).

p. 157 Neanderthal vocal equipment is discussed in B. Arensburg, A. M. Tillier, B. Vandermeersch et al., A Middle Paleolithic hyoid bone, *Nature* 338 (1989): 758–60.

p. 158 Surveys of the new Mount Carmel dates include M. J. Aitken and H. Valladas, Luminescence dating relevant to human origins, *Philosophical Transactions of the Royal Society (London),* ser. B, 337 (1992): 139–44; and R. Grun and C. B. Stringer, Electron spin resonance dating and the evolution of modern humans, *Archaeometry* 33 (1991): 153–99.

p. 159 The possible overlap of modern human and Neanderthal technologies is made more likely by discoveries detailed in N. Mercier, H. Valladas, J. L. Joron et al., Thermoluminescence dating of the late Neanderthal remains from Saint-Césaire, *Nature* 351 (1991): 737–39.

p. 160 The original paper on punctuated equilibrium is N. Eldredge and S. J. Gould, *Punctuated equilibria: An alternative to phyletic gradualism,* in *Models in Paleobiology,* ed. T. J. M. Schopf (San Francisco: Freeman, 1972).

Chapter 7: Two Kinds of Geneticist—Two Kinds of Gene

p. 168 Discussions of the work of H. J. Muller are found in Elof A. Carlson, *Genes, radiation and society: The life and work of H. J. Muller* (Ithaca, N.Y.: Cornell University Press, 1981); James F. Crow, *H. J. Muller's role in evolutionary biology* in *The founders of evolutionary genetics,* ed. S. Sarkar (Amsterdam: Kluwer, 1992); and James F. Crow, H. J. Muller: scientist and humanist, *Wisconsin Academy Review* 37 (1990): 19–22.

p. 169 Chesterton's concerns about eugenics, many of them still very relevant, are found in G. K. Chesterton, *Eugenics and other evils* (London: Cassell, 1922). Daniel J. Kevles, *In the name of eugenics: Genetics and the uses of human heredity* (New York: Knopf, 1985), provides an extensive history of the eugenics movement.

p. 171 Histories of the Columbia fly room include Alfred H. Sturtevant, *A history of genetics* (New York: Harper & Row, 1965); and Leslie C. Dunn, *A short history of genetics: The development of some of the main lines of thought* (New York: McGraw-Hill, 1965).

p. 182 Perhaps the best history of Lysenkoism is David Joravsky, *The Lysenko affair* (Cambridge, Mass.: Harvard University Press, 1970).

p. 183 The little book Muller incautiously touted to Stalin is Hermann J. Muller, *Out of the night: A biologist's view of the future* (New York: Vanguard, 1935).

p. 184–85 Muller's seminal paper on the dangers to our species of genetic damage is H. J. Muller, Our load of mutations, *American Journal of Human Genetics* 2 (1950): 111–76. The recent career of the two sperm banks founded using Muller's ideas can be found in T. Gorman, Measure of success elusive for sperm bank, in *Los Angeles Times,* April 12, 1992, pp. B1–B7.

p. 188 A short history of Dobzhansky's early career is found in William B. Provine, *Origins of "The genetics of natural populations" series* in *Dobzhansky's genetics of natural populations I–XLIII,* ed. R. C. Lewontin et al. (New York: Columbia, 1981), 1–83. Other accounts include T. Dobzhansky, *The reminiscences of Theodosius Dobzhansky* (New York: Columbia University Oral History Collection, 1962–63); and Barbara Land, *Evolution of a scientist: The two worlds of Theodosius Dobzhansky* (New York: Crowell, 1973).

p. 197 Dobzhansky's landmark book went through three editions. The first, which includes much historical material left out of the later editions, is T. Dobzhansky, *Genetics and the origin of species* (New York: Columbia University Press, 1937).

Chapter 8: Pulling the Plug

p. 202 Kimura provides a brief reminiscence of his life in Motoo Kimura, Genes, populations and molecules: A memoir, in *Population genetics and molecular evolution,* eds., Tomoko Ohta and Kenichi Aoki (Tokyo: Japan Scientific Societies Press, 1985), 459–81.

p. 203 The early history of the field of population genetics, and Wright's contributions to it, can be found in William B. Provine, *The origins of theoretical population genetics* (Chicago: University of Chicago Press, 1971); and William B. Provine, *Sewall Wright and evolutionary biology* (Chicago: University of Chicago Press, 1986).

p. 207 The mathematics of some of these discoveries is given in James F. Crow and Motoo Kimura, *An introduction to population genetics theory* (New York: Harper & Row, 1970).

p. 209 The paper that founded molecular evolutionary studies is E.

Zuckerkandl and L. Pauling, *Molecular disease, evolution, and genic heterogeneity* in *Horizons in Biochemistry,* ed. M. Kasha and B. Pullman (New York: Academic Press, 1962).

p. 214 The two papers detailing the extent of enzyme polymorphism are R. C. Lewontin and J. L. Hubby, A molecular approach to the study of genic heterogeneity in natural populations. II. Amount of variation and degree of heterozygosity in natural populations of *Drosophila pseudoobscura, Genetics* 54 (1966): 595–609; and H. Harris, Enzyme polymorphisms in man, *Proceedings of the Royal Society (London),* ser. B, 164 (1966): 298–310.

p. 214 The red flag that infuriated the selectionists was J. L. King and T. H. Jukes, Non-Darwinian evolution, *Science* 164 (1969): 788–98.

p. 215 The paper that drew Kimura's ire was T. L. Blundell and S. P. Wood, Is the evolution of insulin Darwinian or due to selectively neutral mutation? *Nature* 257 (1975): 197–203.

p. 217 N. Proudfoot, Pseudogenes, *Nature* 286 (1980): 840–41, discusses these remarkable disabled genes.

Chapter 9: The Love Song of the Fruit Fly and Other Amazing Gene Stories

p. 222 Some of the landmarks in the study of the period gene are recounted in C. P. Kyriacou and J. C. Hall, Interspecific genetic control of courtship song production and reception in *Drosophila, Science* 232 (1986): 494–97; G. Petersen, J. C. Hall, and M. Rosbash, The period gene of *Drosophila* carries species-specific behavioral instructions, *EMBO Journal* 7 (1988): 3939–47; R. Costa, A. A. Peixoto, J. R. Thackeray et al., Length polymorphism in the threonine-glycine-encoding repeat region of the *period* gene in *Drosophila, Journal of Molecular Evolution* 32 (1991): 238–46; and D. A. Wheeler, C. P. Kyriacou, M. L. Greenacre et al., Molecular transfer of a species-specific behavior from *Drosophila simulans* to *Drosophila melanogaster, Science* 251 (1991): 1082–85.

p. 226 Some of the properties of highly organized genes are set out in Christopher Wills, *The wisdom of the genes* (New York: Basic, 1989).

p. 228 The lack of correlation between vitamin D production and skin pigment is investigated in L. Y. Matsuoka, J. Wortsman, J. G. Haddad et al., Racial pigmentation and the cutaneous synthesis of vitamin D, *Archives of Dermatology* 127 (1991): 536–38.

Stern's simple genetic model is in C. Stern, Model estimates of the number of gene pairs involved in pigmentation variability of the Negro-American, *Human Heredity* 20 (1970): 165–68.

p. 230 Some of the genetic compexities surrounding melanin production are examined in V. J. Hearing and K. Tsukamoto, Enzymatic control of pigmentation in mammals, *FASEB Journal* 5 (1991): 2902–9; J. M. Naeyaert, M. Eller, P. R. Gordon et al., Pigment content of cultured human melanocytes does not correlate with tyrosinase message level, *British Journal of Dermatology* 125 (1991): 297–303; I. J. Jackson, D. M. Chambers, P. S. Budd et al., The tyrosinase-related protein-1 gene has a structure and promoter sequence very different from tyrosinase, *Nucleic Acids Research* 19 (1991): 3799–3804; and M. Yaar and B. A. Gilchrist, Human melanocyte growth and differentiation: A decade of new data, *Journal of Investigative Dermatology* 97 (1991): 612–17.

p. 233 The evolution of haptoglobins was worked out by N. Maeda, S. M. McEvoy, H. F. Harris et al., Polymorphisms in the human haptoglobin gene cluster: Chromosomes with multiple haptoglobin-related (Hpr) genes, *Proceedings of the National Academy of Science (U. S.)* 83 (1986): 7395–99; and S. M. McEvoy and N. Maeda, Complex events in the evolution of the haptoglobin gene cluster in primates, *Journal of Biological Chemistry* 263 (1988): 15740–47.

p. 235 The similarity between the two species is documented in H. L. Carlson, W. E. Johnson, P. S. Nair et al., Allozymic and chromosomal similarity in two *Drosophila* species, *Proceedings of the National Academy of Sciences (U. S.)* 72 (1975): 4521–25.

p. 236 Val's genetic analysis is in F. C. Val, *Evolution* 31 (1976): 611–20.

p. 238 Hardy's book is Alister C. Hardy, *The living stream: Evolution and man* (New York: Harper & Row, 1965).

p. 239 Waddington discussed his findings and theory in Conrad H. Waddington, *New patterns in genetics and development* (New York: Columbia University Press, 1962); and Conrad H. Waddington, *The nature of life* (New York: Atheneum, 1962).

Chapter 10: Escape from Stupidworld

A good introduction to many of the phenomena discussed in this section can be found in the September 1992 issue of *Scientific American*, a single issue devoted to Mind and Brain.

p. 245 Wilson discussed his ideas in J. S. Wyles, J. G. Kunkel and A. C. Wilson, Birds, behavior and anatomical evolution, *Proceedings of the National Academy of Sciences (U. S.)* 80 (1983): 4394–97; and A. C. Wilson, The molecular basis of evolution, *Scientific American* 253 no. 4 (1985): 164–73. Recent work on Darwin's finches is summarized in P. R. Grant, Natural selection and Darwin's finches, *Scientific American* 265 (1991): 82–88.

p. 246 The remarkable habits and evolution of these fishes are discussed in Geoffrey Fryer and T. D. Iles, *The cichlid fishes of the great lakes of Africa: Their biology and evolution* (Edinburgh: Oliver & Boyd, 1972). The work showing their recent origin is in A. Meyer, T. D. Kocher, P. Basasibwaki et al., Monophyletic origin of Lake Victoria cichlid fishes suggested by mitochondrial DNA sequences, *Nature* 347 (1990): 550–53.

p. 249 The evolutionary role of play is discussed in Robert Fagen, *Animal play behavior* (New York: Oxford University Press, 1981).

p. 250 The evolution of the brain and the role of brain size are summarized in Harry J. Jerison, *Evolution of the brain and intelligence* (New York: Academic Press, 1973); and H. J. Jerison, *Brain size and the evolution of the mind,* James Arthur lecture on the evolution of the human brain (New York: American Museum of Natural History, 1991).

p. 255 Recent summaries of the various roles of homeobox genes include W. McGinnis and R. Krumlauf, Homeobox genes and axial patterning, *Cell* 68 (1992): 283–302; and P. Holland, Homeobox genes in vertebrate evolution, *Bioessays* 14 (1992): 267–73.

p. 259 The ancient history of neurons is discussed in M. Reuter and M. Gustaffson, "Neuroendocrine" cells in flatworms—progenitors to metazoan neurons? *Archives of Histology and Cytology* 52 (1989): 253–63. Some of the history of the discovery of neuron structure is G. Pilleri, *Camillo Golgi, 1843–1926, Santiago Ramón y Cajal, 1852–1934, Adelchi Negri, 1876–1912: Biographical sketches for the 50th anniversary of the Bern University Brain Anatomy Institute* (Ostermundigen, Switzerland: Verlag des Hirnanatomischen Institutes, 1984).

p. 263 T. W. Deacon, Rethinking mammalian brain evolution, *American Zoologist* 30 (1990): 629–705; and A. J. Rockel, R. W. Hiorns, and T. P. S. Powell, The basic uniformity in structure of the neocortex, *Brain* 103 (1980): 221–44, discuss the organization of the brain at levels higher than the neuron.

p. 265 An accessible discussion of recent work on dinosaur evolution is John Noble Wilford, *The riddle of the dinosaur* (New York: Knopf, 1985).

Chapter 11: The Master Juggler

p. 270 William F. Allman, *Apprentices of wonder: Inside the neural network revolution* (New York: Bantam, 1989), discusses attempts to imitate the workings of the brain through electronics.

p. 273 The story of Broca and his progenitors is in Francis Schiller, *Paul Broca: Founder of French anthropology, explorer of the brain* (Berkeley, Calif.: University of California Press, 1979).

p. 274 The remarkable phenomenon of blindsight is discussed in Lawrence Weiskrantz, *Blindsight: A case study and implications* (New York: Oxford University Press, 1986), O. Braddick, J. Atkinson, B. Hood et al., Possible blindsight in infants lacking one cerebral hemisphere, *Nature* 360 (1992): 461–63; M. Barinaga, Neuroscience—unraveling the dark paradox of blindsight, *Science* 258 (1992): 1438–39; and A. Cowey and P. Stoerig, The neurobiology of blindsight, *Trends in neurosciences* 14 (1991): 140–45. Dennett's insightful discussion is in Daniel C. Dennett, *Consciousness explained* (Boston: Little, Brown, 1991).

p. 275 This unusual neural map is reported in V. S. Ramachandran, M. Stewart, and D. C. Rogers-Ramachandran, Perceptual correlates of massive cortical reorganization, *Neuroreport* 3 (1992): 583–86.

p. 276 Many of the normal and retarded people with remarkable abilities mentioned here are discussed in L. K. Obler and D. Fein, eds. *The exceptional brain* (New York: Guilford, 1988). M. A. Howe, *Fragments of genius: The strange feats of idiots savants* (New York: Routledge, 1989), provides an introducton to savant syndrome.

p. 279 Clara Claiborne Park, *The siege: The first eight years of an autistic child, with an epilogue, fifteen years later* (Boston: Little, Brown, 1982), gives a vivid account of the difficulty of trying to reach an autistic child.

Chapter 12: A Sponge for Knowledge

p. 285 Helen Keller, *The story of my life* (Garden City, New York: Doubleday, 1954) recounts her early years.

p. 287 The horrifying story of Genie is told in Susan Curtiss, *Genie: A psycholinguistic study of a modern-day "wild child"* (New York: Academic Press, 1977).

p. 292 Some information on Williams syndrome can be found in K. L.

Jones, Williams syndrome: An historical perspective of its evolution, natural history and etiology, *American Journal of Medical Genetics* 6 (supp.) (1990): 89–96; and T. L. Jernigan and U. Bellugi, Anomalous brain morphology on magnetic resonance images in Williams syndrome and Down syndrome, *Archives of Neurology* 47 (1990): 529–33.

p. 294 A little of the literature on chimpanzee language can be sampled in R. A. Gardner, B. T. Gardner, and T. E. Van Cantfort, eds. *Teaching sign language to chimpanzees* (Albany, New York: State University of New York Press, 1989); and E. S. Savage-Rumbaugh, J. Murphy, R. A. Sevcik et al., *Language comprehension in ape and child,* in *Monographs of the Society for Research in Child Development,* ed. W. C. Bronson, 1993).

p. 298 The use of alternative pathways to communication by our resourceful brains is discussed in D. P. Corina, J. Vaid, and U. Bellugi, The lingustic basis of left hemisphere specialization, *Science* 255 (1992): 1258–60; and S. W. Anderson, H. Damasio, A. R. Damasio et al., Acquisition of signs from American Sign Language in hearing individuals following left hemisphere damage and aphasia, *Neuropsychologia* 30 (1992): 329–40.

p. 299 This theory of the recent appearance of language is found in W. Noble and I. Davidson, The evolutionary emergence of modern human behavior—language and its archeology, *Man* 26 (1991): 223–53.

Chapter 13: Reining in the Runaway?

p. 302 Thackeray's examination of recent human evolution is J. F. Thackeray and J. A. Kieser, Variability in shape of the dental arcade of *Homo sapiens* in Late Pleistocene and modern samples from southern Africa, *Paleontologia africana* (in press). The dates of ancient Bushman paintings are given in A. I. Thackeray, The Middle Stone Age south of the Limpopo River, *Journal of World Prehistory* 6 (1992): 385–440.

p. 303 A survey of the ancient peoples of southernmost Africa is H. J. Deacon, Southern Africa and modern human origins, *Philosophical Transactions of the Royal Society (London),* ser. B, 337 (1992): 177–83.

p. 303 J. S. Jones, Is evolution over? If we can be sure about anything, it's that humanity won't become superhuman, *New York Times,* September 22, 1991, p. E17, claims that human evolution has stopped. My devastating reply is C. Wills, Has human evolution ended? *Discover* 13 (1992): 22–25.

p. 304 H. J. Curtis, Biological mechanisms underlying the aging process, *Science* 141 (1963): 686–94, gives some interesting observations on the differences in genetic stability of long-lived and short-lived mice.

p. 304 The twin study is T. J. Bouchard, Jr., D. T. Lykken, M. McGue et al., Sources of human psychological differences: The Minnesota study of twins raised apart, *Science* 250 (1990): 223–28. Illiteracy in the Third World is discussed in B. D. Weiss, G. Hart, and R. E. Pust, The relationship between literacy and health, *Journal of Health Care for the Poor and Underserved* 1 (1991): 351–63.

p. 306 The illiteracy study is F. T. Fleener and J. F. Scholl, Academic characteristics of self-identified illiterates, *Perceptual and Motor Skills* 74 (1992): 739–44. Some of the genetic and other complexities underlying dyslexia are explored in G. W. Stuart and W. J. Lovegrove, Visual processing deficits in dyslexia—receptors or neural mechanisms, *Perceptual and Motor Skills* 74 (1992): 187–92; H. W. Catts, Early identification of dyslexia—evidence from a follow-up study of speech-language impaired children, *Annals of Dyslexia* 41 (1991): 163–77; and S. D. Smith, B. F. Pennington, W. J. Kimberling et al., Family dyslexia: Use of genetic linkage data to define subtypes, *Journal of the American Academy of Child and Adolescent Psychiatry* 29 (1990): 204–13.

p. 307 Julian Jaynes, *The origin of consciousness in the breakdown of the bicameral mind* (Boston: Houghton Mifflin, 1976).

p. 308 Some of the many theories about what has powered our recent evolution are found in Charles Darwin, *The descent of man and selection in relation to sex* (Princeton, N.J.: Princeton University Press, 1981); Clifford J. Jolly, *Physical anthropology and archeology* (New York: Random House, 1976); E. S. Vrba, Mammals as a key to evolutionary theory, *Journal of Mammalogy* 73 (1992): 1–28; and Nancy M. Tanner, *On becoming human* (New York: Cambridge University Press, 1981).

GLOSSARY

Acheulian tools. Advanced stone tools found in Africa and Europe but not in Asia. These tools constitute the first real technological advance over the primitive chipped pebbles of the Oldowan industry.

Adaptive radiation. When an ancestral species or group of species gives rise in the course of evolution to many different species occupying a variety of ecological niches.

Allele. One form of a particular gene. Most alleles of a given gene that are found in a gene pool differ only slightly from each other, perhaps by only one or two base differences that have appeared through mutation. Some, however, show larger differences. In the course of evolution, alleles increase or decrease in frequency in the gene pool. Sometimes, one allele replaces another.

Amino acid. Proteins are made up of long, unbranched chains of amino acids, molecules that have both basic and acidic properties. Amino acids can be quite different from each other chemically, which is why proteins can have such a variety of properties. Twenty different amino acids are coded by the *genetic code.*

Australopithecus afarensis. The oldest of the Australopithecines, small brained but with a fully upright posture and postcranial skeleton with very humanlike characteristics. *A. afarensis* flourished in East Africa between three and three-quarters and three and a quarter million years ago.

Australopithecus africanus. A small-brained hominid, with slightly more humanlike characteristics than *A. afarensis,* that lived in southern and eastern Africa between three and two and a half million years ago. The hand of *A. africanus*, in particular, was strikingly more humanlike than that of *A. afarensis*.

Autism. A group of dysfunctional behaviors marked by withdrawal from normal social contact and sometimes by a pathological concentration on certain aspects of the environment.

Axon. The process of a neuron that sends impulses to other neurons or to muscle fibers. Particularly in the brain, the axon may be multibranched.

Bases. Part of the building blocks of DNA and RNA, which provide the specificity for the genetic code. In the double-stranded DNA, adenine (A) always pairs with thymine (T) on the opposite strand, and guanine (G) always pairs with cytosine (C). But the sequence of bases along the strand of the DNA is not constrained by these pairing rules, so that in theory any message coded in the language of these four bases can be written into the DNA strand by the process of evolution.

Blindsight. A phenomenon in which a large blind spot is produced in one or both visual fields as a result of damage to the striate cortex in the back of the brain. People with blindsight can locate objects in the blind spot, even though they insist they do not see them.

Broca's area. A region in the left frontal lobes of the brain that when damaged interferes with the articulation of speech.

Cerebellum. A highly convoluted region of the hindbrain that until recently was thought to be involved only in very basic brain functions. There is growing evidence that the cerebellum is involved in speech production and comprehension.

Chromosomes. Little structures in the nucleus of the cell that carry the DNA and the other molecules needed to make accurate copies of themselves. Most of the genetic information in the cell is coded in the DNA of the chromosomes of the nucleus.

ClB. The chromosome used by H. J. Muller to detect newly induced mutations in his *Drosophila*. All the genes on the chromosome are locked together, so that crossing-over is prevented and the genetics of the flies is greatly simplified.

Cretaceous period. The latter half of the age of dinosaurs, lasting from 135 to 65 million years ago.

Culturgens. The hypothetical cultural traits suggested by Charles Lumsden and E. O. Wilson that, when they arise, select for genes for behavioral and physical traits that allow their carriers to employ the culturgens most effectively.

Cytoplasm. The region of the cell lying outside the nucleus, where proteins are synthesized and the mitochondria are located.

Dendrites. The thin cellular branches radiating out from a neuron that gather information from a sense organ or other neurons.

DNA. Deoxyribonucleic acid, a double-stranded, long-chain molecule in which the two chains are twisted into the famous double helix. In its sequence of *bases,* DNA codes the genetic information of almost all organisms.

Dominance. An allele that masks the effect of another allele in a heterozygote is *dominant* to it.

Drosophila. A genus made up of about 2,500 different species of two-winged flies.

Drosophila heteroneura. A species from the island of Hawaii that has a very laterally elongated head.

Drosophila melanogaster. The fruit fly of choice for geneticists.

Drosophila persimilis. One of the pair of sibling species that Theodosius Dobzhansky studied intensively in order to understand the process of speciation.

Drosophila pseudoobscura. The other of Dobzhansky's sibling species.

Drosophila silvestris. A species from the island of Hawaii, closely related to *D. heteroneura* but without the laterally elongated head.

Drosophila simulans. A sibling species of *D. melanogaster*, not as closely related to it as the sibling species *D. pseudoobscura* and *D. persimilis*.

E. coli. See *Escherichia coli*.

Electron spin resonance. A technique for dating crystalline material that depends on the number of energetic atoms in the crystals that have been produced by natural sources of radiation.

Encephalization quotient. The brain size of a species relative to the sizes of the brains of an average species that has the same body size.

Epistasis. In its broadest sense, epistasis describes the situation when a gene at one locus influences a gene or genes at other loci.

Escherichia coli. A bacterium found commonly in the guts of a wide variety of vertebrates, including man. More is known about the genes of *E. coli* than about those of any other organism.

Eugenics. Francis Galton's term for the betterment of the human species through controlled or regulated breeding.

First Family. Remains of over a dozen individuals of *A. afarensis* found in the Hadar region of Ethiopia and dated to over three million years ago. These fossils are remarkable both for their numbers, which give a good idea of how variable the population must have been, and for the fact that they were not found in association with other animal bones.

Fixation. The process by which a mutant allele completely replaces other alleles at that genetic locus in a population. If the mutant allele is favorable, it may replace the others rapidly; if it is selectively neutral, it may take many generations to do so.

Gene duplication. The process by which an ancestral gene duplicates so that the descendants carry two or more copies of the gene. Once this process occurs, the different copies of the ancestral gene are free to diverge and eventually to take up new functions.

Gene pool. The collection of genes held in common by a population or a species. Each member of the population possesses a sample of the gene pool. Normally, there is much more genetic variation in the gene pool of the population as a whole than there is in an individual.

Genetic code. The four bases found in *DNA* can be arranged to produce sixty-four code words, each three *bases* long. Sixty-one of the words code for the twenty *amino acids*, with between one and six code words for each amino acid. The other three words code for stop signals and indicate that the end of a protein has been reached. Mitochondrial genes have a different genetic code from genes in the nucleus, but the difference is only a slight one—the code is essentially universal.

Genome. All the genes possessed by an organism, a sample from the *gene pool* of the species.

Genotype. A description of the genes carried by an organism, usually the genes that the experimenter happens to be interested in. Thus, the geneticist knows that somebody of blood type O has the genotype *ii*, regardless of what other genes may be in that person's genome.

Giant chromosomes. Chromosomes that have duplicated many times in certain specialized cells of a fruit fly larva, so that they form bundles of threads that are easily visible under the microscope.

Great Rift Valley. The valley running the length of East Africa, through the Red Sea and into the Middle East, that is the probable site of a future split between the major African continental plate and the Somali plate to its east.

Haptoglobin. A protein molecule involved in salvaging the iron-containing part of hemoglobin in people suffering from severe anemia. Haptoglobin genes have duplicated and fused in our immediate evolutionary past, although the reasons they have done so remain obscure.

Heterozygote. At a given genetic locus (or location) on a chromosome, a heterozygous organism carries one copy of each of two different alleles, one from each parent. It is possible for an organism to be heterozygous at one locus but *homozygous* at another.

Homeobox genes. Genes important in development that all share a short piece of DNA called a homeobox. Homeobox genes regulate some of the most fundamental processes of the early embryonic life of creatures as diverse as ourselves and insects.

Hominid. Members of the genera *Homo*, *Australopithecus*, and *Paranthropus*—organisms very like ourselves.

Hominoid. The hominids together with the great apes.

Homo erectus. A large-brained hominid that first appeared in East Africa a little over one and a half million years ago. The last representatives may have lived in Java a little more than two hundred thousand years ago, although the date is uncertain. Peking man is the *type specimen* of *H. erectus*. Its postcranial skeleton was essentially indistinguishable from our own, and, although the shape of its skull was rather different from ours, the brain sizes of later *H. erectus* slightly overlapped our own.

Homo habilis. A possible link between small-brained early Australopithecines and larger-brained *H. erectus*, living in East Africa between two and a half and one and a half million years ago.

Homozygote. If two copies of the same allele, one from each parent, occupy a particular genetic locus, the organism is homozygous at that locus—although it may be heterozygous at other genetic loci.

Interference. If a maternal and a paternal chromosome cross over, nearby crossovers involving the same pair of chromosomes are interfered with. H. J. Muller, who discovered this phenomenon, also found strains of flies in which crossing-over was completely prevented along the entire chromosome (see *ClB*).

Interneurons. Neurons that form the connection between neurons that collect information from the sense organs and those that send signals to the muscles and other organs of the body. Most of the neurons of the brain are interneurons, and our brain has a very large number of complexly connected interneurons.

Inversions. Regions of chromosomes in which the order of the genes has been reversed. If maternal and paternal chromosomes are heterozygous for inversions, this effectively suppresses *recombination* between them.

Java man. A number of finds of late *Homo erectus* found in Java that may range from as much as five hundred thousand to as little as two hundred thousand years old—although the latter date is very doubtful. Some of these hominids had very large brains, virtually overlapping our own in size.

Khoisan. Hunter-gatherer peoples of southern Africa who have been displaced from much of their range over the last two thousand years by peoples from the north. It has been argued that these peoples may be genetically continuous with older groups that have inhabited the region for almost one hundred thousand years.

Kow Swamp. A region in the state of Victoria, Australia, where the remains of a heavy-boned aboriginal group dating from between nine thousand and fifteen thousand years ago have been found.

Laetoli. The site of fragmentary remains, perhaps of *A. afarensis*, and of hominid footprints that have been dated to about 3.7 million years ago. The footprints indicate that even at that early time hominids of the region were fully upright in posture.

Linkage. The nearer two genes are to each other on a chromosome, the more likely it is that they tend to be passed together to the next generation. This is because the likelihood that they will be broken apart by the process of recombination is less than if they are farther apart. Genes near each other are said to be tightly linked—those further apart are said to be loosely linked.

Locus. The location of a gene on the chromosome. Each gene has its own locus.

Lucy. The virtually complete skeleton of a small female *A. afarensis* from Ethiopia, dated to about 3.2 million years ago. This small-brained creature was fully upright, but her hands still showed resemblances to those of an arboreal monkey.

Marsupials. Mammals such as kangaroos and opossums in which most of fetal development takes place not in the womb but in an external pouch of skin belonging to the mother.

Melanin. Various colored pigments made up of long chains of molecules derived from the amino acid tyrosine. Melanin is responsible for the pigmentation of our skins and the irises of our eyes.

Miocene epoch. The epoch extending from twenty-four to five million years ago, marked by a warm and equable climate, during which a number of ape species such as *Proconsul* flourished in Africa and elsewhere.

Mitochondria. Structures in the cytoplasm of the cell that are descended from free-living bacteria that infected the cells of our ancestors perhaps two billion years ago. Among other things, mitochondria enable our cells to burn sugars completely using oxygen and to extract a great deal of useful energy from each sugar molecule.

Mitochondrial Eve. The woman in our distant ancestry who carried the mitochondrial chromosomes from which all the human mitochondrial chromosomes of the present time are descended. She was not—repeat, not!—the only woman alive at the time. Others, who have passed many of their nuclear genes to us, shared the planet with her. But, largely by chance, all the mitochondrial chromosomes in our species have descended from her.

Molecular clock. Proteins and genes of a given type appear to evolve at a fairly constant clocklike rate. Neutralists argue that this is because most of the mutations that arise at these genetic levels have little effect on the organisms carrying them.

Multiple-origins model. The model of human origins that suggests that the present diversity of human races arose relatively early in our evolution, perhaps during the time of *H. erectus*, so that an approach to modern humans has taken place at various times and perhaps at various speeds in different parts of the Old World. This set of apparently parallel evolutionary events was aided by an unknown amount of gene flow from one part of the species to another.

Mutation. In the broadest sense, any inherited alteration in the genetic information of an organism. Mutations can be small, such as single-base changes in the DNA of a gene, or very large, such as the sudden doubling of an entire set of chromosomes.

Neanderthal man. Hominids inhabiting the Middle East and Europe between as early as one hundred thousand years ago and thirty-five thousand years ago, when they were apparently supplanted by modern humans. Neanderthals were very similar in appearance and behavior to ourselves, although opinion has lurched back and forth several times since their discovery 140 years ago about just how like us they were. Current opinion is that they were distinguishably different in many ways, but different or not they certainly had an elaborate tool-using culture that included ritual burial.

Neutral mutations. Mutations that have little or no effect on the organisms carrying them.

Neutralists. Scientists who believe that the great majority of mutations arising in natural populations are neutral, so that their behavior can be described and predicted by simple mathematical models.

Noah's Ark model. The model of human origins that suggests that modern humans arose relatively recently, probably in Africa, and spread rapidly through the Old World, replacing the less advanced hominids that were living there.

Nucleus. A globular structure found in most cells, surrounded by a membrane, that contains the chromosomes.

Oldowan tools. The simplest and most primitive stone-tool technology, the oldest examples of which are from East Africa and have been dated to two and a half million years ago.

Olduvai Gorge. A river gorge in Tanzania that has carved its way through sediments revealing that at least three different kinds of hominid lived there at various times during the last two million years.

Outgroup. A group of organisms that can serve to calibrate a family tree, particularly one based on molecular data. Chimpanzees, for example, can serve as an outgroup for human data—the ratio of the amount of divergence seen among humans to the amount that has accumulated since the ancestors of humans and chimpanzees parted company can give an estimate of the time since the mitochondrial Eve.

Parallel evolution. Evolution in which two groups of organisms descended from an ancestor have gone through very similar evolutionary changes. If *H. habilis* and the robust Australopithecines really were genetically separate, the many similarities between them must have arisen through parallel evolution, because they resemble each other more closely than they do their most likely common ancestor.

Paranthropus. A genus name for some of the more extreme robust Australopithecines, first suggested by Robert Broom, that is growing in popularity. This separate name implies that the robust Australopithecines were clearly different from the evolutionary line leading to humans, a matter that has not been settled.

PCR. See *polymerase chain reaction*.

Period gene. A gene in *Drosophila* that is involved in regulating rhythmic behavior.

Peripheral neurons. Neurons that are involved in gathering information from the environment or sending instructions to the body. Only about one-hundred-thousandth of our neurons are peripheral neurons.

PET. See *positron emission tomography*.

Phenotype. The appearance and behavior of an organism. Our phenotype is influenced by both our genotype and our environment.

Pleistocene epoch. Epoch that lasted from about two million to ten thousand years ago, and ended with the appearance of extensive human cultures. It is marked by many periods of glaciation.

Pliocene epoch. The epoch from five to two million years ago, marked by a warm climate but with a slow cooling trend, during which the mammals reached their greatest diversity.

Polymerase chain reaction. An ingenious technique that lets the experimenter make unlimited quantities of a gene starting with as little as a single copy.

Polymorphism. Variation among the individuals in a population. Polymorphism can be obvious, in appearance or behavior, or less obvious, at the level of the genes or the chromosomes.

Positron emission tomography. A way of measuring activity in the brain by monitoring blood flow.

Potassium-argon dating. A method for dating volcanic materials that are as young as fifty thousand years old or as ancient as several billion years old.

Potential-altering mutations. Mutations that alter the structure or organization of a gene and give rise to an enhanced possibility that potential-realizing mutations can then take place that fine-tune the gene's function.

Potential-realizing mutations. Mutations that fine-tune a gene's function and that are more likely to occur because a potential-altering mutation has already taken place.

Precambrian era. The period of time beginning with the origin of the earth some 4.6 million years ago and extending to the appearance of the majority of the present-day groups of multicellular organisms 550 million years ago. The ancestors of mitochondria probably entered the cells of our own ancestors a little more than halfway through the Precambrian era.

Proconsul. An ape living in the Miocene epoch, some eighteen to twenty million years ago, that is thought to be an ancestor of the hominids.

Pseudogenes. Genes that have been damaged so that they no longer function. Now invisible to natural selection, these genes are free to accumulate mutational changes at a high rate.

Punctuated equilibrium. Name given by Niles Eldredge and Stephen Jay Gould to the phenomenon seen commonly in the fossil record, in which periods of stasis are punctuated by periods of swift change. So swift are the changes, and so limited the sizes of the populations in which they take place, that fossils documenting these changes cannot usually be found.

Quantitative (multigenic) variation. If several genes contribute to a character, the phenotypes that result can be described only in quantitative terms rather than the qualitative terms Mendel used to describe his short and tall pea plants. If several genes had contributed to the character Mendel used in his crosses, his pea plants would have ranged from short through intermediate to tall without falling into discrete classes. Quantitative genetic variation is very common in natural populations.

Recessive. An allele of a gene is recessive to another allele if its effects are masked by that other allele in a heterozygote.

Recombination, genetic. By this process, segments are exchanged between the maternally derived and paternally derived chromosomes each generation. Were it not for this process, all the genes on any given chromosome would always be inherited as a unit.

Regional continuity model. See *multiple-origins model.*

Restriction enzymes. Enzymes that recognize and cut DNA molecules at very specific sequences, usually four or six bases long. These enzymes are essential for gene manipulation.

RNA. This DNA-like molecule, ribonucleic acid, consists of a single chain.

It carries a copy of the DNA code to the cytoplasm of the cell, where the coded information is used to manufacture proteins.

Robust Australopithecines. A group of hominids that lived between two and a half and one million years ago in South and East Africa and that are marked by large teeth and heavy jaw musculature. They seem to have used only primitive tools.

Runaway sexual selection. An idea suggested by R. A. Fisher in 1915. Organisms competing for mates might evolve extreme structures or behaviors that put them in jeopardy—increasing their mating fitness at the same time as their overall fitness is lowered. Such selection can continue as long as their overall fitness is not lowered too much.

Savant syndrome. People with mental retardation who nonetheless can develop particular skills to a high level.

Selectionists. Scientists who believe that most of the genetic variation in natural populations affects its carriers to some degree.

Skhul and Qafzeh. Caves in northern Israel containing remains of quite modern-appearing humans that have been dated to as long as ninety thousand years ago.

Sterkfontein. A site near Johannesburg at which gracile Australopithecines were found.

Swartkrans. A site near Sterkfontein at which robust Australopithecines and *Homo erectus* have been found.

Symbiosis. Living together with mutual benefit. We and our mitochondria live together in a long-continued and highly beneficial symbiosis.

Synapse. A connection between neurons.

Tabun and Kebara. Caves in northern Israel where Neanderthal remains dating to as much as sixty thousand years ago have been found.

Teleology. From the Greek *telos* (end). Evolution that proceeds toward a goal is said to be teleological; such a process is not a part of modern evolutionary theory.

Therapsids. These mammal-like creatures were the dominant large animals for one hundred million years before the age of dinosaurs. Largely supplanted by the dinosaurs, a few of the therapsid lineages survived and evolved into the mammals.

Thermoluminescence. The light emitted by heated materials as their excited atoms fall to ground state. The amount of thermoluminescence is correlated with the amount of ambient radioactivity and the age of the sample.

Transitions. Mutational changes that substitute one base for another of a similar size in the DNA.

Transversions. Changes in which a base of a different size is substituted.

Type specimen. The specimen, usually the first found or the best preserved, to which all other specimens of a species can be compared.

Tyrosinase. An enzyme that helps to produce the large pigment molecules of melanin from many small molecules of the amino acid tyrosine.

Waddington effect. Usually, confusingly, called genetic assimilation. A group of organisms developing in a stressful environment is phenotypically more variable than a similar group developing under more relaxed circumstances, and this increased phenotypic variation increases the effectiveness of natural selection.

Wernicke's area. An area at the side of the brain that regulates speech comprehension and meaningful speech production. People suffering damage to this area have difficulty understanding speech, and, although they can talk fluently, what they say is largely nonsense.

Willandra Lakes. A series of now-dry lakes in southern Australia inhabited by a light-boned people who lived there as long as 30,000 years ago.

Williams syndrome. An inherited condition marked by mental retardation but almost normal production and comprehension of speech.

X chromosome. The female sex-determining chromosome. A female carries two X chromosomes and a male carries an X and a Y.

X-linkage. An X-linked gene is one that is located on the X chromosome, like the Bar-eyed gene in *Drosophila*. The Y chromosome has few genes in common with the X, so genes on the Y cannot mask the effects of genes on the X.

Y chromosome. The male sex-determining chromosome.

Zhoukoudian. A site near Beijing at which remains of *Homo erectus* (Peking man), dated to about half a million years ago, were found.

PICTURE CREDITS

Grateful acknowledgment is made to the following for their permission to reprint figures and photographs:

Figures 1.4 and 1.5 from R. Cann et al., "Mitochondrial DNA and human evolution," *Nature* 325 (1987): 31–36. Copyright © 1987 Macmillan Magazines Limited.

Figure 2.1 from Coon and Carleton, *The origin of races* (New York: Knopf, 1962), p. 29.

Figure 5.1 from Richard Leakey, *One life: An autobiography* (Topsfield, Mass.: Salem House Publishers, 1984), p. 13.

Figure 5.2 from S. L. Washburn, "Tools and human evolution," *Scientific American* (September 1960): 70.

Figure 5.3 from A. Walker and R. E. F. Leakey, "The hominids of East Turkana," *Scientific American* (August 1978): 60.

Figure 5.4 from M. D. Leakey, "Tracks and tools," *Philosophical transactions of the royal society, London* B 292 (1981): 95–102, fig. 1.

Figure 10.1 from G. Fryer and T. D. Iles, *The cichlid fishes of the great lakes of Africa* (Harlow, Essex [UK]: Oliver and Boyd, 1972), fig. 18.

Figure 10.2 (top) from H. P. Whiting and L. B. Halstead Tarlo, "The brain of the heterostraci (agnatha)," *Nature* 207 (1965): 829–31, fig. 2. Copyright © 1965 Macmillan Magazines Limited.

Figure 10.2 (bottom) from J. Nolte, *The human brain* (St. Louis: Mosby-Year Book, 1988), fig. 2.3.

Figure 10.5 from W. McGinnis and R. Krumlauf, "Homeobox genes and axial patterning," *Cell* 68 (1992): 283–302, fig. 2. Copyright © 1992 by Cell Press.

Plate 5 from B. Latimer et al., "Talocrural joint in African hominids: Implications for Australopithecus Afarensis," *American Journal of Physical Anthropology* 74 (1978): 155–75, fig. 3.

Plate 6 from P. Mold et al., "Fission-track annealing characteristics of meteoric phosphates," *Nuclear Tracks* 9 (1984): 119–28. With permission from Pergamon Press Ltd., Headington Hill Hall, Oxford.

Plate 7 courtesy of the Institute of Human Origins.

Plate 8 copyright © John Reader, Science Source/Photo Researchers.

Plate 9 photo by A. Walker. Copyright © by National Museum of Kenya.

Plate 10 courtesy of Wistar Institute.

Plates 11 and 13 copyright © E. Trinkaus.

Plate 14 from G. Lefevre, "A Photographic representation and interpretation of the polytene chromosomes of Drosophila melanogaster," in *Genetics and Biology of Drosophila*, vol. 1A, ed. M. Ashburner and E. Novitski (London: Academic Press, 1976), fig. 3.

Plate 15 from T. Dobzhansky and C. Epling, *Contributions to the genetics, taxonomy, and ecology of Drosophila pseudoobscura* (Washington, D.C.: Carnegie Institution, 1944), plate 1.

Plate 18 from J. Nolte, *The human brain* (St. Louis: Mosby-Year Book, 1988), plate 8 A-E.

Index

Note: Page numbers in **bold** type indicate glossary entries; *n* indicates material in footnotes.